# METHODS IN MOLECULAR BIOLOGY

*Series Editor*
**John M. Walker**
**School of Life and Medical Sciences**
**University of Hertfordshire**
**Hatfield, Hertfordshire, AL10 9AB, UK**

For further volumes:
http://www.springer.com/series/7651

# Biomedical Nanotechnology

## Methods and Protocols

## Second Edition

Edited by

## Sarah Hurst Petrosko

*Department of Chemistry and International
Institute for Nanotechnology
Northwestern University
Evanston, IL, USA*

## Emily S. Day

*Department of Biomedical Engineering
University of Delaware
Newark, DE, USA*

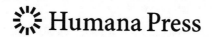

*Editors*
Sarah Hurst Petrosko
Department of Chemistry and International
   Institute for Nanotechnology
Northwestern University
Evanston, IL, USA

Emily S. Day
Department of Biomedical Engineering
University of Delaware
Newark, DE, USA

ISSN 1064-3745                    ISSN 1940-6029   (electronic)
Methods in Molecular Biology
ISBN 978-1-4939-8314-8           ISBN 978-1-4939-6840-4   (eBook)
DOI 10.1007/978-1-4939-6840-4

Printed on acid-free paper

This Humana Press imprint is published by Springer Nature
The registered company is Springer Science+Business Media LLC
The registered company address is: 233 Spring Street, New York, NY 10013, U.S.A.

# Preface

Nanoscience and technology focuses on the synthesis and application of structures that have at least one dimension on the sub-100 nm length scale. It deals with investigating the fundamental properties of such structures, which usually differ significantly from that of the bulk material, and taking advantage of these qualities to construct novel materials and devices or develop unique applications that can address global challenges in health, energy, the environment, and beyond. Owing to widespread interest and investment, biomedical nanotechnology, geared toward the ultimate use of nanostructures in medicinal and clinical applications, is an area of intense research that is growing and progressing at an extraordinary pace. This rapid development is fueled by the fact that nanomaterials often offer superior capabilities when compared to materials that are conventionally used for the detection, diagnosis, and treatment of disease. Nanomaterials are enabling promising advances toward the attainment of long-standing goals within biomedicine, such as the ability to track and treat disease in real time and the capability to provide personalized or precision medicine approaches to each individual patient, among others.

The goal of this volume is to provide an overview of some of the types of nanostructures commonly utilized in nanobiomedicine; many of these nanostructures possess both inorganic and organic or biological components and are of a size that allows them to interface with biological systems. The majority of this volume consists of protocol chapters, which provide practical information on the synthesis and characterization of a variety of solution-phase and surface-bound nanomaterials and show how they can be used in sensing, imaging, therapeutics, or more than one of these capacities simultaneously. Most chapters provide step-by-step instructions and insight into overcoming possible pitfalls and challenges associated with the completion of the protocol. Most chapters also offer the reader insight into how the protocol can be changed with the reader's own research goals in mind (e.g., to target a different gene or to detect a different biomarker).

The ability to reliably and reproducibly synthesize highly uniform nanomaterials that can be fully characterized is important in many areas of nanoscience, including biomedicine. Chapters 1–6 discuss protocols for synthesizing and characterizing molecule and biomolecule-functionalized nanoconjugates with gold, iron oxide, or polymeric cores, which are often utilized in biomedical nanotechnology. Here, novel fluorescence (Chapter 1), $^1$H nuclear magnetic resonance (NMR) (Chapter 2), nanoparticle tracking (Chapter 3), and superconducting quantum interference device (SQUID) magnetometer-based (Chapter 4) methodologies are discussed. Emphasis is placed on using such knowledge to uncover structure–function relationships that can be used to control how nanomaterials interact with biological systems. Chapter 5 describes the synthesis of dumbbell-like gold-iron oxide nanoparticles as well as other types of magnetic particles; many nanomaterials are designed with multiple components that lend them multifunctionality. Then, Chapter 6 shows the reader how to synthesize non-biodegradable polystyrene-based, polyethylene glycol functionalized nanoparticles useful for analyzing the brain microenvironment. Many researchers are currently interested in understanding and treating diseases of the brain; indeed, this topic represents a highly pertinent research area within biomedical nanotechnology.

A realm of biomedical nanotechnology that has experienced early success lies in the use of nanomaterials for biosensing and imaging applications. One example of a biosensing platform that utilizes well-characterized nanoparticles is presented in Chapter 7. The volumetric bar-chart chip, or V-chip, can be used for the instant visual quantitation of target proteins in a multiplexed manner. Flow cytometry-based methods have also emerged as powerful tools within biomedical nanotechnology. Researchers are devising ways to make this technique amenable to high-throughput, quantitative measurements (Chapter 8), and they are applying it to gain knowledge of how nanomaterials interact with and influence the behaviors of cells, especially immune cells, an important consideration for any nanomaterial that will be placed into a living system (Chapter 9). In fact, researchers in the field of biomedical nanotechnology are placing special attention on figuring out how to keep immune systems from negatively reacting to nanoparticles (eliciting toxicity), but also on understanding how nanoparticles can be used to upregulate or downregulate immune responses in the context of immunotherapy. Nanostructures that can be used to detect and track biomolecules intracellularly are also being developed. Chapter 10 presents a nanostructure that can track messenger RNA inside differentiating stem cells, using a method that does not result in cell death, and hence allows for downstream processing or long-term analysis. Nanoparticle tracking is also being used to understand the motion of membrane neurotransmitter transporters on the surface of cells (Chapter 11), and photoacoustic imaging is being used to map nanoparticle distribution in vivo (Chapter 12).

In addition to enabling applications in biosensing or imaging, there are various ways in which nanomaterials can function as therapeutic agents within the field of biomedical nanotechnology. A desirable feature of many nanomaterials is that their modular architectures allow them to perform multiple functions simultaneously. For example, in Chapter 13, conjugated gold nanorod structures that can utilize both drug delivery and photothermal therapy to treat disease are discussed. In Chapter 14, a nanostructure is presented that is designed to collapse upon irradiation into free oligonucleotides, drug molecules, and small molecule payloads (all with therapeutic potential), and Chapter 15 discusses structures composed of amphiphilic block copolymers containing the appropriate cargo (drug or diagnostic agent) and enzyme-responsive peptides that fall apart when they encounter specific enzymes. These chapters highlight an important push within the field of biomedical nanotechnology—to prepare dynamic, bioresponsive structures. Another thrust in the field is toward developing structures that mimic natural biological systems. The Nanoscript, which mimics the structure and function of transcription factors (TFs), is designed to interact directly with genes to provide a therapeutic benefit through the regulation of gene expression (Chapter 16). These chapters focus primarily on the synthesis of the nanoconjugates themselves, but allude to their downstream therapeutic applications.

Some types of nanostructures utilized in biomedical nanotechnology are effective when coupled to substrates comprised of nano- and microstructures, as demonstrated in Chaps. 17–19. In Chapter 17, nanovesicles are loaded into microneedle-array patches to aid in the delivery of insulin. Chapter 18 focuses on the synthesis of nanofiber configurations on surfaces that can be used as biological grafts and implants to engineer and regenerate soft tissues, and Chapter 19 shows how one can use nanopatterned surfaces to treat diseases of the eye. Nanoparticle toxicity is also discussed in this volume because it is particularly relevant when introducing nanomaterials to biological environments, especially the human body. Chapter 20 discusses how to assess toxicity related to graphene oxide nanomaterial exposure; importantly, this chapter highlights another important use of a flow cytometry-based method (imaging flow cytometry). Chapter 21 expounds on how to analyze the toxicity of nanoparticle aerosols using a special air-liquid exposure system; such studies are important to bolster the translation of nanoparticle treatments that employ this delivery method to the market. The final chapter in this volume provides a case study on patent landscapes within nanoscience and technology, following up on ideas presented in the first edition of this volume (Chapter 22).

The chapters are written by leading researchers from all over the world, several of whom also participated in the first edition of this volume, working within nanoscience and technology, biology, chemistry, physics, engineering, medicine, and the law. Together, these chapters demonstrate the potential of nanotechnology to revolutionize medical care. Moreover, they beautifully illustrate how the fundamental property differences associated with nanomaterials (relative to bulk materials) and their potential for use as multicomponent/multifunctional structures can be exploited to transform the study, detection, and treatment of disease. This volume is a useful reference for scientists and researchers at all levels who are interested in working in a new area of nanoscience and technology or in expanding their knowledge base in their current field. Chapter 22, in particular, will be of interest to the social scientist, lawyer, or businessperson, who wants to learn about how the patent process applies to the field of nanotechnology. We are optimistic that advances in nanotechnology, and related fields, will lead to solutions to key issues within biomedicine, as evidenced by the exciting work featured in the second edition of this volume.

*Evanston, IL*
*Newark, DE*

*Sarah Hurst Petrosko*
*Emily S. Day*

# Contents

# Contributors

GHANASHYAM ACHARYA • *Ocular Nanomedicine Research Laboratory, Department of Ophthalmology, Baylor College of Medicine, Houston, TX, USA*

AKRIVI ASIMAKOPOULOU • *Aerosol and Particle Technology Laboratory, CPERI/CERTH, Thessaloniki, Greece*

DANIELLE M. BAILEY • *Department of Chemistry, Vanderbilt University, Nashville, TN, USA; Interdisciplinary Materials Science Program, Vanderbilt University, Nashville, TN, USA; Department of Pharmacology, Vanderbilt University, Nashville, TN, USA*

LIMING BIAN • *Department of Mechanical and Automation Engineering (Biomedical Engineering), Shun Hing Institute of Advanced Engineering, The Chinese University of Hong Kong, Shatin, New Territories, Hong Kong, China*

MARGARET M. BILLINGSLEY • *Department of Biomedical Engineering, University of Delaware, Newark, DE, USA*

DIANA M. BOWMAN • *Center for the Study of Law, Science and Technology, Sandra Day O'Connor College of Law, Arizona State University, Tempe, AZ, USA*

CASSANDRA E. CALLMANN • *Department of Chemistry and Biochemistry, University of California–San Diego, La Jolla, CA, USA*

SI CHEN • *Bioengineering Department, University of Illinois at Urbana-Champaign, Urbana, IL, USA*

CHUN KIT K. CHOI • *Department of Electronic Engineering (Biomedical Engineering), The Chinese University of Hong Kong, Shatin, New Territories, Hong Kong, China*

CHUNG HANG J. CHOI • *Department of Electronic Engineering (Biomedical Engineering), Shun Hing Institute of Advanced Engineering, The Chinese University of Hong Kong, Shatin, New Territories, Hong Kong, China*

PAIGE COLLINS • *Department of Biomedical Engineering, University of Connecticut, Storrs, CT, USA*

GRACE CONARD • *Indiana University School of Medicine, Indianapolis, IN, USA*

KHOLUD DARDIR • *Department of Chemistry and Chemical Biology, Rutgers University, Piscataway, NJ, USA*

EMILY S. DAY • *Department of Biomedical Engineering, University of Delaware, Newark, DE, USA; Helen F. Graham Cancer Center and Research Institute, Newark, DE, USA; Department of Materials Science and Engineering, University of Delaware, Newark, DE, USA*

NATHAN C. GIANNESCHI • *Department of Chemistry and Biochemistry, University of California–San Diego, La Jolla, CA, USA*

JORDAN J. GREEN • *Department of Biomedical Engineering, Translational Tissue Engineering Center, and Institute for Nanobiotechnology, Johns Hopkins University School of Medicine, Baltimore, MD, USA; Department of Materials Science and Engineering, Johns Hopkins University, Baltimore, MD, USA; Department of Ophthalmology, and Neurosurgery, Johns Hopkins University, Baltimore, MD, USA*

ZHEN GU • *Joint Department of Biomedical Engineering, University of North Carolina at Chapel Hill, Chapel Hill, NC, USA; Department of Biomedical Engineering, North Carolina State University, Raleigh, NC, USA; Center for Nanotechnology in Drug Delivery and Division of Molecular Pharmaceutics, UNC Eshelman School of Pharmacy,*

*University of North Carolina at Chapel Hill, Chapel Hill, NC, USA; Department of Medicine, University of North Carolina at Chapel Hill, Chapel Hill, NC, USA*

PAVAN GUPTA • *Chicago College of Osteopathic Medicine, Downers Grove, IL, USA*

PRINCESS I. IMOUKHUEDE • *Bioengineering Department, University of Illinois at Urbana-Champaign, Urbana, IL, USA*

ATHANASIOS KONSTANDOPOULOS • *Aerosol and Particle Technology Laboratory, CPERI/CERTH, Thessaloniki, Greece; Department of Chemical Engineering, Aristotle University, Thessaloniki, Greece*

KOSTAS KOSTARELOS • *Nanomedicine Laboratory, School of Health Sciences, Faculty of Biology, Medicine and Health and National Graphene Institute, The University of Manchester, Manchester, UK*

OLEG KOVTUN • *Department of Chemistry, Vanderbilt University, Nashville, TN, USA*

NICOLE L. KREUZBERGER • *Department of Biomedical Engineering, University of Delaware, Newark, DE, USA*

SANGAMESH G. KUMBAR • *Department of Orthopaedic Surgery, University of Connecticut Health, Farmington, CT, USA; Department of Biomedical Engineering, University of Connecticut, Storrs, CT, USA*

KI-BUM LEE • *Department of Chemistry and Chemical Biology, Rutgers University, Piscataway, NJ, USA*

JUSTIN LETENDRE • *Department of Biomedical Engineering, University of Connecticut, Storrs, CT, USA*

NASTASSJA A. LEWINSKI • *Institute for Work and Health (IST), University of Lausanne, Lausanne, Switzerland; Department of Chemical and Life Science Engineering, Virginia Commonwealth University, Richmond, VA, USA*

YING LI • *Department of Nanomedicine, Houston Methodist Research Institute, Houston, TX, USA; Department of Cell and Developmental Biology, Weill Medical College of Cornell University, New York, NY, USA*

FRANCES S. LIGLER • *Joint Department of Biomedical Engineering, University of North Carolina at Chapel Hill, Chapel Hill, NC, USA; Department of Biomedical Engineering, North Carolina State University, Raleigh, NC, USA*

NATHAN J. LIU • *Institute for Work and Health (IST), University of Lausanne, Lausanne, Switzerland; Department of Medicine, Weill Cornell Medical College, Cornell University, New York, NY, USA*

GEOFFREY P. LUKE • *Thayer School of Engineering, Dartmouth College, Hanover, NH, USA*

LORENA MALDONADO-CAMARGO • *Department of Chemical Engineering, University of Florida, Gainesville, FL, USA*

OHAN S. MANOUKIAN • *Department of Orthopaedic Surgery, University of Connecticut Health, Farmington, CT, USA; Department of Biomedical Engineering, University of Connecticut, Storrs, CT, USA*

DANIELA C. MARCANO • *Ocular Nanomedicine Research Laboratory, Department of Ophthalmology, Baylor College of Medicine, Houston, TX, USA*

ANTHONY D. MARINO • *Center for the Study of Law, Science and Technology, Sandra Day O'Connor of Law, Arizona State University, Tempe, AZ, USA*

RITA MATTA • *Department of Biomedical Engineering, University of Connecticut, Storrs, CT, USA*

AUGUSTUS D. MAZZOCCA • *Department of Orthopaedic Surgery, University of Connecticut Health, Farmington, CT, USA*

JILIAN R. MELAMED • *Department of Biomedical Engineering, University of Delaware, Newark, DE, USA*

JILL E. MILLSTONE • *Department of Chemistry, University of Pittsburgh, Pittsburgh, PA, USA*

AUSTIN VAN NAMEN • *Thayer School of Engineering, Dartmouth College, Hanover, NH, USA*

ELIZABETH NANCE • *Department of Chemical Engineering, University of Washington, Seattle, WA, USA*

DENNIS B. PACARDO • *Joint Department of Biomedical Engineering, University of North Carolina at Chapel Hill, Chapel Hill, NC, USA; Department of Biomedical Engineering, North Carolina State University, Raleigh, NC, USA; Center for Nanotechnology in Drug Delivery and Division of Molecular Pharmaceutics, UNC Eshelman School of Pharmacy, University of North Carolina at Chapel Hill, Chapel Hill, NC, USA*

ELENI PAPAIOANNOU • *Aerosol and Particle Technology Laboratory, CPERI/CERTH, Thessaloniki, Greece*

KINAM PARK • *Industrial and Physical Pharmacy and Weldon School of Biomedical Engineering, Purdue University, West Lafayette, IN, USA*

JAMES PARKIN • *Bioengineering Department, California Institute of Technology, Pasadena, CA, USA*

SARAH HURST PETROSKO • *Department of Chemistry and International Institute for Nanotechnology, Northwestern University, Evanston, IL, USA*

LIDONG QIN • *Department of Nanomedicine, Houston Methodist Research Institute, Houston, TX, USA; Department of Cell and Developmental Biology, Weill Medical College of Cornell University, New York, NY, USA*

CHRISTOPHER RATHNAM • *Department of Chemistry and Chemical Biology, Rutgers University, Piscataway, NJ, USA*

MICHAEL RIEDIKER • *Institute for Work and Health (IST), University of Lausanne, Lausanne, Switzerland; SAFENANO, IOM Singapore, Singapore, Singapore*

RACHEL S. RILEY • *Department of Biomedical Engineering, University of Delaware, Newark, DE, USA*

CARLOS RINALDI • *Department of Chemical Engineering, University of Florida, Gainesville, FL, USA; J. Crayton Pruitt Family Department of Biomedical Engineering, Gainesville, FL, USA*

SANDRA J. ROSENTHAL • *Department of Chemistry, Vanderbilt University, Nashville, TN, USA; Department of Pharmacology, Vanderbilt University, Nashville, TN, USA; Vanderbilt Institute for Nanoscale Science and Engineering, Vanderbilt University, Nashville, TN, USA; Department of Physics and Astronomy, Vanderbilt University, Nashville, TN, USA; Department of Chemical and Biomolecular Engineering, Vanderbilt University, Nashville, TN, USA; Materials Science and Technology Division, Oak Ridge National Laboratory, Oak Ridge, TN, USA*

CRYSTAL S. SHIN • *Ocular Nanomedicine Research Laboratory, Department of Ophthalmology, Baylor College of Medicine, Houston, TX, USA*

ASHLEY M. SMITH • *Department of Chemistry, University of Pittsburgh, Pittsburgh, PA, USA*

YUJUN SONG • *Department of Nanomedicine, Houston Methodist Research Institute, Houston, TX, USA; Department of Cell and Developmental Biology, Weill Medical College of Cornell University, New York, NY, USA*

XIAOLIAN SUN • *State Key Laboratory of Molecular Vaccinology and Molecular Diagnostics & Center for Molecular Imaging and Translational Medicine, School of Public Health, Xiamen University, Xiamen, China*

SHOUHENG SUN • *Department of Chemistry, Brown University, Providence, RI, USA*

DOUGLAS J. SYLVESTER • *Center for the Study of Law, Science and Technology, Sandra Day O'Connor College of Law, Arizona State University, Tempe, AZ, USA*

XUYU TAN • *Department of Chemistry and Chemical Biology, Northeastern University, Boston, MA, USA*

MYTHREYI UNNI • *Department of Chemical Engineering, University of Florida, Gainesville, FL, USA*

DANIELLE M. VALCOURT • *Department of Biomedical Engineering, University of Delaware, Newark, DE, USA*

SANDRA VRANIC • *Nanomedicine Laboratory, School of Health Sciences, Faculty of Biology, Medicine and Health and National Graphene Institute, The University of Manchester, Manchester, UK*

JARED WEDDELL • *Bioengineering Department, University of Illinois at Urbana-Champaign, Urbana, IL, USA*

DAVID R. WILSON • *Department of Biomedical Engineering, Translational Tissue Engineering Center, Institute for Nanobiotechnology, Johns Hopkins University School of Medicine, Baltimore, MD, USA*

JICHENG YU • *Joint Department of Biomedical Engineering, University of North Carolina at Chapel Hill, Chapel Hill, NC, USA; North Carolina State University, Raleigh, NC, USA; Center for Nanotechnology in Drug Delivery and Division of Molecular Pharmaceutics, UNC Eshelman School of Pharmacy, University of North Carolina at Chapel Hill, Chapel Hill, NC, USA*

SHANN S. YU • *Swiss Institute for Experimental Cancer Research (ISREC) and Institute for Bioengineering (IBI), École Polytechnique Fédérale de Lausanne, Lausanne, Switzerland*

KE ZHANG • *Department of Chemistry and Chemical Biology, Northeastern University, Boston, MA, USA*

YUQI ZHANG • *Joint Department of Biomedical Engineering, University of North Carolina at Chapel Hill, Chapel Hill, NC, USA; North Carolina State University, Raleigh, NC, USA; Center for Nanotechnology in Drug Delivery and Division of Molecular Pharmaceutics, UNC Eshelman School of Pharmacy, University of North Carolina at Chapel Hill, Chapel Hill, NC, USA*

# Chapter 1

# Quantification of siRNA Duplexes Bound to Gold Nanoparticle Surfaces

## Jilian R. Melamed, Rachel S. Riley, Danielle M. Valcourt, Margaret M. Billingsley, Nicole L. Kreuzberger, and Emily S. Day

## Abstract

RNA interference (RNAi)-based gene regulation has recently emerged as a promising strategy to silence genes that drive disease progression. RNAi is typically mediated by small interfering ribonucleic acids (siRNAs), which, upon delivery into the cell cytoplasm, trigger degradation of complementary messenger RNA molecules to halt production of their encoded proteins. While RNAi has enormous clinical potential, its in vivo utility has been hindered because siRNAs are rapidly degraded by nucleases, cannot passively enter cells, and are quickly cleared from the bloodstream. To overcome these delivery barriers, siRNAs can be conjugated to nanoparticles (NPs), which increase their stability and circulation time to enable in vivo gene regulation. Here, we present methods to conjugate siRNA duplexes to NPs with gold surfaces. Further, we describe how to quantify the resultant amount of siRNA sense and antisense strands loaded onto the NPs using a fluorescence-based assay. This method focuses on the attachment of siRNAs to 13 nm gold NPs, but it is adaptable to other types of nucleic acids and nanoparticles as discussed throughout the protocol.

**Key words** Nanoparticles, siRNA, Nucleic acids, Gene regulation, Conjugation, Loading

## 1 Introduction

RNA interference (RNAi) is a potent method to regulate gene expression that is under intense investigation as a therapy for a variety of diseases including cancer, hepatitis C, Alzheimer's, and Parkinson's [1]. In RNAi, exogenous small interfering RNAs (siRNAs) delivered into cells initiate the degradation of complementary messenger RNA (mRNA) molecules by the cells' internal machinery; this halts production of the proteins encoded by the mRNAs, resulting in reduced gene expression [1]. While RNAi has potential to transform our ability to treat disease, there are several challenges associated with delivering siRNAs to diseased sites for gene therapy. For example, siRNAs rapidly degrade in the presence of nucleases, have a poor biodistribution profile, and cannot passively enter cells due to their negative charge [1, 2]. To facilitate

Sarah Hurst Petrosko and Emily S. Day (eds.), *Biomedical Nanotechnology: Methods and Protocols*, Methods in Molecular Biology, vol. 1570, DOI 10.1007/978-1-4939-6840-4_1, © Springer Science+Business Media LLC 2017

passage across negatively charged cell membranes, siRNAs are typically complexed with cationic transfection agents. These cationic agents are useful for in vitro studies, but their high toxicity precludes in vivo use [3, 4]. Accordingly, researchers are developing new strategies to enable in vivo siRNA delivery, and the most common approach is to use nanoparticles as siRNA carriers [3–5]. Nanoparticles are advantageous as siRNA delivery vehicles because they can overcome several of the aforementioned challenges related to siRNA delivery [3–5]. The two main methods of siRNA delivery using nanoparticles are encapsulation, wherein siRNAs are entrapped inside porous nanoparticles or within layers of positively charged materials surrounding nanoparticles [4, 6–10], and conjugation, wherein siRNAs are bound to nanoparticle surfaces and exposed as the outer layer [5, 11–14]. Both of these methods have been shown to protect siRNAs from degradation, promote their cellular uptake, and improve gene regulation both in vitro and in vivo [6–15]. Therefore, there is substantial evidence to support continued development of nanoparticles for siRNA delivery. In this chapter, we describe the synthesis and characterization of siRNA nanocarriers prepared by the conjugation method.

Thorough and consistent characterization of siRNA nanocarriers is critical for their successful implementation as mediators of RNAi. Researchers must accurately quantify the amount of siRNA loaded within or on nanoparticle carriers to precisely dose therapies. Further, loading density is known to influence the cell uptake of nanocarriers coated with siRNA (or other nucleic acids), with high density favoring increased cell uptake [16–18], so quantitative characterization is essential to understand and enhance the interactions between siRNA-coated nanoparticles and cells. Several methods exist that could be used to qualitatively or quantitatively measure siRNA bound to or entrapped within nanoparticles. For example, zeta potential and dynamic light scattering measurements can confirm siRNA loading, as the addition of siRNA changes nanoparticles' surface charge and hydrodynamic diameter [12]. While qualitatively confirming the presence of siRNA is useful, quantification of siRNA loading provides more valuable information. One common method to quantify siRNA loading involves determining the siRNA remaining in solution after nanoparticle conjugation and purification by recording the absorbance at 260 nm (the peak absorbance of siRNA) with a spectrophotometer and calculating the amount of siRNA present using the Beer-Lambert law [19]. The strength of this method is its simplicity, but it is not ideal because it does not directly measure siRNAs on nanoparticles, assumes no losses during processing, has low sensitivity, and cannot distinguish sense- and antisense strand loading. Some alternative approaches have been developed to directly and quantitatively measure siRNAs and other nucleic acids bound to nanoparticles, including methods based on real-time polymerase chain reaction (PCR) [20] and methods that use fluorophore-labeled nucleic acids to quantify loading via fluorescence measurements [21, 22]. The main limitation of the PCR approach is its complexity, and the fluorescence-based approaches are limited by the high cost of fluorophore-labeled siRNA and potential errors in quantification that may result from fluorophores being quenched due to their close proximity to each other and the nanoparticles' surfaces. Additionally, fluorophore modifications may alter siRNA loading onto nanoparticles and, unless both strands are fluorophore labeled, fluorescence readout cannot quantify both antisense and sense RNA strands. Given that the number of sense and antisense oligonucleotides on nanoparticles may not be equal [23], it is imperative to measure both strands individually when quantifying loading.

Here, we describe methods to coat nanoparticles containing gold surfaces with thiolated siRNAs (Fig. 1) and to quantify the amount of conjugated siRNAs using a simple, fluorescence-based approach that does not require the siRNAs to be modified with fluorophores (Fig. 2). We focus on gold-based nanoparticles since these are the most extensively studied nanocarriers that utilize a surface conjugation strategy [5]. In this protocol, antisense RNA strands are first hybridized to sense RNA strands containing a 3′ thiol (*see* **Note 1**), which facilitates siRNA duplex loading on the nanoparticles' surfaces via gold–thiol bond formation (Fig. 1). Next, the conjugated siRNA is quantified as shown in Fig. 2. Briefly, the antisense strands are dehybridized from the nanoparticle-bound sense strands then the nanoparticles are

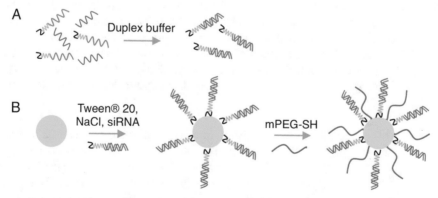

**Fig. 1** Schematic representing siRNA conjugation to gold nanoparticle (AuNP) surfaces. (**a**) siRNA duplexes are prepared by mixing antisense (*blue*) and thiolated sense (*red*) oligonucleotides in duplex buffer at a 1:1 ratio, heating to 95 °C, and cooling to 37 °C slowly over 1 h. The *black* portion of the sense strand indicates a thiol group, and the gray portion indicates a PEG spacer. (**b**) Freshly prepared siRNA duplexes are conjugated to AuNP surfaces by adding excess siRNA to AuNPs (*yellow*) suspended in dilute Tween® 20 and NaCl. siRNA loading is maximized by slowly increasing the concentration of NaCl, which screens charges between siRNA duplexes. Last, any exposed surface area is passivated with methoxy-PEG-thiol (mPEG-SH, *gray*) to stabilize the siRNA-coated AuNPs

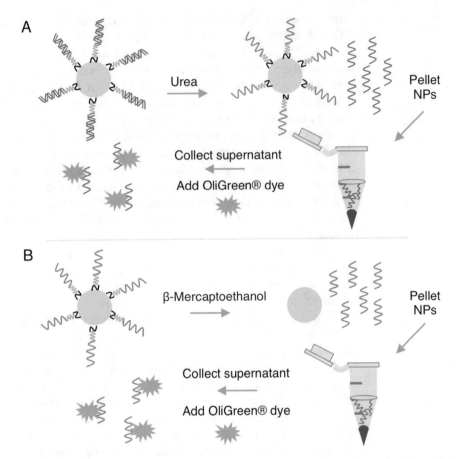

**Fig. 2** Schematic representing the procedure to quantify siRNA duplexes on nanoparticles (NPs). (**a**) Antisense RNA strands (*blue*) are dehybridized from sense RNA strands (*red*) on nanoparticles by incubating in 8 M urea at 45 °C. The sense-loaded nanoparticles are pelleted by centrifugation, and the supernatant containing the antisense strands is collected. Antisense RNA strands within the supernatant are measured using components of the Quant-iT™ OliGreen® kit. (**b**) Sense RNA strands (*red*) are removed from the nanoparticles' surfaces by breaking the gold–thiol bond with β-Mercaptoethanol. The nanoparticles are pelleted by centrifugation, and the supernatant containing the sense RNA strands is collected for analysis of RNA content with the Quant-iT™ OliGreen® kit

pelleted by centrifugation. The antisense-containing supernatant is collected and incubated with the components of the Quant-IT OliGreen® kit to produce a fluorescent signal (*see* **Note 2**). The OliGreen® dye is weakly fluorescent in solution, but produces a strong fluorescent signal that is easily detected with a fluorescent plate reader upon binding to single-stranded nucleic acids. The concentration of antisense RNA in the sample is determined by comparing the sample fluorescence intensity to that of a standard curve of known antisense RNA concentration. Next, thiolated sense RNA strands are released from the nanoparticles by breaking the gold–thiol bonds in β-mercaptoethanol. The released sense RNA strands are collected and quantified using the Quant-iT™

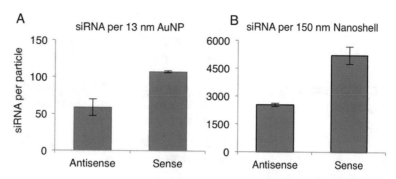

**Fig. 3** siRNA loading on nanoparticle surfaces increases with nanoparticle surface area. Representative antisense and sense RNA loading data is shown for 13 nm diameter AuNPs (**a**) and 150 nm diameter silica core/gold shell nanoshells (**b**). Data indicate mean ± standard deviation of three batches of siRNA-coated nanoparticles

OliGreen® kit. This method for quantifying siRNA duplexes loaded on nanoparticles is advantageous because it does not require siRNA that has been fluorophore-labeled, can be adapted to suit various nanoparticle core materials, and allows for individual measurement of both antisense and sense RNA strands. We have found that approximately twice as many sense strands bind to nanoparticles' surfaces as antisense strands (Fig. 3), so it is useful to obtain measurements for each sequence.

We describe this protocol using 13 nm diameter spherical gold nanoparticles (AuNPs) as the siRNA nanocarrier, but it is translatable to other types of nanoparticles and oligonucleotides as well. For example, we and other researchers have previously described how this assay can be adapted to load and measure DNA (deoxyribonucleic acid) or microRNA on 13 nm AuNPs or larger AuNPs [24–26]. To demonstrate the versatility of this protocol for different nanocarriers, we have added notes to describe how it may be adapted to the synthesis and characterization of siRNA-coated nanoshells, which are nanoparticles that contain 120 nm silica cores and 15 nm-thick gold shells (Fig. 3). The 13 nm AuNPs and 150 nm nanoshells used in this protocol are meant to serve as model systems; readers should optimize the instructions provided as necessary for their specific nanoparticle formulation and application.

# 2  Materials

Prepare all solutions using ultrapure, RNase-free water (*see* **Note 3**) and store at appropriate temperatures as indicated below. In addition to the reagents listed in this section, readers should have access to the following: −80 °C freezer, 4 °C refrigerator, microcentrifuge tubes, pipettes, black-walled 96-well plates, thermomixer, sonicator bath, rocking platform, ice, ice bucket, vortex, microcentrifuge, spectrophotometer, fume hood, and a fluorescent plate reader.

**2.1  siRNA Duplex Preparation**

1. Sense- and antisense RNA oligonucleotides (*see* **Note 4**): Dried or reconstituted single-stranded RNA should be stored at −80 °C.

2. RNase-free duplex buffer (*see* **Note 5**): Store at room temperature.

**2.2  Oligonucleotide Conjugation to AuNPs**

1. AuNPs, made RNase-free by treatment with DEPC (diethylpyrocarbonate) (*see* **Note 6**): Store at room temperature.

2. Tween® 20: Prepare a 10% solution of Tween® 20 by diluting 1 mL Tween® 20 in 9 mL water (*see* **Note 7**). Store at room temperature.

3. Sodium chloride (NaCl): Prepare 5 M NaCl solution by dissolving 58.44 g NaCl in 200 mL ultrapure water (this water does not have to be RNase-free since the final solution will be treated with DEPC). DEPC-treat the entire solution to deactivate RNases (*see* **Note 3**). Store at room temperature.

4. Methoxy PEG-thiol (mPEG-SH) at desired molecular weight (*see* **Note 8**): Store mPEG-SH as a lyophilized powder at −80 °C under argon. Fresh 1 mM solution should be made in ultrapure RNase-free water prior to use.

5. Phosphate buffered saline (PBS), 1×, RNase-free: Store at room temperature.

**2.3  Quantifying Antisense Oligonucleotides Bound to AuNPs**

1. siRNA-coated AuNPs: Prepare as described in the Methods (Subheading 3.2). Store at 4 °C.

2. 8 M Urea, RNase-free: Prepare by dissolving 12.01 g urea with 25 mL RNase-free water. Store at room temperature.

3. Tween® 20: Prepare a 0.1% solution in Ultrapure RNase-free water by diluting the previously prepared 10% stock solution 1:100.

4. Quant-iT™ OliGreen® ssDNA Assay Kit: Store at 4 °C (*see* **Note 2**).

5. Antisense oligonucleotides: Store at −80 °C. These antisense RNA strands will be used to generate a standard curve.

**2.4  Quantifying Sense Oligonucleotides Bound to AuNPs**

1. Sense strand-coated AuNPs: Prepared as described in the Methods (Subheading 3.3) and stored at 4 °C.

2. PBS, 1×, RNase-free: Store at room temperature.

3. 2 M β-Mercaptoethanol: Prepare a 2 M β-Mercaptoethanol solution by combining 0.7 μL of 14.3 M β-Mercaptoethanol with 4.3 mL TE buffer diluted 1:20 in water. (TE buffer is

10 mM Tris–HCl, 1 mM EDTA, pH 7.5 and is provided in the OliGreen® kit.)

4. Tween® 20: Prepare a 0.1% solution in ultrapure RNase-free water by diluting the previously prepared 10% stock solution 1:100. Store at room temperature.

5. Quant-iT™ OliGreen® ssDNA Assay Kit: Store at 4 °C (*see* **Note 2**).

6. Sense oligonucleotides: Store at −80 °C. These sense RNA strands will be used to generate a standard curve.

# 3   Methods

Subheadings 3.1 and 3.2 describe how to attach siRNA duplexes to AuNPs and are depicted in Fig. 1. Subheadings 3.3 and 3.4 describe how to quantify siRNA sense and antisense strands bound to AuNPs and are depicted in Fig. 2.

## 3.1   siRNA Duplex Preparation

1. Thaw sense and antisense oligonucleotides to room temperature and keep on ice throughout the procedure. Dilute oligonucleotides in RNase-free duplex buffer to the desired concentration (*see* **Note 9**).

2. Mix equal molar amounts of sense and antisense oligonucleotides in an RNase-free microcentrifuge tube (*see* **Note 10**).

3. Heat the combined oligonucleotides to 95 °C for 5 min, shaking at 400 rpm in a thermomixer.

4. Cool the oligonucleotides to 37 °C slowly over 1 h, shaking at 400 rpm (*see* **Note 11**).

5. Duplexed siRNA should be used immediately for conjugation or stored at −80 °C.

## 3.2   Oligonucleotide Conjugation to AuNPs

1. Add Tween® 20 to a final concentration of 0.2% Tween® 20 to a solution of 10 nM RNase-free 13 nm AuNPs.

2. Add NaCl to the Tween-stabilized AuNPs to a final concentration of 150 mM. Incubate 5 min at room temperature.

3. Add the previously prepared siRNA duplexes at a concentration of 1 nmole siRNA per mL of 10 nM AuNPs (*see* **Note 12**). Sonicate for 30 s using a sonicator bath, and incubate the particles at room temperature on a rocking platform at a moderate speed for 4 h.

4. After 4 h, increase the NaCl concentration of the solution to 350 mM. Sonicate again for 30 s in the sonicator bath, and incubate the particles overnight at room temperature on a rocking platform at a moderate speed (*see* **Note 13**).

5. After overnight incubation, backfill any remaining empty space on the AuNPs' surfaces with mPEG-SH by adding mPEG-SH to a final concentration of 10 μM. Incubate at room temperature on a rocking platform at a moderate speed for 4 h (*see* **Note 14**).

6. Purify the siRNA-coated AuNPs by sequential centrifugation. For 13 nm AuNPs, centrifuge at $21,000 \times g$ for 30 min to form a pellet. Remove the supernatant and dilute the AuNP pellet to half of the original volume with RNase-free 1× PBS. Repeat this procedure three times, resuspending to the desired concentration after the final wash (*see* **Note 15**).

7. To determine the concentration of the siRNA-coated AuNPs, dilute a small volume of the sample 1:100 in water and measure the extinction at 520 nm in a spectrophotometer. For 13 nm AuNPs, an extinction coefficient of $2.70 \times 10^8$ L/(mol·cm) can be used to calculate the concentration using Beer's law (*see* **Note 16**).

8. Store siRNA-coated AuNPs at desired concentration at 4 °C. We typically store siRNA-coated AuNPs at or above 100 nM.

### 3.3 Quantification of Antisense Oligonucleotides Bound to AuNPs

1. Dilute siRNA-coated AuNPs (prepared as in Subheading 3.2 and stored at 100 nM) to 6.67 nM in 8 M urea to a final volume of 300 μL in an RNase-free microcentrifuge tube. Incubate at 45 °C for 20 min in a thermomixer shaking at 400 rpm to dehybridize the antisense strands from the sense strands, which will remain bound to the AuNPs (Fig. 2).

2. Add 300 μL of 0.1% Tween® 20 to the urea-treated AuNPs, then centrifuge the solution at $21,000 \times g$ for 30 min.

3. Collect and transfer the supernatant, which contains the dehybridized antisense RNA strands, to an RNase-free tube. The AuNP pellet containing surface-bound sense strands can be stored at 4 °C until sense strand quantification.

4. Prepare a standard curve of known antisense concentration by diluting single-stranded antisense oligonucleotides in RNase-free water. The following antisense concentrations are recommended: 80, 40, 20, 10, 5, 0 nM antisense RNA (*see* **Note 17**).

5. In a black-walled 96-well plate, add 76.6 μL of each antisense RNA standard to individual wells. Add 23.4 μL of 8 M urea to each standard to match the concentration of urea in the samples prepared from siRNA-coated AuNPs. It is recommended to prepare the standard curve in duplicate or triplicate wells to minimize error.

6. In the 96-well plate, add 50 μL of each antisense sample obtained from **steps 1–3** of Subheading 3.3 to three separate wells. Dilute the samples 1:2 by adding 50 μL of RNase-free water to each well.

7. Prepare the Quant-iT™ OliGreen® reagent. First, prepare 1× TE Buffer by adding 250 μL 20× TE Buffer to 4.75 mL RNase-free water. Dilute the OliGreen® reagent 1:200 by adding 25 μL OliGreen® reagent to the 1× TE Buffer.

8. Add 100 μL of the diluted OliGreen® reagent to each well containing antisense samples obtained from the AuNPs or containing antisense RNA standards.

9. On a microplate reader, measure the fluorescence intensity of the samples using an excitation wavelength of ~480 nm and an emission wavelength of ~520 nm (*see* **Note 18**).

10. Use the fluorescence intensities for each point in the standard curve to calculate the concentration of antisense RNA in the samples prepared from siRNA-coated AuNPs (*see* **Note 19**). The number of antisense strands loaded per AuNP can be calculated by dividing the sample antisense concentration by the AuNP concentration represented in the plate, taking into account the sample dilutions throughout the procedure. Typical antisense siRNA loading on 13 nm AuNPs and on 150 nm nanoshells is provided in Fig. 3.

*3.4 Quantification of Sense Oligonucleotides Bound to AuNPs*

1. Remove the AuNP pellet obtained from **step 3** in Subheading 3.3 from the refrigerator. Gently pipette to remove all of the remaining supernatant, which contains antisense strands that have been dehybridized from the AuNPs. Be careful not to remove any of the AuNP pellet, which contains surface-bound sense RNA strands (Fig. 2).

2. Dilute the AuNP pellet in 500 μL RNase-free 1× PBS and vortex. The nanoparticles should disperse into the solution.

3. Dilute the AuNPs 1:2 by combining 250 μL sample and 250 μL RNase-free water and read the extinction at 520 nm. Calculate the AuNP concentration as previously described (Subheading 3.2, **step 7**).

4. In a fume hood, combine 100 μL of the AuNP solution from **step 2** with 100 μL of 2 M β-Mercaptoethanol diluted in 1× TE buffer in an RNase-free microcentrifuge tube (Fig. 2) (*see* **Note 20**).

5. Wrap the microcentrifuge tube in aluminum foil and incubate on a rocking platform at a moderate speed at room temperature for 24 h (*see* **Note 20**).

6. In a fume hood, remove the aluminum foil wrapper from the sample. Add 200 μL 0.1% Tween® 20 to the AuNP sample, and then centrifuge at $21,000 \times g$ for 30 min to form a pellet.

7. Working in a fume hood, collect the supernatant, which contains the sense strands that have been displaced from the AuNPs' surfaces, and place it into a new RNase-free microcentrifuge tube.

8. Prepare a standard curve of known sense oligonucleotide concentrations by diluting sense oligonucleotides in RNase-free water. The following sense concentrations are recommended: 40, 20, 10, 5, 1, 0 nM sense RNA (*see* **Note 21**).

9. Prepare the Quant-iT™ OliGreen® reagent as previously described (Subheading 3.3, **step** 7).

10. In a black-walled 96-well plate, place 100 μL of each sense standard into individual wells in a 96-well plate. It is recommended to run each standard in duplicate or triplicate wells.

11. Place 50 μL of the sense-containing supernatant obtained in **step** 7 into three individual wells. Dilute the samples 1:2 by adding 50 μL RNase-free water to each well.

12. Add 100 μL of diluted OliGreen® reagent to each well containing sense RNA samples obtained from the AuNPs or containing sense RNA standards.

13. On a microplate reader, measure the fluorescence intensity of the samples using an excitation wavelength of ~480 nm and an emission wavelength of ~520 nm (*see* **Note 18**). Compare the fluorescence intensity of the samples obtained from the AuNPs to that of the standard curve to determine the amount of sense strands as described previously (**step 10** of Subheading 3.3). Calculate the number of sense RNA strands per AuNP by dividing the sample sense concentration by the AuNP concentration determined in **step 3** of Subheading 3.4, taking into account the sample dilutions throughout the procedure. Typical sense RNA loading on 13 nm AuNPs and on 150 nm nanoshells is provided in Fig. 3.

## 4  Notes

1. We use siRNA that is thiolated at the 3′ end of the sense strand, but other positions for the thiol group would also be suitable.

2. We describe the use of the Quant-iT™ OliGreen® kit for this assay, but other dyes that are weakly fluorescent until they bind nucleic acids, which then amplifies their fluorescence, would be suitable as well. The OliGreen® dye is advantageous in that it is highly sensitive with the ability to accurately detect nucleic acids as dilute as 100 pg/mL.

3. It is imperative to use water that is nuclease-free. Ultrapure RNase/DNase-free water can be purchased from a variety of commercial vendors. Alternatively, purified water such as Milli-Q water can be made nuclease-free by adding diethylpyrocarbonate (DEPC) to the water at 0.1% (v/v), heating the solution to 37 °C for several hours, and autoclaving. We autoclave at 121 °C for 40 min for 500 mL of solution, but the time and

temperature should be adjusted based on the volume of liquid to be autoclaved.

4. We use siRNAs with a sense sequence that contains, at the 3′ end, an overhang (such as dithymidine) to increase stability against intracellular nucleases, a short polyethylene glycol (PEG) spacer to enhance loading onto nanoparticles, and a thiol group to bind the gold surface. An example siRNA sense sequence against yellow fluorescence protein (YFP) mRNA is as follows: 5′–UGA CAG UCC AAC UAC AAC AGC TT–PEG$_{36}$–SH–3′. We typically also use antisense sequences that contain a 3′ dithymidine overhang, but no additional modifications. This protocol may be adapted for use with other siRNA sense and antisense sequences, or for use with other nucleic acids (such as DNA and microRNA).

5. Upon thawing, RNA sense and antisense strands should be reconstituted in nuclease-free duplex buffer. Nuclease-free duplex buffer at pH 7.5 contains 30 mM 4-(2-hydroxyethyl)-1-piperazineethanesulfonic acid (HEPES) and 100 mM potassium acetate. Excess RNA molecules may be stored in this duplex buffer.

6. This protocol uses 13 nm AuNPs produced by the Frens method [27] and stored at a concentration of 10 nM. The protocol can easily be adapted for particles of different sizes or concentrations by scaling the reagents up or down according to the available particle surface area. To render 13 nm AuNPs RNase-free, add 0.1% v/v DEPC, shake at 37 °C for several hours and autoclave. We have observed that some types of nanoparticles, including nanoshells, become unstable upon autoclaving. In this case, we have found that treating the particles with DEPC at 37 °C for 48 h allows sufficient time for DEPC to deactivate RNases and then completely hydrolyze to prevent future side reactions.

7. We use Tween® 20 in this protocol, but other surfactants such as sodium dodecyl sulfate would also be suitable.

8. The purpose of the mPEG-SH is to stabilize the nanoparticles. We have tested mPEG-SH molecular weights ranging from 400–5000 Da. In general, higher molecular weight PEG that extends beyond the length of the siRNA duplex will increase nuclease resistance, but may prevent the siRNA from interacting with cells in downstream applications, and lower molecular weight PEG that is much shorter than the siRNA duplex may not be sufficient to prevent the nanoparticles from aggregating. We typically use either 2000 or 5000 Da mPEG-SH for backfilling nanoparticles coating with siRNAs containing ~21 nucleotides per strand. The mPEG-SH utilized should be optimized for specific nucleic acid sequences and intended applications.

9. Typical resuspension concentration for individual RNA sense and antisense oligonucleotides is ~200 μM in RNase-free duplex buffer. Store lyophilized and resuspended oligonucleotides at −80 °C.

10. All methods should be completed using RNase-free materials including microcentrifuge tubes and pipette tips. These can be purchased RNase-free (recommended), or rendered RNase-free by DEPC treatment. We recommend spraying and wiping any materials, including pipettes and the lab bench, with an RNase-removal agent to remove any RNases.

11. To cool siRNA duplexes slowly, set the thermomixer to 37 °C and mix at 400 rpm for 1 h. The siRNA duplexes will anneal as they cool to 37 °C.

12. The amount of siRNA used during synthesis will require optimization for individual siRNA sequences. We have found that 1–2 nmole siRNA per mL of 13 nm AuNPs at a concentration of 10 nM works well for most sequences. For 150 nm diameter nanoshells, 0.1–0.2 nmole siRNA per mL of nanoshells at a concentration of 0.0045 nM works well.

13. Generally, increasing the concentration of NaCl added during synthesis will increase the amount of siRNA that loads onto the AuNPs since the NaCl screens charges between duplexes. Additionally, increasing the number of NaCl additions over several hours may improve siRNA loading. This should be optimized for individual sequences and types of nanoparticles.

14. The amount of mPEG-SH added and duration of incubation should be optimized for individual siRNA sequences and types of nanoparticles. mPEG-SH concentrations have been tested in the range of 5–30 μM, while recommended incubation times range from 1–4 h. Further, incubating mPEG-SH with nanoparticles at 4 °C may improve siRNA loading.

15. After washing the AuNPs three times, suspend the AuNP pellet in a small volume for storage at your desired concentration. For example, an original volume of 5 mL 10 nM AuNPs should be resuspended in 500 μL 1× PBS after the final wash for storage at ~100 nM. We have stored siRNA-conjugated AuNPs at 100 nM for several weeks without adversely impacting the functionality of the siRNA.

16. Beer's Law states that $A = \varepsilon c l$ where $A$ is the nanoparticle absorbance at its peak resonance wavelength as measured by the spectrophotometer, $\varepsilon$ is the extinction coefficient, $c$ is the concentration of the nanoparticles, and $l$ is the path length of the sample. The extinction coefficient for 13 nm AuNPs is provided, but readers should confirm the extinction coefficient for other types of nanoparticles as they adapt this protocol to their needs.

17. The antisense standard curve will be further diluted in 8 M urea to mimic the urea concentration in the samples prepared from siRNA-coated AuNPs. It is recommended to prepare the standard curve such that the desired antisense concentrations (for example: 80, 40, 20, 10, 5, 0 nM antisense) are reached after adding 8 M urea to the standards. For example, to prepare the standard curve to reach the previously stated final concentrations, dilute antisense RNA to the following concentrations: 104.44, 52.22, 26.11, 13.05, 6.53, 0 nM antisense RNA. While these standard curve concentrations work well for the concentration and loading of siRNA on 13 nm AuNPs presented here, they may need to be optimized for different nanoparticles, nanoparticle concentrations, or loading densities. For example, the final antisense standard curve concentrations for 150 nm diameter nanoshells should be 20, 10, 5, 2, 0.5, 0 nM antisense. It is crucial for the sample fluorescence readings to be within the range of the standard curve.

18. It is recommended to use excitation ~480 nm and emission ~520 nm, but the exact wavelengths can be altered as long as the excitation and emission wavelengths do not overlap. We typically use excitation at 485 nm and emission at 515 nm.

19. If the standard curve is linear, the concentration of antisense RNA from the AuNPs can be calculated using $y = mx + b$ where $y$ is the measured fluorescence reading, $m$ is the slope of the standard curve plot, $x$ is the antisense concentration to be calculated, and $b$ is the intercept of the standard curve plot.

20. We perform this step and all other steps involving β-Mercaptoethanol in a chemical fume hood because the β-Mercaptoethanol is odorous. To minimize odor when transferring the tube containing β-Mercaptoethanol outside the hood, we wrap it in aluminum foil. Alternatively, the rocking platform could be placed in the chemical fume hood.

21. For nanoshells, we recommend using the following sense standard curve concentrations: 10, 5, 2.5, 1, 0.5, and 0 nM sense RNA.

## Acknowledgement

The authors acknowledge support from the University of Delaware Research Foundation, the W.M. Keck Foundation, Grant IRG14-251-07-IRG from the American Cancer Society, and an Institutional Development Award (IDeA) from the National Institutes of General Medical Sciences (NIGMS) of the National Institutes of Health (NIH) under grant number U54-GM104941. J.R.M. received support from a National Defense Science and Engineering Graduate Fellowship from the Department of Defense.

## References

1. Deng Y, Wang CC, Choy K et al (2014) Therapeutic potentials of gene silencing by RNA interference: Principles, challenges, and new strategies. Gene 538:217–227
2. Hill AB, Chen M, Chen C-K et al (2016) Overcoming gene-delivery hurdles: Physiological considerations for nonviral vectors. Trends Biotechnol 34(2):91–105
3. Chen J, Guo Z, Chen X (2016) Production and clinical development of nanoparticles for gene delivery. Mol Ther Methods Clin Dev 3:16023
4. Kim HJ, Kim A, Miyata K et al (2016) Recent progress in development of siRNA delivery vehicles for cancer therapy. Adv Drug Deliv Rev 104:61–77
5. Ding Y, Jiang Z, Saha K et al (2014) Gold nanoparticles for nucleic acid delivery. Mol Ther 22(6):1075–1083
6. Deng ZJ, Morton SW, Ben-Akiva E et al (2013) Layer-by-layer nanoparticles for systemic codelivery of an anticancer drug and siRNA for potential triple-negative breast cancer treatment. ACS Nano 7(11):9571–9584
7. Elbakry A, Zaky A, Liebl R et al (2009) Layer-by-layer assembled gold nanoparticles for siRNA delivery. Nano Lett 9(5):2059–2064
8. Lee J-S, Green JJ, Love KT et al (2009) Gold, poly(β-amino ester) nanoparticles for small interfering RNA delivery. Nano Lett 9(6):2402–2406
9. Dahlman JE, Barnes C, Khan OF et al (2014) In vivo endothelial siRNA delivery using polymeric nanoparticles with low molecular weight. Nat Nanotechnol 9:648–655
10. Lee JB, Hong J, Bonner DK et al (2012) Self-assembled RNA interference microsponges for efficient siRNA delivery. Nat Mater 11:316–322
11. Giljohann DA, Seferos DS, Prigodich AE et al (2009) Gene regulation with polyvalent siRNA-nanoparticle conjugates. J Am Chem Soc 131:2072–2073
12. Jensen SA, Day ES, Ko CH et al (2013) Spherical nucleic acid nanoparticle conjugates as an RNAi-based therapy for glioblastoma. Sci Transl Med 5(209):209ra152
13. Randeria PS, Seeger MA, Wang X-Q et al (2015) siRNA-based spherical nucleic acids reverse impaired wound healing in diabetic mice by ganglioside GM3 synthase knockdown. Proc Natl Acad Sci U S A 112(18):5573–5578
14. Zheng D, Giljohann DA, Chenc DL et al (2012) Topical delivery of siRNA-based spherical nucleic acid nanoparticle conjugates for gene regulation. Proc Natl Acad Sci U S A 109(30):11975–11980
15. Barnaby SN, Lee A, Mirkin CA (2014) Probing the inherent stability of siRNA immobilized on nanoparticle constructs. Proc Natl Acad Sci U S A 111(27):9739–9744
16. Choi CHJ, Hao L, Narayan SP et al (2013) Mechanism for the endocytosis of spherical nucleic acid nanoparticle conjugates. Proc Natl Acad Sci U S A 110(19):7625–7630
17. Patel PC, Giljohann DA, Daniel WL et al (2010) Scavenger receptors mediate cellular uptake of polyvalent oligonucleotide-functionalized gold nanoparticles. Bioconjug Chem 21(12):2250–2256
18. Giljohann DA, Seferos DS, Patel PC et al (2007) Oligonucleotide loading determines cellular uptake of DNA-modified gold nanoparticles. Nano Lett 7(12):3818–3821
19. Raja MAG, Katas H, Wen TH (2015) Stability, intracellular delivery, and release of siRNA from chitosan nanoparticles using different cross-linkers. PLoS One 10(6):e1028963
20. Kim E-Y, Stanton J, Vega RA et al (2006) A real-time PCR-based method for determining the surface coverage of thiol-capped oligonucleotides bound onto gold nanoparticles. Nucleic Acids Res 34(7):e54
21. Demers LM, Mirkin CA, Mucic RC et al (2000) A fluorescence-based method for determining the surface coverage and hybridization efficiency of thiol-capped oligonucleotides bound to gold thin films and nanoparticles. Anal Chem 72:5535–5541
22. McKenzie F, Steven V, Ingram A et al (2009) Quantitation of biomolecules conjugated to nanoparticles by enzyme hydrolysis. Chem Commun 2009:2872–2874
23. Randeria PS, Jones MR, Kohlsted KL et al (2015) What controls the hybridization thermodynamics of spherical nucleic acids? J Am Chem Soc 137(10):3486–3489
24. Kouri FM, Hurley LA, Daniel WL et al (2015) miR-182 integrates apoptosis, growth, and differentiation programs in glioblastoma. Genes Dev 29(7):732–745

25. Rosi NL, Giljohann DA, Thaxton CS et al (2006) Oligonucleotide-modified gold nanoparticles for intracellular gene regulation. Science 312(5776):1027–1030

26. Hill HD, Millstone JE, Banholzer MJ et al (2009) The role radius of curvature plays in thiolated oligonucleotide loading on gold nanoparticles. ACS Nano 3(2):418–424

27. Frens G (1973) Controlled nucleation for the regulation of the particle size in monodisperse gold suspensions. Nat Phys Sci 241:20–22

# Chapter 2

# Ligand Exchange and $^1$H NMR Quantification of Single- and Mixed-Moiety Thiolated Ligand Shells on Gold Nanoparticles

Ashley M. Smith and Jill E. Millstone

## Abstract

The use of nanoparticles in biomedicine critically depends on their surface chemistry. For metal nanoparticles, a common way to tune this surface chemistry is through mass action ligand exchange, where ligand exchange can be used to expand the functionality of the resulting nanoparticle conjugates. Specifically, the quantity, identity, and arrangement of the molecules in the resulting ligand shell each can be tuned significantly. Here, we describe methods to exchange and quantify thiolated and non-thiolated ligands on gold nanoparticle surfaces. Importantly, these strategies allow the quantification of multiple ligand types within a single ligand shell, simultaneously providing ligand composition and ligand density information. These results are crucial for both designing and assigning structure-function relationships in bio-functionalized nanoparticles, and these methods can be applied to a broad range of nanoparticle cores and ligand types including peptides, small molecule drugs, and oligonucleotides.

Key words Gold nanoparticles, Nanoparticle functionalization, Ligand exchange, Ligand quantification, $^1$H NMR

## 1 Introduction

Gold nanoparticles (AuNPs) are versatile materials that display ever-growing potential in fields that range from bioimaging [1, 2] to drug delivery [3, 4]. As with all NPs, their surface chemistry has a strong impact on their behavior in these applications and dictates many of their interactions in biological environments. Therefore, it is important to be able to understand and tailor this surface chemistry (e.g., control the quantity and composition of the appended ligands). Ligand quantification has been achieved in various ways, including thermogravimetric analysis [5–7], optical spectroscopy methods [8–10], and nuclear magnetic resonance spectroscopy (NMR) [11–13]. Yet, only NMR has demonstrated the necessary chemical resolution to yield information about ligand identity [12, 14], quantity [15–17], and arrangement [18, 19], even within a

Sarah Hurst Petrosko and Emily S. Day (eds.), *Biomedical Nanotechnology: Methods and Protocols*, Methods in Molecular Biology, vol. 1570, DOI 10.1007/978-1-4939-6840-4_2, © Springer Science+Business Media LLC 2017

single experiment. Moreover, postsynthetic modification of the ligand shell is typically not necessary for [1]H NMR quantification analysis. Therefore, one can analyze the *active* molecule, peptide, or protein of interest without modifications such as fluorophore labeling, which can alter the behavior of the conjugate within the system of interest and prevent correlation with the un-labeled conjugate. Here, we will describe the quantification of 1 kDa poly(ethylene glycol) methyl ether thiol (PEGSH) and 8-mercaptooctanoic acid (MOA) ligands, both as single-ligand and mixed-ligand shells. PEG-based ligands are of particular interest because they are widely used in nanobiomedicine for solubilization, stability, particle size control, and anti-biofouling. [20–23]. This method is remarkably general and may be applied to any ligand shell on a NP of interest, provided that the ligands composing the shell each have at least one spectroscopically distinct chemical shift.

## 2    Materials

Prepare all solutions using ultrapure water (resistivity 18.2 M$\Omega$·cm), unless otherwise noted. Before use, wash all glassware and Teflon-coated stir bars with aqua regia (3:1 ratio of concentrated hydrochloric acid (HCl) and nitric acid ($HNO_3$) by volume) and subsequently rinse thoroughly with water. *Caution: Aqua regia is highly toxic and corrosive and requires proper personal protective equipment. Aqua regia should be handled in a fume hood only.*

### 2.1  Gold Nanoparticle (AuNP) Synthesis [24]

1. $HAuCl_4$ solution: 1.0 mM. Weigh 0.197 grams of hydrogen tetrachloroaurate (III) trihydrate ($HAuCl_4$·3 $H_2O$) (*see* **Note 1**, chemical purity), and dissolve it in 500 mL of water in a 1-L three-neck round-bottom flask to form a pale yellow solution. Add a 1-inch Teflon-coated stir bar.

2. Trisodium citrate solution: 33.0 mM. Weigh 0.493 grams of sodium citrate tribasic dihydrate (citrate) (*see* **Note 2**, NP size control), and dissolve it in 50 mL of water to form a clear solution.

3. Reflux condenser.

4. 500 mL media bottle.

### 2.2  Citrate to Thiol Ligand Exchanges

1. 0.45 μm disposable poly(vinylidene fluoride) (PVDF) filters (25 mm GD/XP filters, Whatman, Inc.).

2. 1.5 mL centrifuge tubes.

3. Poly(ethylene glycol) methyl ether thiol (PEGSH), average MW = 1000 Da (Laysan Bio, Inc., Arab, AL, USA) solution: 5.0 mM in water (*see* **Note 3**, potential ligands). Store at 4 °C.

4. 8-mercaptooctanoic acid (MOA) (>95%, Sigma-Aldrich, St. Louis, MO, USA) solution: 1.0 mM in water. Store at 4 °C.

5. Base solution: 10 mM sodium hydroxide in water (*see* **Note 4**, role of base).

6. Deuterium oxide ($D_2O$): D, 99.9%.

*2.3 Inductively Coupled Plasma Mass Spectrometry (ICP-MS) for Au Concentration Determination*

1. Ultrapure aqua regia solution: In a 3:1 ratio, combine HCl (37 wt.% in $H_2O$, 99.999% trace metal basis) and $HNO_3$ (70%, purified by redistillation, ≥99.999% trace metal basis) to form a red, fuming solution (*see* **Note 5**, aqua regia cautionary statement).

2. 5% aqua regia matrix: Add 25 mL of concentrated ultrapure aqua regia to 475 mL of water.

3. 200 ppb Au stock solution: Dilute 2 μL of a gold standard for ICP (Fluka, TraceCERT 1001 ± 2 mg/L Au in HCl) with the 5% aqua regia matrix in a 10 mL volumetric flask.

4. 15 mL centrifuge tubes.

5. 10 mL volumetric flasks.

*2.4 ¹H NMR for Ligand Concentration Determination*

1. 5 mm borosilicate glass NMR tubes.

2. Acetonitrile (ACN) standard solution: 0.25% v/v, 15 μL of ACN in 6 mL of $D_2O$.

3. PEGSH solution: 1 mM in $D_2O$. Store at 4 °C.

4. MOA solution: 1 mM in $D_2O$. Store at 4 °C.

---

# 3    Methods

*3.1 Synthesis of 13 nm AuNPs*

1. While stirring at a rate of at least 800 rpm, bring the $HAuCl_4$ solution to a rapid reflux, with a drip rate of approximately 1 drop/second.

2. Rapidly add the citrate solution to the refluxing solution. Allow to mix for 5 min before removing from heat. Within these 5 min, the solution will change from yellow to clear to black to purple to ruby red ($\lambda_{max} = 519$ nm, *see* **Note 6**, NP colors). Cool the NP solution to room temperature before transferring to a clean 500-mL media bottle. Store at 4 °C.

*3.2 Preliminary Controls for Citrate to Thiol Ligand Exchanges*

*3.2.1 Time of Ligand Exchange*

1. Before use, filter 50 mL of the citrate-capped AuNPs through a PVDF filter. After filtration, concentrate the NPs by transferring 1.5 mL of NPs into a 1.5 mL centrifuge tube and centrifuging the solution at 20,000 rcf for 5 min. Remove the supernatant and add another 1.5 mL of NPs to the same centrifuge tube. Centrifuge again, and remove the supernatant to yield a concentrated pellet of citrate-capped NPs (*see* **Note 7**,

washing citrate-capped NPs). This yields a single tube with a concentrated particle pellet that will be used for subsequent ligand exchange and analysis. Prepare a total of 16 tubes with concentrated pellets (*see* **Note 8**, tube count).

2. Sequentially add 900 μL of water, 50 μL of PEGSH solution, and 50 μL of base solution to resuspend the concentrated pellets. Place the resulting mixtures on a temperature-controlled mixer at 1000 rpm and 25 °C. To test the time necessary for ligand exchange to proceed to completion while at a high excess (*see* **Note 9**, ligand excess), remove two tubes at each time point (here, 1, 2, 3, 4, 6, 8, 20, and 24 h after initiating ligand exchange) (*see* **Note 10**, time controls).

3. After the determined incubation time, wash the NPs by centrifuging the tubes for 5 min at 20,000 rcf. Remove the supernatant, and resuspend the pellet in 1 mL of water. Repeat this washing process for a total of two washes in water and two washes in $D_2O$. After the final centrifuge cycle in the second wash with $D_2O$, remove the supernatant to yield the concentrated pellet of PEGSH-capped AuNPs.

4. These experiments establish the necessary time for ligand exchange to proceed to a state consistent with equilibrium (Fig. 1a). Once the plot of ligand density vs. time reaches a plateau, it can be assumed that the ligand exchange has proceeded to completion. Here, a steady state is reached within 1 h, and all further experiments are conducted with an incubation time of 4 h to ensure that time is not a limiting factor in the ligand exchanges (*see* **Note 11**, completion of ligand exchange).

*3.2.2 Determining Necessary Excess Ligand Concentrations*

1. Repeat **step 1** from Subheading 3.2.1 for ten tubes with concentrated pellets of citrate-capped AuNPs.

2. After the necessary time for ligand exchange to reach a steady state is confirmed, the ligand excess with respect to NP surface area must be tested so that experiments are conducted where the ligand concentration is not the limiting factor for ligand coverage on the NP surface. Resuspend the concentrated pellets in various amounts of PEGSH to test different ligand excess amounts, all with 50 μL of base solution, added sequentially after the water and PEGSH. Here, for example, we will use an excess of 5× (944.75 μL water/5.25 μL PEGSH), 10× (939.5 μL water/10.5 μL PEGSH), 20× (929.0 μL water/21.0 μL PEGSH), 30× (918.5 μL water/31.5 μL PEGSH), and 50× (897.5 μL water/52.5 μL PEGSH). Place these tubes on a temperature-controlled mixer for 4 h.

3. After 4 h, wash the NPs as described in **step 3** in Subheading 3.2.1. After the final centrifuge cycle in the second wash with

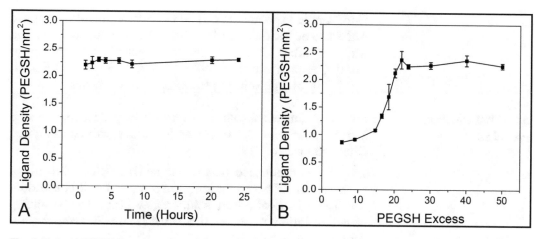

**Fig. 1** Plots of PEGSH ligand density on the AuNP as a function of time in excess (50×) PEGSH (**a**) and as a function of PEGSH excess after an incubation time of 4 h (**b**). Results indicate that maximum loading reaches a steady state on the timescale of minutes and at an excess of at least 20×. Adapted with permission from ref. 11. Copyright 2015 American Chemical Society

$D_2O$, remove the supernatant to yield the concentrated pellet of PEGSH-capped AuNPs from the various ligand excesses.

4. These samples will establish the necessary ligand excess with respect to surface area for ligand exchange to proceed to a steady state (Fig. 1b). Once the plot of ligand density vs. ligand excess reaches a plateau, a steady state of ligand density can be assumed. Here, a steady state is reached at approximately a 20× excess, and all further experiments are conducted with an excess of 50× to ensure ligand excess is not a limiting factor in the ligand exchanges (*see* **Note 12**, ligand exchange completion).

5. These experiments were completed to establish the necessary time and ligand excess for ligand exchange for the MOA ligand as well. These controls yielded results consistent with those obtained with PEGSH, and all subsequent experiments for the MOA ligand will also be conducted at a 50× excess for 4 h.

**3.3 Thiol-To-Thiol Backfilling Procedure**

1. Working with the pellet of PEGSH-capped AuNPs generated after **step 3** of Subheading 3.2.1, sequentially add 687.5 μL of water, 50 μL of base solution, and 262.5 μL of MOA solution to resuspend each pellet (*see* **Note 13**, MOA concentration). Place these tubes on a temperature-controlled mixer for 4 h (*see* **Note 14**, ligand exchange time).

2. After this time, wash the NPs as described in **step 3** in Subheading 3.2.1. After the final centrifuge cycle in the second wash with $D_2O$, remove the supernatant to yield the concentrated pellet of AuNP conjugates (*see* **Note 15**, extent of ligand exchange).

3. Comparable experiments can be conducted for MOA-capped AuNPs to be backfilled with PEGSH. In this situation, working with ten tubes with a concentrated pellet of MOA-capped AuNPs, sequentially add 897.5 μL water, 52.5 μL PEGSH, and 50 μL of base solution to resuspend each pellet.

### 3.4 Thiol Co-Loading Procedure

1. Filter and concentrate citrate-capped NPs as described in **step 1** of Subheading 3.2.1 to obtain five tubes with concentrated citrate-capped AuNPs.

2. To these concentrated pellets, sequentially add 792.5 μL of water, 50 μL of base solution, 26.25 μL of PEGSH solution, and 131.25 μL of MOA solution (*see* **Note 16**, premixing ligand solutions). Place these tubes on a temperature-controlled mixer for 4 h.

3. After this time, wash the NPs as described in **step 3** in Subheading 3.2.1. After the final centrifuge cycle in the second wash with $D_2O$, remove the supernatant to yield the concentrated pellet of mixed-moiety AuNPs capped with a mixture of PEGSH and MOA ligands (*see* **Note 17**, ligand stoichiometry).

### 3.5 ICP-MS Preparations and Method

1. Prepare 5 Au standards from the 200 ppb Au stock solution by diluting in the 5% aqua regia matrix (*see* **Note 18**, matrix considerations). Specifically, prepare the five different standards of 1, 5, 10, 20, and 30 ppb by diluting 50, 250, 500, 1000, and 1500 μL of the 200 ppb Au stock to 10 mL with the 5% aqua regia matrix in 10 mL volumetric flasks. Transfer the standards to 15 mL centrifuge tubes for storage (*see* **Note 19**, storage of standards).

2. Digest the washed pellets of thiol-capped AuNPs formed above in the ligand exchange steps with ~5 μL of fresh, concentrated aqua regia (*see* **Note 20**, digestion considerations). Allow digestion to proceed overnight (*see* **Note 21**, extent of digestion). After digestion, dilute the digested pellet to a volume of 500 μL. Then, remove 1 μL of this solution and dilute in 10 mL of the 5% aqua regia matrix.

3. Analyze the five standards by ICP-MS, measuring each standard five times and averaging to build a 5-point calibration curve (*see* **Note 22**, multi-element calibration standards). Next, analyze the unknown, digested thiol-functionalized AuNP samples, measuring each in triplicate and averaging. Use a five-minute flush time with the 5% aqua regia matrix between each run, and analyze a blank sample consisting of only 5% aqua regia between samples to confirm residual Au has been removed (*see* **Note 23**, "sticky" elements).

4. In conjunction with the AuNP size (which can be determined using transmission electron microscopy), the concentration of

Au reported by the ICP-MS will allow for calculation of the number of AuNPs in the sample. First, calculate the number of Au atoms in the particle size used (e.g., 13 nm AuNPs contain ~71,970 atoms; *see* **Note 24**, number of atoms calculation). Then, divide the number of Au atoms determined by ICP-MS by the number of Au atoms in the AuNP to find the number of AuNPs in the sample, taking into account any dilution factors.

### 3.6  $^1$H NMR Preparations and Method

1. Prepare five PEGSH standards from the 1 mM PEGSH stock solution by diluting in $D_2O$. Specifically, dilute this stock for the five different standards of 0.1, 0.25, 0.50, 0.75, and 1.0 mM by diluting 50, 125, 250, 375, and 500 μL of the 1 mM PEGSH stock to 500 μL with $D_2O$. To each standard, add 5 μL of the ACN standard solution. Mix well. Repeat these dilutions for the MOA ligand to prepare five additional MOA standards. Transfer each standard to an NMR tube.

2. Use the remainder of the digested and diluted AuNP samples from **step 2** in Subheading 3.5. Add 5 μL of the ACN standard solution to each sample and mix thoroughly (*see* **Note 25**, standard concentration). Transfer each sample to an NMR tube.

3. For all $^1$H NMR spectra to be obtained, it is recommended to apply water suppression (*see* **Note 26**, water suppression). Run the first standard sample on an NMR to obtain the $^1$H spectrum (*see* **Note 27**, number of scans and signal-to-noise ratio). Integrate the ACN standard peak and the most prominent ligand peak (Fig. 2a). Repeat for the remaining standards to generate a calibration curve for both PEGSH and MOA, where ligand concentration is plotted against the integrated ratio of ligand/ACN (Fig. 2b). From this plot, a linear equation can be

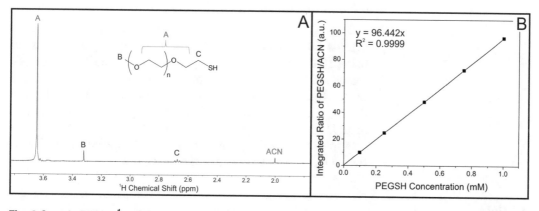

**Fig. 2** Sample PEGSH $^1$H NMR spectrum displaying integrated peaks labeled in *red* (peak A from the PEGSH and ACN) (**a**), and a sample calibration curve for PEGSH generated by plotting the PEGSH concentration against the ratio of PEGSH to ACN integrated peak intensities (where "peak" refers to the selected peaks shown in (**a**)) (**b**). The equation of the line obtained from the calibration curve allows for the calculation of unknown ligand concentrations. Adapted with permission from ref. 11. Copyright 2015 American Chemical Society

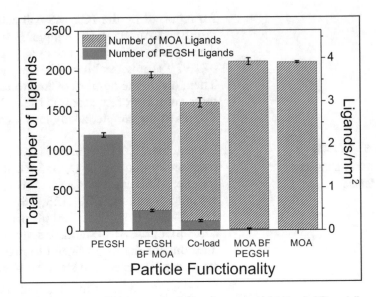

**Fig. 3** Comparison of the amount of ligands appended to a AuNP and ligand density for both single moiety (PEGSH and MOA) and mixed moiety (backfilled (BF) and co-loaded) ligand exchanges. In both backfilling and co-loading, the MOA dominates the ligand shell. Adapted with permission from ref. 11. Copyright 2015 American Chemical Society

obtained that will allow for the evaluation of the ligand concentration in the unknown samples.

4. After obtaining the $^1$H spectrum of each unknown, thiol-functionalized AuNP sample, integrate the ACN standard peak and the same ligand peak as was integrated for the relevant standards above. Using the equation from the calibration curve, insert the ratio of ligand/ACN integrated peak intensities to solve for the unknown ligand concentration in the sample.

5. Calculate ligand/particle numbers by dividing the number of ligands found using NMR by the number of particles found using ICP-MS. Ligand density in units of ligands/nm$^2$ can be calculated by dividing the number of ligands per particle found using NMR by the surface area of a single NP (Fig. 3). Ligand density can also be calculated by dividing the total number of ligands found using NMR by the total amount of NP surface area in the sample as determined by NP concentration. These two density calculations give similar values.

## 4    Notes

1. Using pure reagents can help to improve both the yield and monodispersity of the resulting NPs. Thus, it is recommended to use the highest purity reagents available for both ligand

exchange and NP synthesis. Polydispersity in the ligand MW (chain length) can impact both ligand exchange and NP synthesis [25].

2. The AuNP size can be controlled by altering the Au:citrate molar ratio used in the particle synthesis. For 13 nm AuNPs, a 1:3.35 Au:citrate molar ratio was used. By increasing this ratio, larger NPs can be synthesized. Mole ratios of Au:citrate higher than 1 are not recommended as they do not produce stable, monodisperse NPs.

3. Here, 1 kDa PEGSH and MOA are used in the ligand exchanges. Other water-soluble thiolated ligands can also be appended to the NPs by following the same procedures described here.

4. The base solution is added to the ligand exchanges to prevent multilayer formation due to hydrogen bonding between the terminal carboxylic acids on the MOA. While it is not necessary to use the base with the PEGSH samples, it is added to all ligand exchanges for consistency.

5. Aqua regia is highly toxic and corrosive. Wear proper personal protective equipment and handle only in a fume hood. Do not mix with organic solvents and always allow the solution to vent—gasses evolve continuously upon mixing the HCl and $HNO_3$ solutions.

6. While not an absolute indicator, visual color inspection can be useful in monitoring the AuNPs. Citrate-capped 13 nm AuNPs will appear ruby red; as the particle size increases, the red will darken to purple. Additionally, if at any time during subsequent ligand exchanges the color dramatically changes (e.g., from red to purple or black), the NPs have irreversibly aggregated and will not be suitable for either further use or robust analysis.

7. Avoid over-washing the citrate-capped AuNPs. Because citrate is weakly bound to the AuNPs, washing more than three times will cause the NPs to begin to aggregate.

8. The number of tubes necessary for each sample will vary depending on NP concentration, NP size, and ligand identity. The number of tubes contained herein is appropriate for the PEGSH and MOA system; the number can be increased or decreased as necessary.

9. These concentrations of PEGSH are based on the ligand excess with respect to NP surface area. See ref. 11 for a detailed explanation and sample calculations.

10. The time necessary for ligand exchange can vary depending on the ligand identity. It is therefore recommended to complete a time analysis for all ligand types studied.

11. If the ligand exchanges are not given enough time to proceed to completion, the final ligand shell will be more variable in terms of both composition and ligand quantity from batch to batch. To ensure accurate and consistent particle preparation, allow enough time such that the plot of ligand density vs. time of ligand exchange has reached a plateau before continuing with further ligand studies.

12. Similar to time of exchange being a limiting factor, a limiting ligand excess will also produce variable particle conjugates with inconsistent numbers of ligands.

13. Using this concentration of MOA will yield roughly a $50\times$ ligand excess with respect to particle surface area (depending on exact NP concentrations). Here, we used a $50\times$ excess out of an abundance of caution to remain on the plateau region in Fig. 1b, however a $50\times$ excess is not necessary to achieve full ligand loading as indicated by the surface coverage plot in Fig. 1b. The ligand concentration mentioned in this step can be adjusted depending on desired excess.

14. The time for ligand exchange can and should be altered, depending upon the necessary time for the ligand exchange to reach completion. As shown in Fig. 1a, the ligand loading using the ligands described here reaches a steady state within an hour, so any time frame longer than this period can be used. We use 4 h as a compromise between experiment expediency and an abundance of caution.

15. The ligand exchange of PEGSH for MOA is unlikely to completely displace all of the original ligand present on the NP surface (here, PEGSH). This incomplete exchange is exceptionally likely in cases where the particle binding moiety is the same for both the ligand on the particle and the ligand that is being added (as is the case here). In this case, where we seek to exchange PEGSH appended to the particle with a new ligand, MOA, we have shown that the displacement is not complete, and under the conditions listed produced a particle with a ligand shell that is 13% PEGSH and 87% MOA. The degree of ligand displacement depends strongly on the particle binding moiety of the two ligands, as well as the total ligand architecture. The degree of displacement must be determined for every ligand combination studied. Luckily, the composition of the ligand shell is readily extracted from the NMR experiment described herein and does not require any additional analytical steps or procedures. Indeed, there is no intrinsic limit on the number of different ligands that can be identified in a single NMR analysis, provided each ligand has at least one spectroscopically distinct chemical shift and is present in a quantity above the detection limit of NMR.

16. It is recommended to premix the ligand solutions before addition to the AuNPs to ensure that neither one has additional time for ligand exchange with respect to the other. Additionally, these concentrations are to co-load the ligands at a $50\times$ total ligand excess, with a $25\times$ excess for each ligand and at a 50:50 ratio. This ratio can be altered to yield final ligand shells with different compositions [26] (*see* also **Note 17**).

17. When attempting to functionalization a Au NP with two ligands simultaneously (i.e., "co-loading"), it is natural to expect that the stoichiometry of the ligands added will be reflected in their final composition ratios on the NP. Unfortunately, the relationship is much more complex and depends on both ligand–ligand interactions in solution (e.g., ligands may form small aggregates or "rafts," as in the case of lipids), as well as the affinity of each ligand for the NP surface. The relationship between the stoichiometry of ligands added and the stoichiometry appended to the particle after ligand introduction must be determined for every new ligand combination to predict the relationship between the amount of ligand added and the amount of ligand that ultimately binds to the particle. However, two points are important to note, despite these challenges. First, on-particle ligand compositions *can* be tuned by changing the stoichiometry of the ligands added, it just may not directly match the ratio added. Second, the NMR analysis method can be used, in all cases, to elucidate what the relationship is between ligand composition added and ligand composition appended.

18. An aqua regia matrix is used since it is effective for digesting Au. However, some metals (e.g., silver) are incompatible with this matrix (since silver chloride will precipitate out in the presence of the chloride ions). In these situations, the matrix can be altered. For example, in the case of silver, a 5% nitric acid matrix can be used.

19. For best results, prepare fresh standards for ICP analysis every day, as any matrix evaporation or metal adsorption on containment vessels will alter the standard concentrations.

20. Use only a small ($<10~\mu L$) amount of aqua regia in the digestion, as samples with a high ionic strength are difficult to tune on the NMR.

21. After digestion, the Au samples will be a pale yellow color. Black or purple specks indicate that the digestion is not complete. Sonicating the tubes or placing them on the temperature-controlled mixer at ~35 °C can aid in the digestion.

22. Calibration curves will be needed for all elements being analyzed on the ICP-MS. As long as there are no inter-element interferences, multi-element standards can be used.

23. Certain elements, including Au and to lesser degree Ag, can be "sticky" in the ICP-MS, leaving residual Au in the tubing and internal components that will be reflected in subsequent measurements. Long flush times and analyzing a blank will help to confirm removal of the residual metals. Analyzing samples at lower concentrations will also help to reduce the problem of sticking, but if longer (>5 min) flush times do not yield clean blank samples, a flush with 2% Triton detergent solution can aid in the removal of sticky metals.

24. The number of atoms in a pseudospherical AuNP can be estimated by using the diameter of the NP to find the NP volume and dividing this volume by the volume of a Au unit cell [11]. Then, knowing that there are 4 Au atoms/unit cell, the number of atoms/particle is obtained.

25. While here we add 5 μL of ACN standard solution to both each ligand standard as well as each ligand sample, this concentration can be adjusted so that the internal standard peak is proportional in intensity to ligand peaks of interest.

26. Water suppression for the $^1$H NMR spectra is useful when the AuNPs are synthesized and undergo ligand exchange in water. Even after multiple washes in $D_2O$, residual water will remain; this water will contribute to a peak at 4.7 ppm. Depending on the amount of water, the peak can be large, obscuring the baseline for the relevant ligand peaks. A water suppression pulse sequence minimizes this peak, allowing for a smoother baseline and more accurate integration of the ligand peaks.

27. Depending on sample concentrations, more or less scans can be run on each sample. For the concentrations contained herein, 64 scans should give sufficient resolution for a minimum signal-to-noise ratio of 10.

# Acknowledgment

This work was supported by the National Science Foundation (CHE-1253143) and the University of Pittsburgh.

# References

1. Huang X, Jain PK, El-Sayed IH, El-Sayed MA (2007) Gold nanoparticles: interesting optical properties and recent applications in cancer diagnostics and therapy. Nanomedicine 2:681–693

2. Murphy CJ, Gole AM, Stone JW, Sisco PN, Alkilany AM, Goldsmith EC, Baxter SC (2008) Gold nanoparticles in biology: beyond toxicity to cellular imaging. Acc Chem Res 41:1721–1730

3. Ghosh P, Han G, De M, Kim CK, Rotello VM (2008) Gold nanoparticles in delivery applications. Adv Drug Deliv Rev 60:1307–1315

4. Giljohann DA, Seferos DS, Daniel WL, Massich MD, Patel PC, Mirkin CA (2010) Gold nanoparticles for biology and medicine. Angew Chem Int Ed 49:3280–3294

5. Hostetler MJ, Wingate JE, Zhong C-J, Harris JE, Vachet RW, Clark MR, Londono JD, Green SJ, Stokes JJ, Wignall GD (1998) Alkanethiolate gold cluster molecules with core diameters from 1.5 to 5.2 nm: core and monolayer properties as a function of core size. Langmuir 14:17–30

6. Takae S, Akiyama Y, Otsuka H, Nakamura T, Nagasaki Y, Kataoka K (2005) Ligand density effect on biorecognition by PEGylated gold nanoparticles: regulated interaction of RCA120 lectin with lactose installed to the distal end of tethered PEG strands on gold surface. Biomacromolecules 6:818–824

7. Corbierre MK, Cameron NS, Sutton M, Laaziri K, Lennox RB (2005) Gold nanoparticle/polymer nanocomposites: dispersion of nanoparticles as a function of capping agent molecular weight and grafting density. Langmuir 21:6063–6072

8. Demers LM, Mirkin CA, Mucic RC, Reynolds RA, Letsinger RL, Elghanian R, Viswanadham G (2000) A fluorescence-based method for determining the surface coverage and hybridization efficiency of thiol-capped oligonucleotides bound to gold thin films and nanoparticles. Anal Chem 72:5535–5541

9. Xia X, Yang M, Wang Y, Zheng Y, Li Q, Chen J, Xia Y (2012) Quantifying the coverage density of poly(ethylene glycol) chains on the surface of gold nanostructures. ACS Nano 6:512–522

10. Hurst SJ, Lytton-Jean AK, Mirkin CA (2006) Maximizing DNA loading on a range of gold nanoparticle sizes. Anal Chem 78:8313–8318

11. Smith AM, Marbella LE, Johnston KA, Hartmann MJ, Crawford SE, Kozycz LM, Seferos DS, Millstone JE (2015) Quantitative analysis of thiolated ligand exchange on gold nanoparticles monitored by 1H NMR spectroscopy. Anal Chem 87:2771–2778

12. Anderson NC, Hendricks MP, Choi JJ, Owen JS (2013) Ligand exchange and the stoichiometry of metal chalcogenide nanocrystals: Spectroscopic observation of facile metal-carboxylate displacement and binding. J Am Chem Soc 135:18536–18548

13. Valdez CN, Schimpf AM, Gamelin DR, Mayer JM (2014) Low capping group surface density on zinc oxide nanocrystals. ACS Nano 8:9463–9470

14. Hens Z, Martins JC (2013) A solution NMR toolbox for characterizing the surface chemistry of colloidal nanocrystals. Chem Mater 25:1211–1221

15. Crawford SE, Andolina CM, Smith AM, Marbella LE, Johnston KA, Straney PJ, Hartmann MJ, Millstone JE (2015) Ligand-mediated "turn on," high quantum yield near-infrared emission in small gold nanoparticles. J Am Chem Soc 137:14423–14429

16. Becerra L, Murray C, Griffin R, Bawendi M (1994) Investigation of the surface morphology of capped CdSe nanocrystallites by 31P nuclear magnetic resonance. J Chem Phys 100:3297–3300

17. Sachleben JR, Wooten EW, Emsley L, Pines A, Colvin VL, Alivisatos AP (1992) NMR studies of the surface structure and dynamics of semiconductor nanocrystals. Chem Phys Lett 198:431–436

18. Liu X, Yu M, Kim H, Mameli M, Stellacci F (2012) Determination of monolayer-protected gold nanoparticle ligand–shell morphology using NMR. Nat Commun 3:1182

19. Guarino G, Rastrelli F, Scrimin P, Mancin F (2012) Lanthanide-based NMR: A tool to investigate component distribution in mixed-monolayer-protected nanoparticles. J Am Chem Soc 134:7200–7203

20. Jokerst JV, Lobovkina T, Zare RN, Gambhir SS (2011) Nanoparticle PEGylation for imaging and therapy. Nanomedicine 6:715–728

21. De Jong WH, Borm PJ (2008) Drug delivery and nanoparticles: applications and hazards. Int J Nanomedicine 3:133

22. Boisselier E, Astruc D (2009) Gold nanoparticles in nanomedicine: preparations, imaging, diagnostics, therapies and toxicity. Chem Soc Rev 38:1759–1782

23. Soppimath KS, Aminabhavi TM, Kulkarni AR, Rudzinski WE (2001) Biodegradable polymeric nanoparticles as drug delivery devices. J Control Release 70:1–20

24. Frens G (1973) Controlled nucleation for the regulation of the particle size in monodisperse gold suspensions. Nature 241:20–22

25. Marbella LE, Andolina CM, Smith AM, Hartmann MJ, Dewar AC, Johnston KA, Daly OH, Millstone JE (2014) Gold-cobalt nanoparticle alloys exhibiting tunable compositions, near-infrared emission, and high T2 relaxivity. Adv Funct Mater 24:6532–6539

26. Smith, A.M., Johnston, K.A., Marbella, L.E., Millstone, J.E. (2016) unpublished results

# Chapter 3

# Nanoparticle Tracking Analysis for Determination of Hydrodynamic Diameter, Concentration, and Zeta-Potential of Polyplex Nanoparticles

David R. Wilson and Jordan J. Green

## Abstract

Nanoparticle tracking analysis (NTA) is a recently developed nanoparticle characterization technique that offers certain advantages over dynamic light scattering for characterizing polyplex nanoparticles in particular. Dynamic light scattering results in intensity-weighted average measurements of nanoparticle characteristics. In contrast, NTA directly tracks individual particles, enabling concentration measurements as well as the direct determination of number-weighted particle size and zeta-potential. A direct number-weighted assessment of nanoparticle characteristics is particularly useful for polydisperse samples of particles, including many varieties of gene delivery particles that can be prone to aggregation. Here, we describe the synthesis of poly(beta-amino ester)/deoxyribonucleic acid (PBAE/DNA) polyplex nanoparticles and their characterization using NTA to determine hydrodynamic diameter, zeta-potential, and concentration. Additionally, we detail methods of labeling nucleic acids with fluorophores to assess only those polyplex nanoparticles containing plasmids via NTA. Polymeric gene delivery of exogenous plasmid DNA has great potential for treating a wide variety of diseases by inducing cells to express a gene of interest.

**Key words** Nanoparticle tracking analysis, Hydrodynamic diameter, Zeta-potential, Concentration, DNA labeling

## 1 Introduction

Nanoparticle-based therapies have the potential to cure many difficult-to-treat or undruggable diseases by directly modulating cell expression of genes via delivery of deoxyribonucleic acid (DNA) or small interfering ribonucleic acid (siRNA) molecules. In particular, polymeric nanoparticles have great promise as gene regulatory agents for their ability to efficiently transfect a wide variety of cell types and their ability to be specifically targeted to certain cell populations, offering increased specificity for treating diseases such as cancer [1]. Polymeric nanoparticles for gene delivery, termed polyplexes, generally form via self-assembly due to electrostatic interactions between cationic polymers and anionic nucleic

Sarah Hurst Petrosko and Emily S. Day (eds.), *Biomedical Nanotechnology: Methods and Protocols*, Methods in Molecular Biology, vol. 1570, DOI 10.1007/978-1-4939-6840-4_3, © Springer Science+Business Media LLC 2017

acids, with condensation to discrete particles further driven by the hydrophobic effect [2]. Consequently, polyplex particle distributions can be polydisperse in diameter, making accurate characterization a challenge [3].

Traditionally, characterization methods for polyplex nanoparticles have included dynamic light scattering (DLS), transmission electron microscopy (TEM), atomic force microscopy (AFM), and fluorescence correlation spectroscopy [4]. Of these techniques, dynamic light scattering has been the most widely applied technique for determining hydrated nanoparticles' properties. DLS utilizes photon correlation spectroscopy to yield intensity-weighted average results. For monodisperse nanoparticle solutions, DLS produces an accurate estimate of the nanoparticles' hydrodynamic diameter; however, for polydisperse nanoparticle populations the hydrodynamic diameter determined by DLS is less accurate because DLS is biased towards larger particles within the sample [5, 6]. Nanoparticle tracking analysis (NTA), which was developed in the mid-2000s, offers an alternative nanoparticle characterization technique in which individual particles are tracked to yield number-weighted average hydrodynamic diameter values. These number-weighted averages do not disproportionately represent larger nanoparticles in the solution, rendering NTA more accurate than DLS for characterizing heterogeneously sized materials such as polyplex nanoparticles. The difference between number-weighted and intensity-weighted measurements of nanoparticles' properties can be substantial, as shown in Fig. 1 [4, 5]. Therefore, it is advantageous to use NTA rather than DLS when analyzing polydisperse nanoparticle solutions. NTA also enables zeta-potential and nanoparticle concentration to be determined, allowing for unique applications such as determining the number of plasmids per polymeric nanoparticle [8, 9]. This information is important for gene delivery applications, as it allows researchers to correlate loading with gene regulation efficiency.

In this chapter, the methods to characterize polymeric polyplex gene delivery nanoparticles by NTA are described. As an example, we detail how to synthesize a poly(beta-amino ester) (PBAE) that has been shown to be effective for the transfection of a variety of cell types, and how to use NTA to analyze the polyplex nanoparticles formed by complexation of PBAE with plasmid DNA [10–13]. Considerations for determining polyplex nanoparticle size and concentration as well as zeta-potential are presented. Methods for labeling plasmid DNA with a fluorophore in order to differentiate nanoparticles containing DNA from nanoparticles that do not contain DNA are also presented. It is important to specifically analyze plasmid-containing nanoparticles as these are the nanoparticles responsible for successful gene delivery and they may have different properties from the polymeric nanoparticles that do not contain DNA [8, 14].

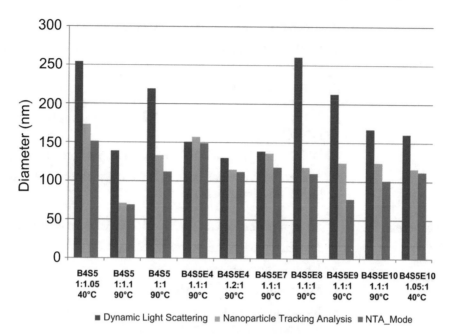

■ Dynamic Light Scattering  ■ Nanoparticle Tracking Analysis  ■ NTA_Mode

**Fig. 1** A comparison of DLS and NTA shows the difference in measured nanoparticle sizes resulting from intensity-weighted versus number-weighted measurements obtained by DLS and NTA, respectively. Here, each grouping represents a different type of nanoparticle prepared from different polymer precursors. The measurements shown from left to right for each polymeric nanoparticle formulation are intensity-weighted z-average particle size by DLS, mean number-weighted particle size by NTA, and the mode number-weighted particle size by NTA; the NTA mode measurement refers to the peak of the particle size histogram in a number-weighted size distribution. For polydisperse nanoparticle populations, larger aggregate particles are disproportionally represented in DLS, but not in NTA. Reproduced with permission from [7]

# 2  Materials

Readers are assumed to have access to basic wet lab equipment and materials such as micropipettes and microcentrifuge tubes. Please follow all waste-disposal guidelines at your institution for proper disposal of used materials.

*2.1  Polymer Synthesis*

1. 1,4-Butanediol diacrylate (B4).

2. 4-amino-1-butanol (S4).

3. 1-(3-aminopropyl)-4-methylpiperazine (E7).

4. Anhydrous tetrahydrofuran (THF).

5. Anhydrous diethyl ether.

6. Anhydrous Dimethyl sulfoxide (DMSO).

7. Teflon-lined screw cap glass scintillation vials, 5 mL.

8. Teflon-coated magnetic micro stir bars.

9. Magnetic stir plate.

10. Incubator/oven capable of reaching 90 °C.

11. Centrifuge capable of reaching at least 3000 rcf for 15 mL centrifuge tubes.

12. Vacuum chamber.

**2.2 Polyplex Nanoparticle Formation**

1. Polymer from Subheading 3.1 at 100 μg/μL in anhydrous DMSO.

2. DNA at 1 μg/μL in DNAse- and RNAse-free $H_2O$.

3. 3 M sodium acetate buffer (NaAc) (*see* **Note 1** to make NaAc buffer from salt form).

**2.3 Hydrodynamic Diameter and Concentration Measurements**

1. Nanoparticles formed in 25 mM NaAc from Subheading 3.2.

2. 150 mM phosphate buffered saline (PBS).

3. 1.5 mL microcentrifuge tubes.

4. Micropipettes and pipette tips

5. NanoSight NS500, NS300 or LM10 (*see* **Notes 1** and **2** regarding hardware and software version).

6. NanoSight syringe pump (optional; *see* **Note 3** regarding benefit).

**2.4 Zeta-Potential Measurements**

1. Nanoparticles formed in 25 mM NaAc from Subheading 3.2.

2. 10 mM NaCl.

**2.5 Fluorescent Particle Analysis**

1. Plasmid DNA.

2. Cy3-$NH_2$.

3. NHS Psoralen (succinimidyl-[4-(psoralen-8-yloxy)]-butyrate).

4. 1 M HEPES buffer, pH 8.5 (*see* **Note 4** for making buffer from salt form (required sodium hydroxide pellets as well)).

5. UV Lamp (365 nm).

6. Microcentrifuge.

7. 95% Ethanol.

8. 3 M sodium acetate buffer (NaAc) (*see* **Note 1** for making NaAc buffer from salt form).

9. Small volume spectrophotometer.

10. NanoSight syringe pump.

11. 1 mL disposable plastic syringe.

# 3 Methods

**3.1 Polymer Synthesis**

The following section specifically details the synthesis of 100 mg of a poly(beta-amino ester) (PBAE) demonstrated in previous

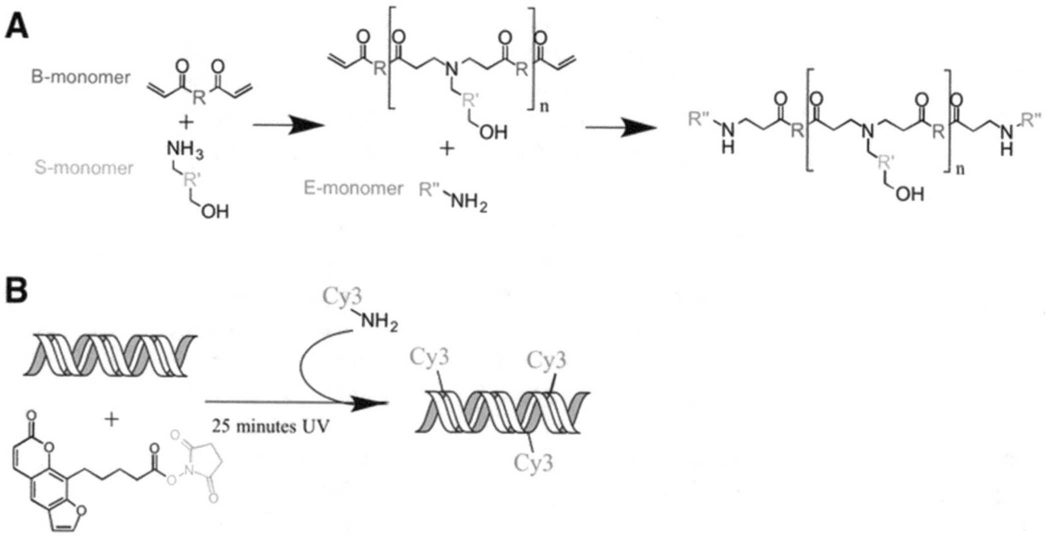

Fig. 2 (a) PBAE synthesis from B and S monomers with E monomer endcapping. (b) Nucleic acid labeling via NHS-psoralen UV crosslinking with DNA followed by reaction of the NHS moiety with a primary-amine containing fluorophore

publications to be highly effective for the transfection of a variety of cell types [10–13]. Figure 2a shows a schematic that describes the reaction to form an end-capped PBAE. The amount of polymer synthesized can be increased by linearly scaling the amounts of the reagents used.

1. **Base polymer synthesis**. Weigh 71 mg of monomer B4 into a tared 5 mL screw cap glass scintillation vial via micropipette. Re-tare and add 29 mg of monomer S4 for a stoichiometric ratio between acrylate/amine monomers of 1.1:1 (*see* **Note 5** regarding molar ratios when synthesizing PBAEs). Add a Teflon-coated magnetic stir bar to the vial and stir at approximately 500 rpm for 24 h at 90 °C.

2. **Base polymer endcapping**. Dissolve base polymer to 166.7 μg/μL by adding 600 μL anhydrous THF to the vial from **step 1** of Subheading 3.1 and vortex with the cap on (*see* **Note 6** regarding endcapping solvent and ether precipitation). Weigh 31.5 mg of monomer E7 into a tared 5 mL screw cap glass scintillation vial and dissolve in 400 μL THF to make a 0.5 M solution. Add the E7 solution in THF to the base polymer solution and stir for 1 h at room temperature.

3. **Purification via ether precipitation**. Record mass of empty 15 mL centrifuge tube then transfer the 1 mL solution of endcapped polymer and add at least 5× volume of anhydrous diethyl ether. Vortex vigorously to allow excess endcapping molecules to be removed from polymer. Centrifuge at

minimum 3000 rcf for 5 min to precipitate the polymer. Remove supernatant ether/THF solution and repeat **step 3** for a total of two ether precipitations. Place the centrifuge tube containing the polymer in a vacuum chamber at room temperature for 24–48 h to allow for residual diethyl ether to evaporate.

4. **Polymer storage**. When residual ether can no longer be detected in the polymer, determine the final mass of the polymer and dissolve to 100 μg/μL in anhydrous DMSO by adding the necessary volume of DMSO and vortexing. Aliquot polymer into microcentrifuge tubes and store at −20 °C with desiccant (*see* **Note 7** regarding storage).

**3.2 Polyplex Nanoparticle Formation**

This section describes the preparation of PBAE polyplex nanoparticle formulations. While PBAE/DNA polyplexes are prepared in sodium acetate buffer (NaAc), other polyplexes are to be prepared for transfection in their respective buffers (i.e. 150 mM NaCl for polyethylenimine/DNA or 10 mM HEPES for polylysine/DNA). Buffers ensure that the pH of the nanoparticle formation solution is appropriate to enable the amine-containing polymer to be sufficiently positively charged to complex with the anionic DNA.

1. **Reagent preparation**. Dilute 3 M NaAc to 25 mM pH 5.0 using ultrapure water. Prepare DNA to a concentration of 1 μg/μL in unbuffered ultrapure $H_2O$. *See* **Note 1** for preparation of NaAc buffer from salt form.

2. **Reagent dilution**. Dilute DNA to 0.06 μg/μL in 25 mM NaAc pH 5.0 using micropipettes for a minimum volume of 10 μL for each intended measurement to be taken. Thaw a polymer aliquot and dilute to 3.6 μg/μL or 1.8 μg/μL in 25 mM NaAc for 60 w/w or 30 w/w ratio of polymer/DNA polyplex nanoparticles respectively. Ensure polymer is entirely dissolved by vortexing vigorously.

3. **Polyplex formation**. Mix 10 μL each of polymer:DNA solutions 1:1 and incubate for 10 min at room temperature to allow for polyplex nanoparticle formation. Polyplex nanoparticles are utilized directly after self-assembly without a purification step.

**3.3 Hydrodynamic Diameter and Concentration Measurements**

These methods describe how to characterize polyplex nanoparticles using a NanoSight NS500 with version 3.0+ software. For general information on how to operate NS500 software and hardware or how to use other versions refer to Malvern published guides [15]. Users unfamiliar with NanoSight software are encouraged to refer to Appendix Fig. A.1, which outlines rudimentary software features.

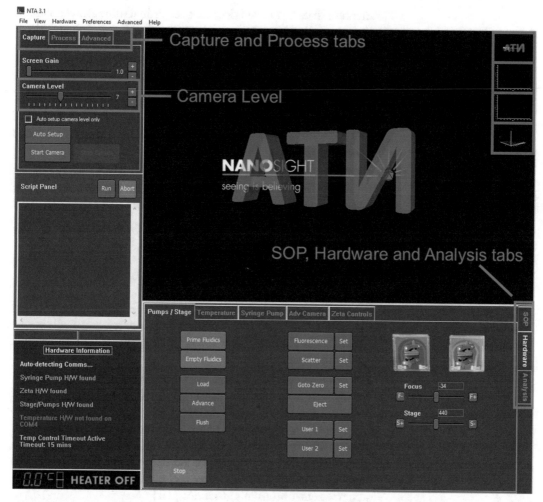

**Fig. A.1** NanoSight software overview showing the main tabs for Capture and Process in the *top right* with the camera level highlighted and the main tabs for SOP, Hardware and Analysis in the *bottom left*

1. **NanoSight preparation**. Prime the NanoSight with Millipore water and ensure that the sample chamber is clean. Move the sample stage to scatter position.

2. **Polyplex preparation**. Dilute polyplex nanoparticles in 150 mM PBS 1:1000 to make 1 mL of diluted nanoparticle solution. A two-step dilution is recommended to ensure dilution volumes are accurate (*See* **Note 8** for recommended dilutions to try for various polyplex nanoparticles).

3. **Dilution factor determination**. Load the diluted polyplex nanoparticles into the NanoSight sample chamber and start the camera. Increase the camera level and assess the number of particles per frame. (*See* **Note 9** regarding finding optimal dilution and camera level to avoid biasing particle tracking to larger particles.) To do this, increase dilution factor until an

increase in camera level does not result in a dramatic increase in the number of particles able to be visualized. Ideally, there should 20–80 particles per frame of the video as this enables enough particles to be tracked in a 60 s capture and is not beyond the tracking capability of the computer hardware.

4. **Video capture**. Under the SOP Tab take a standard measurement of 3 captures of 60 s each, which should be sufficient if you have the required number of particles per frame (*see* **Note 10** regarding optimal capture length).

5. **Analysis**. After the capture completes, scroll through the video using the frame slider and under the Process tab adjust the detection threshold to avoid false positives and false negatives.

6. **Data export**. Following processing, export results to a PDF and optionally .csv files for *\*particledata* and *\*experimentsummary* (*see* **Note 11** regarding .csv file content and methods of processing to yield raw particle data for analysis).

*3.4  Zeta-Potential Measurements*

1. **NanoSight and polyplex preparation**. Prime the NanoSight using the standard setup protocol and prepare polyplexes as detailed in Subheading 3.2 by mixing diluted polymer and DNA in 25 mM NaAc.

2. **NanoSight focusing**. In the zeta-potential tab, go to the zero position and reduce the camera level enough to ensure that background scatter does not saturate the detector, visible as large, colored circles with halos from the nanoparticles you are characterizing. Bring the background glass scatter into focus as best as possible and assess the centering of the thumbprint then return to zeta position 1. Refer to Malvern's software guide for general zeta-potential guidelines [15].

3. **Polyplex dilution**. Dilute polyplex nanoparticles in 10 mM NaCl approximately 1:1000 to yield 20–80 particles per frame, as recommended by Malvern. Load particles into the Nano-Sight as done in **step 2** of Subheading 3.3 (*see* **Note 12** regarding diluents for zeta-potential measurements).

4. **Focus adjustment**. Go to zeta position 1 and assess focus level. Adjust focus at position 1 if needed. Then move through positions 2–5 to assess that the beam of focus forms a horizontal or near-horizontal line across the screen. The focus of individual particles will worsen with an increase in position number but adjusting of focus level is not recommended. A slightly higher than normal camera level may be required to still be able to see particles at position 5.

5. **Video capture**. Take a zeta-potential measurement with 90 s duration and 30 s secondary duration with 24 V applied and the temperature held constant at 25 or 37 °C. For nanoparticles, 37 °C is generally more physiologically relevant.

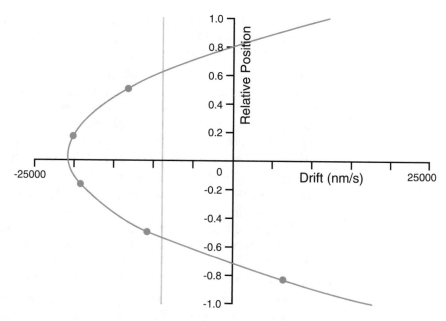

**Fig. 3** The electrophoretic curve should have a parabolic shape as shown opening either to the right or left sides for negative or positive zeta-potentials respectively

6. **Analysis**. For each capture, adjust detection threshold as needed, scrolling through the video to assess multiple frames.

7. **Data export and assessment**. Following data analysis, assess the quality of the zeta-potential measurement by examining the parabolic character of electrophoretic curve on the second page of the PDF export comparing to the electrophoretic curve shown on page 35 of the Malvern software guide and reproduced below (Fig. 3) [15].

*3.5 Measurements of Fluorescent Particles*

The following section details the conjugation of Cy3 fluorophores to plasmid DNA for analysis of polyplex nanoparticles intended for use in gene therapy applications with a NS500 equipped with a 532 nm green laser. These methods enable the properties of the particles specifically containing plasmid DNA to be ascertained in addition to characterization of all polyplex nanoparticles in the population. For alternative fluorophore/laser pairings and details regarding conditions for fluorescent particle analysis, *see* **Note 13**. The detailed method for labeling plasmid DNA described here is more cost-effective for labeling bulk amounts of DNA than using standard labeling kits. The plasmid labeling protocol is based off of that described by Akinc et al. with slight reagent modifications and is shown as a schematic in Fig. 2b [16].

1. **Reagent preparation**. Dissolve NHS-Psoralen and Cy3-NH$_2$ in anhydrous DMSO at 1 μg/μL and 10 μg/μL respectively which should be stored in small aliquots at −20 °C. Prepare

plasmid DNA at 1 μg/μL in ultrapure water; if necessary, ethanol precipitate DNA and resuspend in ultrapure water as detailed via **step 4** of Subheading 3.5 to remove salts.

2. **Psoralen crosslinking**. Add 100 μg of plasmid DNA and 12.5 μg NHS-psoralen to a well of a round bottom 96-well plate. Then place the round bottom plate on ice and expose directly to a 365 nm UV lamp for 25 min to crosslink psoralen to plasmid DNA. A round bottom plate ensures good contact with ice to keep the solution cold under UV light, thus helping to prevent the NHS moiety degradation.

3. **Reaction with fluorophore**. Transfer the solution from **step 2** of Subheading 3.5 to a 1.5 mL microcentrifuge tube. Add 13 μL (1/10th volume) 1 M HEPES buffer pH 8.5 and 40 μg Cy3-NH$_2$. Mix and allow the solution to react in the tube for 1 h protected from light at room temperature (*see* **Note 14** regarding buffer selection and DMSO content of the solution).

4. **Ethanol precipitation**. Add 13 μL (1/10th total solution volume) of 3 M NaAc and 285 μL (2× volume) of ice-cold 95% ethanol to the microcentrifuge tube to reduce DNA solubility. Mix well and place the tube at −80 °C for 10 min, then centrifuge at 14,000 rcf for 15 min. Look for a red pellet, approximately 1 mm in diameter of Cy3-labeled-DNA, then aspirate the supernatant containing excess fluorophore and NHS-psoralen. Resuspend the DNA pellet in 70% ethanol, ensuring that the pellet is broken to help remove any excess fluorophore then centrifuge again, aspirate to dryness and resuspend in 50 μL ultrapure water. Use a small volume spectrophotometer to assess the DNA concentration and labeling density via absorbance at 260 nm and 550 nm, respectively. Add additional ultrapure water to yield labeled plasmid DNA at 1 μg/μL which should be stored in small aliquots at −20 °C.

5. **Polyplex formation**. Form polyplexes as detailed previously in Subheading 3.2 using a PBAE, such as B4S4E7, and Cy3-labeled plasmid DNA. Dilute accordingly in 150 mM PBS to yield 1 mL of diluted polyplexes.

6. **NanoSight setup for fluorescent measurement**. Prime the NanoSight as previously detailed in **step 1** of Subheading 3.3. Load the diluted polyplex sample into a 1 mL disposable plastic syringe and attach it to the NanoSight syringe pump. Start the camera using the software and move to the fluorescence location using the software button under the hardware tab. With the fluorescence filter removed, use the syringe pump to infuse the sample at a rate of 1000 (unitless) until particles clearly appear in the camera field of view then set the infusion rate to 30 and insert the fluorescence filter to assess if the rate of perfusion is sufficient to avoid rapid photobleaching.

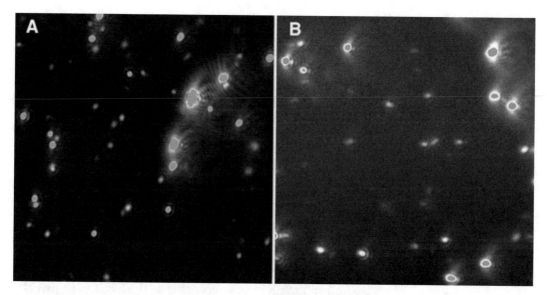

**Fig. 4** PBAE nanoparticles containing Cy3-labeled plasmid DNA can be imaged via NanoSight to determine the properties of those particles specifically containing plasmids. (**a**) NTA without using the 565 nm longpass filter with a camera level of 11. (**b**) NTA to detect only fluorescent particles using the filter with a camera level of 15

Nanoparticles are recommended to contain an average of 100 organic fluorophores for effective detection with NanoSight [17].

7. **Taking a fluorescent measurement**. Once the particle concentration in the field of view appears to have stabilized, insert the fluorescence filter and set the camera level to 15 or 16, adjusting the screen gain if it appears to improve the contrast of the particles to the background. Refer to Fig. 4 regarding the appearance of particles under normal and fluorescence conditions. Then set up to take a measurement in the SOP tab using the advanced button to set the infusion rate during capture to 30 (unitless). Take 3 captures of 60 seconds each with a goal of having 20–80 particles per frame.

8. **Fluorescent Particle Analysis**. Analyze the video captures in the same manner as before and export data as described in **step 6** of Subheading 3.3. To directly compare a sample with and without fluorescence measurements it is recommended to separately dilute the prepared particles in their respective buffer for each individual NanoSight measurement; this ensures that the length of time in buffer was not varied between the sample to avoid confounding results due to potential aggregation or degradation (*see* **Note 15** regarding comparing the properties of polymer only particles to particles specifically containing plasmid DNA).

# 4    Notes

1. These methods would generally work with other NanoSight instruments and other implementations of Nanoparticle Tracking Analysis. Sodium acetate buffer set to pH 5.0 can be purchased at a stock concentration of 3.0 M in water. Alternatively, proportionally to make 3 M sodium acetate buffer add 2.46 g of sodium acetate salt to 10 mL of water and adjust to pH 5.0 by adding 10 N NaOH.

2. Refer to Malvern NanoSight software guides for versions of software and hardware other than those referenced in this chapter [15].

3. Using a syringe pump will generally result in more accurate measurements, as sample perfusion through the chamber can allow an order of magnitude greater number of particles to be tracked to increase the sample size of the number of particles characterized. Use of a syringe pump has been demonstrated to result in both more accurate size and concentration measurements [15].

4. HEPES buffer (1 M) at pH 8.5 can be made by adding 2.38 g of HEPES salt to 8 mL of water. Then add NaOH pellets while mixing until the pH is approximately 8.5. Add distilled water to bring the volume to 10 mL and adjusted the pH as needed using 1 N NaOH or HCl to approximately 8.5.

5. Stoichiometric ratios between acrylate/amine monomers affect the resulting molecular weight of polymers. $M_N$ and $M_W$ will increase as the ratio between B and S monomers approaches unity. The role of polymer molecular weight has been studied in relation to PBAEs [18].

6. Polymers can be endcapped in anhydrous DMSO or THF. Ether precipitation is preferred to isolate the polymer from unreacted monomer.

7. Single-use polymer aliquots are recommended as PBAEs can degrade by hydrolysis and DMSO is hydroscopic. Repeated uses of the same aliquot may result in water entering polymer aliquots and lead to differing results due to polymer degradation. If polymers will be used for cell culture transfection aliquoting to sterile microcentrifuge tubes is recommended.

8. Particles per frame, referring to the mean number of particles visible in the field of view over the entire capture, scales roughly with the concentration of polymer. Optimal dilution typically results in a polymer concentration between 1–5 μg/mL.

9. Polyplex nanoparticles are "soft" polymeric particles, relatively diffuse and have reduced refractive properties compared to more solid polymeric nanoparticles such as poly(lactic-co-glycolic acid) or inorganic gold nanoparticles; as a result, polyplex

nanoparticles are more difficult to detect via NTA [4]. Coupled with the possibility of a heterogeneous particle distribution and the fact that light refraction is partially a function of particle size, dilution factor and camera level must be balanced in order to ensure that all particles in the sample are being accounted for. Increasing the camera level (unitless) allows detection of smaller polyplexes that may be difficult to track at low camera levels.

10. Capture length must be sufficient to track a representative sample of the particles. Longer capture lengths will in general result in more uniformly measured particle size distributions, which is particularly true for low particle-per-frame concentrations. Use of a syringe pump to perfuse the sample at a flow rate between 20–50 (unitless) will reduce the need to use long capture times as more particles will pass through the chamber and be tracked as a result of the convection.

11. NanoSight software allows for export of raw particle data in the form of *.csv* files with the end-name *ParticleData.csv*. Additionally *ExperimentSummary.csv* files contain all information from each capture with machine conditions as well as temperature and user notes. The data in these files are useful for additional analysis or to make figures directly from the raw data but only the particles considered "true" for purposes of the distribution must be considered. A sample Matlab script for parsing the *.csv* files for the diameter and diffusion coefficient data for individual particles included in the distribution is provided below.

```
%######### Import all ParticleData ############
% Put all *ParticeData.csv files for one sample measurement in a
folder.
% Navigate to that folder within Matlab, set the temperature below
% Run this script

% For all particles included in the distribution from all capture
*ParticleData.csv files in directory:
% diameter = column vector of diameters
T = 22;      % Deg. C
diameter = [];
int = [];
CSV = dir(''*Data.csv'');
CSV = {CSV.name};
for i = 1:size(CSV,2) % Runs for loop for total number of files ending
in *Data.csv
  fid = fopen(CSV{i},'r'');
```

```
   DD = textscan(fid, repmat(''%s'',1,10), ''delimiter'','','',
''CollectOutput'',true);
 DD = DD{1};
 fclose(fid);
 snm = DD(strcmpi(''True'', DD(:,7)), 2);
 diameter = [diameter; str2double(snm)]; %column of all tracked
particle diameters
end
```

12. The current visible under the zeta-potential hardware tab with an applied voltage of 24 V must be below 80 μA to be able to take an accurate zeta-potential measurement as per-Malvern's recommendations. Higher molarity salt solutions will result in the current being too high to take an accurate measurement. For samples that result in high current readings, a lower voltage down to 8 V can be used but may result in less optimal measurements. Additionally, because particle drift under an applied voltage is necessary for zeta-potential measurements the syringe pump cannot be used.

13. Malvern recommends a minimum of ten fluorophores per nanoparticle for tracking, with 100 fluorophores per particle preferable to make tracking more effective. Perfusion of the fluorescent particles through the chamber at flow rates between 20–50 (unitless) allows them to be refreshed as the fluorophores are photobleached by the laser. Please refer to the cited Malvern technical note regarding recommended fluorophores for other NanoSight lasers [17].

14. Alternative buffers between pH 7.5–9 that do not contain primary or secondary amines can be used to buffer the reaction between NHS-psoralen and primary amine fluorophores. Some fluorophores that are more hydrophobic may require a higher DMSO content in the solution (up to 30%) to successfully react with the NHS-psoralen pre-crosslinked to plasmid DNA.

15. Fluorescent analysis of polyplex nanoparticles containing fluorescently labeled DNA enables the properties of plasmid-containing particles to be determined. The properties of the entire particle population can likewise be determined by doing a nonfluorescent measurement. The properties of polymer only nanoparticles can be ascertained by doing a nonfluorescent measurement of sample of particles at the same polymer concentration without any DNA (Appendix Fig. A.1).

## Acknowledgements

This work was supported in part by the National Institutes of Health (1R01EB016721) and a Johns Hopkins University Catalyst Award. D.R.W. thanks the National Science Foundation for fellowship support (DGE-0707427).

## References

1. Kim J, Wilson DR, Zamboni CG, Green JJ (2015) Targeted polymeric nanoparticles for cancer gene therapy. J Drug Target 23:627–641. doi:10.3109/1061186X.2015. 1048519

2. Liu Z, Zhang Z, Zhou C, Jiao Y (2010) Hydrophobic modifications of cationic polymers for gene delivery. Prog Polym Sci 35:1144–1162. doi:10.1016/j.progpolymsci. 2010.04.007

3. Tzeng SY, Yang PH, Grayson WL, Green JJ (2011) Synthetic poly(ester amine) and poly (amido amine) nanoparticles for efficient DNA and siRNA delivery to human endothelial cells. Int J Nanomedicine 6:3309–3322. doi:10.2147/IJN.S27269

4. Troiber C, Kasper JC, Milani S et al (2013) Comparison of four different particle sizing methods for siRNA polyplex characterization. Eur J Pharm Biopharm 84:255–264. doi:10. 1016/j.ejpb.2012.08.014

5. Boyd RD, Pichaimuthu SK, Cuenat A (2011) New approach to inter-technique comparisons for nanoparticle size measurements; using atomic force microscopy, nanoparticle tracking analysis and dynamic light scattering. Colloids Surfaces Asp A Physicochem Eng 387:35–42 doi: http://dx.doi.org/10.1016/j.colsurfa. 2011.07.020

6. Filipe V, Hawe A, Jiskoot W (2010) Critical evaluation of Nanoparticle Tracking Analysis (NTA) by NanoSight for the measurement of nanoparticles and protein aggregates. Pharm Res 27:796–810

7. Bhise NS, Gray RS, Sunshine JC et al (2010) The relationship between terminal functionalization and molecular weight of a gene delivery polymer and transfection efficacy in mammary epithelial 2-D cultures and 3-D organotypic cultures. Biomaterials 31:8088–8096. doi:10. 1016/j.biomaterials.2010.07.023

8. Bhise NS, Shmueli RB, Gonzalez J, Green JJ (2012) A novel assay for quantifying the number of plasmids encapsulated by polymer nanoparticles. Small 8:367–373. doi:10.1002/smll. 201101718

9. Shmueli RB, Bhise NS, Green JJ (2013) Evaluation of polymeric gene delivery nanoparticles by nanoparticle tracking analysis and high-throughput flow cytometry. J Vis Exp: e50176. doi:10.3791/50176

10. Bhise NS, Wahlin KJ, Zack DJ, Green JJ (2013) Evaluating the potential of poly(beta-amino ester) nanoparticles for reprogramming human fibroblasts to become induced pluripotent stem cells. Int J Nanomedicine 8:4641–4658. doi:10.2147/IJN.S53830

11. Kim J, Sunshine JC, Green JJ (2014) Differential polymer structure tunes mechanism of cellular uptake and transfection routes of poly(β-amino ester) polyplexes in human breast cancer cells. Bioconjug Chem 25:43–51. doi:10. 1021/bc4002322

12. Tzeng SY, Higgins LJ, Pomper MG, Green JJ (2013) Student award winner in the Ph.D. category for the 2013 society for biomaterials annual meeting and exposition, april 10–13, 2013, Boston, Massachusetts : biomaterial-mediated cancer-specific DNA delivery to liver cell cultures using synthetic poly(beta-amino ester)s. J Biomed Mater Res A 101:1837–1845. doi: 10.1002/jbm.a.34616

13. Mangraviti A, Tzeng SY, Kozielski KL et al (2015) Polymeric nanoparticles for nonviral gene therapy extend brain tumor survival in vivo. ACS Nano 9:1236–1249. doi:10. 1021/nn504905q

14. Beh CW, Pan D, Lee J et al (2014) Direct interrogation of DNA content distribution in nanoparticles by a novel microfluidics-based single-particle analysis. Nano Lett 14:4729–4735. doi:10.1021/nl5018404

15. Malvern-Instruments (2014) NanoSight NS500 NTA Software Guide. 1–24.

16. Akinc A, Langer R (2002) Measuring the pH environment of DNA delivered using nonviral vectors: implications for lysosomal trafficking. Biotechnol Bioeng 78:503–508. doi:10. 1002/bit.20215

17. Malvern-Instruments (2015) Fluorescence nanoparticle detection using NTA http:// www.malvern.com/en/pdf/secure/TN14030-3FluorescentlyLabeledNanoparticles.pdf

18. Bishop CJ, Ketola T, Tzeng SY et al (2013) The effect and role of carbon atoms in poly(β-amino ester)s for DNA binding and gene delivery. J Am Chem Soc 135:6951–6957. doi:10. 1021/ja4002376

# Chapter 4

# Magnetic Characterization of Iron Oxide Nanoparticles for Biomedical Applications

## Lorena Maldonado-Camargo, Mythreyi Unni, and Carlos Rinaldi

## Abstract

Iron oxide nanoparticles are of interest in a wide range of biomedical applications due to their response to applied magnetic fields and their unique magnetic properties. Magnetization measurements in constant and time-varying magnetic field are often carried out to quantify key properties of iron oxide nanoparticles. This chapter describes the importance of thorough magnetic characterization of iron oxide nanoparticles intended for use in biomedical applications. A basic introduction to relevant magnetic properties of iron oxide nanoparticles is given, followed by protocols and conditions used for measurement of magnetic properties, along with examples of data obtained from each measurement, and methods of data analysis.

**Key words** Magnetic nanoparticles, Anisotropy constant, Blocking temperature, Magnetic relaxation, Saturation magnetization, Remanent magnetization, Coercivity

## 1 Introduction

Iron oxide magnetic nanoparticles have been widely used due to the capability of manipulating particle motion, causing energy dissipation, or providing imaging contrast in the presence of an external magnetic field [1, 2]. Their use in biomedical applications such as cancer therapy [3, 4], magnetically triggered drug release [5, 6], magnetofection [7, 8], magnetic resonance imaging [9, 10], and magnetic particle imaging [11, 12] has been widely researched. Examples of important magnetic properties of nanoparticles include the saturation magnetization, remanence and coercivity, magnetic diameter, magnetocrystalline anisotropy constant, mechanism of magnetic relaxation, and blocking temperature, all of which can be material specific and may be influenced by the method of synthesizing and coating the nanoparticles and by the method of sample preparation for magnetic measurements. Accurately quantifying these properties is vital to enable reproducibility in research and to achieve the maximum potential of iron oxide nanoparticles in specific applications. Unfortunately, many publications

Sarah Hurst Petrosko and Emily S. Day (eds.), *Biomedical Nanotechnology: Methods and Protocols*, Methods in Molecular Biology, vol. 1570, DOI 10.1007/978-1-4939-6840-4_4, © Springer Science+Business Media LLC 2017

describing the preparation and use of magnetic nanoparticles lack even basic magnetic characterization, making it difficult to evaluate and compare the work. The methods described in this chapter present detailed procedures to determine the magnetic properties of iron oxide nanoparticles that most significantly impact their biomedical applications.

The magnetic response of iron oxide nanoparticles to an external field depends mainly on the degree of magnetic ordering and on the temperature of the sample. The magnetic moment per unit volume of particle, i.e., the magnetization, may be defined depending on the spin or the orbital energy possessed by the dipole. Particles with large crystallite sizes have dipoles arranged in multiple domains separated by a domain wall so as to maintain the lowest energy state. There exists a critical size (typically less than 100 nm) below which it is energetically unfavorable for domain walls to form, resulting in single domain nanoparticles [13]. As predicted by Louis Néel, nanoparticles in the single domain regime no longer exhibit hysteresis behavior in an applied magnetic field, a condition that is referred to as superparamagnetism [14]. The most commonly used magnetic nanoparticles are ferrites $MFe_2O_4$ that exhibit a spinel or inverse spinel structure. The distribution of the metal ions relative to the oxygen ions in the crystal lattice results in the formation of dipoles and determines the overall magnetization of the material. A measure of this ordering and the strength of the dipoles in single domain particles may be obtained from the magnetic diameter obtained from fitting the Langevin equation (*see* Subheading 3.5) to an experimental equilibrium magnetization curve [15, 16].

In the presence of a magnetic field, magnetic spins tend to align in the direction of the field, resulting in an induced magnetization [17]. The maximum induced magnetization is termed the saturation magnetization of the sample. This induced magnetization may remain even after the field is removed, in which case it is referred to as the remanent magnetization. In those cases, the coercive field corresponds to the magnetic field required to revert the magnetization to zero. Nanoparticles with ferro- and ferri-magnetic behavior often exhibit hysteresis. In nanoparticles that exhibit superparamagnetism, remanence and coercive field becomes negligible.

Nanoparticles show a certain preference for the direction along which their magnetic dipole tends to align, referred to as magnetic anisotropy, which can arise due to the shape and inherent crystalline structure of the nanoparticles. This anisotropy can be intrinsic to the material such as magnetocrystalline, shape, and exchange anisotropy, or induced by an external process. Magnetocrystalline anisotropy refers to the tendency of the magnetization to align along a preferred crystallographic direction. In contrast, polycrystalline samples with no preferred crystal orientation tend to magnetize along a long axis, in what is known as shape anisotropy. Finally, exchange anisotropy arises from interaction between anti-ferromagnetic and ferromagnetic materials [13, 17]. The rate at which the magnetic dipole within a particle will align in a given direction of applied magnetic field is influenced by the temperature of the system (thermal energy) and the magnitude of the magnetic anisotropy energy barrier between easy axes for magnetization.

The magnitude of the magnetocrystalline anisotropy can be determined by temperature-dependent magnetization measurements such as zero-field-cooled (ZFC) and Dynamic Magnetic Susceptibility (DMS) curves in samples wherein nanoparticles' physical rotation is suppressed, either by freezing the sample or dispersing it in a solid matrix (*see* Subheadings 3.7 and 3.8).

There are two mechanisms by which the magnetization of a colloidal suspension of nanoparticles responds after the removal of an external applied field [18–20]. In the first mechanism, the relaxation of the magnetic dipole occurs by physical particle rotation in the liquid. The corresponding characteristic rotational diffusion time $\tau_B$, referred to as the Brownian relaxation time, is given by

$$\tau_B = \frac{3 V_h \eta}{kT} \tag{1}$$

where $\eta$ is the viscosity of the carrier liquid, $V_h$ is the hydrodynamic volume of the particle, $k$ is the Boltzmann constant, and $T$ is the temperature. In the second mechanism, the magnetic dipole rotates within the particle. The corresponding characteristic time $\tau_N$ for dipole rotation is termed the Néel relaxation time and is given by

$$\tau_N = \tau_0 \exp\left(\frac{K V_m}{kT}\right) \tag{2}$$

where $\tau_0$ is a characteristic time with an approximate value of $10^{-9}$s, $V_m$ is the magnetic core volume, and $K$ is the anisotropy constant. In a colloidal suspension of nanoparticles, both relaxation mechanisms are present but the faster mechanism dominates. The Brownian relaxation time is proportional to the viscosity of the carrier liquid and the particle hydrodynamic diameter, whereas the Néel mechanism is solely related to the volume of the magnetic core and the anisotropy constant of the material. Dynamic Magnetic Susceptibility (DMS) measurements (*see* Subheadings 3.9) can be used to obtain information of the magnetic relaxation properties of nanoparticles in suspension, including measurements of their characteristic magnetic relaxation time.

Here, we describe in detail how superconducting quantum interference device (SQUID) magnetometers and dynamic magnetic susceptometers can be applied to quantify magnetic properties of iron oxide nanoparticles in a liquid or solid matrix.

Sample preparation methods and data analysis are also explained. We remark that determining properties of nanomaterials is often difficult on the basis of a single technique. The methods described in this chapter should provide a fairly complete assessment of the magnetic properties of iron oxide nanoparticles that most significantly impact their biomedical applications including low and high field magnetization curves, Zero-Field-Cooled/Field-Cooled (ZFC/FC) magnetization curves, and Dynamic Magnetic Susceptibility (DMS) measurements. However, there are other magnetic measurements, such as isothermal reversibility measurements, First Order Reversal Curves (FORC), and Verwey transition analysis that we have left out of this chapter [21–25]. The reader may adapt this protocol to characterize their specific nanoparticle of interest by taking into consideration the compatibility of the particle surface coating with the polymer or solvent selected.

## 2  Materials

### 2.1  Immobilization of Hydrophobic Nanoparticles in a Solid Matrix (See Note 1)

1. Magnetic nanoparticles synthesized or obtained commercially and coated with organic molecules such as oleic acid, hydrophobic polymers, and oleylamine.
2. Styrene ReagentPlus®, containing 4-tert-butylcatechol as a stabilizer, ≥99%. Store at 4 °C.
3. Divinylbenzene technical grade, 80% (DVB). Store at 4 °C.
4. 2,2'-Azobis(2-methylpropionitrile) (AIBN). Store at 4 °C.
5. 3 mL glass vial.
6. 0.7 mL glass test tubes with screw cap (6 mm OD, 50 mm length).
7. Ultrasonicator probe fitted with a tapered microtip (3/6 in).
8. Oil bath.

### 2.2  Immobilization of Hydrophilic Nanoparticles in a Solid Matrix (See Note 2)

1. Magnetic nanoparticles synthesized or obtained commercially and coated with hydrophilic molecules, such as hydrophilic polymers, amines, carboxylic groups or peptized.
2. Tetraethylene glycol dimethacrylate (TEGDMA), technical grade, ≥90%. Store at 4 °C.
3. 2,2'-Azobis(2-methylpropionitrile) (AIBN). Store at 4 °C.
4. 3 mL glass vial.
5. 0.7 mL glass test tubes with screw cap (6 mm OD, 50 mm length).
6. Ultrasonicator probe fitted with a tapered microtip (3/6 in).
7. Oil bath.

**2.3  Nanoparticle Suspensions**

1. Magnetic nanoparticles: Iron oxide, cobalt ferrite, manganese ferrite, etc., synthesized or obtained commercially. This protocol uses iron oxide nanoparticles.

2. Toluene, hexane, chloroform, 1-octadecene, tetrahydrofuran, or any organic solvent suitable to suspend particles coated with organic molecules, such as oleic acid or oleylamine. This protocol uses iron oxide nanoparticles coated with oleic acid suspended in 1-octadecene.

3. Water or other suitable polar solvents to suspend particles coated with hydrophilic molecules, such as hydrophilic polymer, amines, carboxylic groups, or peptized. This protocol uses water.

4. Filters: Nylon filters for filtration of aqueous solutions. Polytetrafluoroethylene (PTFE) filters for filtration of organic solvents. Polyvinylidene fluoride (PVDF) filters for filtration of nonaggressive aqueous and mild organic solutions (see **Note 3**).

5. Mechanical ultrasonic bath.

**2.4  Equipment**

1. Dynamic magnetic susceptometers operating at low amplitude fields ~0.5 mT with an ideal excitation frequency ranging from 1 to 100 kHz (such as Acreo DynoMag).

2. Vibrating sample magnetometer (VSM) (such as 7400-S from Lake shore Cryotronics, or VSM from Quantum Design).

3. Superconducting Quantum Interference Device (SQUID) magnetometer (such as MPMS3 from Quantum Design or S700X from Cryogenic Ltd) (see **Note 4**).

# 3   Methods: Sample Preparation

As seen in the introduction, magnetic nanoparticles may relax by either of two mechanisms; Néel and Brownian relaxation. Distinguishing between these two mechanisms of magnetic relaxation is critical in many biomedical applications, as physical particle rotation may be significantly impaired once nanoparticles accumulate in tissues or inside cells. Measuring magnetic properties for nanoparticles fixed in a solid matrix allows one to abrogate the effects of Brownian relaxation in the magnetic response of the nanoparticles during a measurement, whereas when the nanoparticles are in suspension both mechanisms contribute to the response. As such, characterization of samples in liquid and solid matrices can provide useful insights into their relaxation-dependent properties.

**3.1  Immobilization of Hydrophobic Nanoparticles in a Solid Matrix**

1. Prepare a stock solution of the monomers and initiator by mixing 3 mL of styrene, 450 µL of DVB, and 10 mg of AIBN in a 10 mL glass vial (see **Note 5**).

2. Weigh ~10 mg of the magnetic nanoparticle sample into a glass vial. The weight percentage of the magnetic core in the sample must be measured prior to immobilization in a solid matrix, so as to properly estimate the amounts for the various components (*see* **Note 6**).

3. Add appropriate amount of the monomer/initiator solution to the nanoparticles in the glass vial to obtain a concentration of magnetic core of ~0.1 wt% (*see* **Note 7**). For example, if the nanoparticle sample (i.e., magnetic core and the ligands on the particles) is 15 wt% of magnetic core and 85 wt% ligands on the particle, add 1.5 g of the monomer/initiator solution (1.5 mL assuming density ~1 g/mL) to 0.01 g of magnetic nanoparticle sample, such as the final magnetic core concentration in the solution is 0.1 wt%.

4. Shake the solution vigorously. If necessary, use an ultrasonicator probe fitted with a tapered microtip (3/16 in.) to disperse the particles in the polymer.

5. Place 500 μL of the solution in glass test tubes with screw-cap (Fig. 1).

6. Place the tube in an oil bath and increase the temperature to 70 °C.

7. Allow the reaction to proceed for 4 h. The monomer should be completely polymerized, forming a solid matrix (Fig. 1).

8. To release the polymer from the tube, tap the sides of the tube. Do this lightly and be careful not to break the sample or the glass tube.

9. Weigh the sample to account for any evaporation losses and recalculate the nanoparticle concentration if needed.

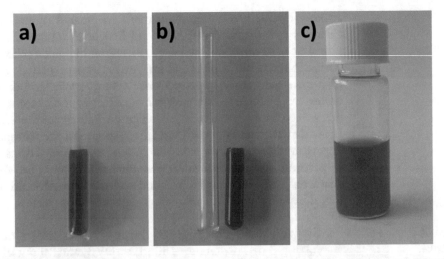

**Fig. 1** Sample preparation. (**a**) 500 μL of the mixture iron oxide nanoparticles and poly(styrene-divinylbenzene) in glass test tubes. (**b**) Iron oxide nanoparticles immobilized in a poly(styrene-divinylbenzene). (**c**) Iron oxide nanoparticles in water solution

**3.2 Immobilization of Hydrophilic Nanoparticles in a Solid Matrix**

1. Weigh ~10 mg of the magnetic nanoparticle sample into a glass vial. The weight percentage of the magnetic core in the sample must be measured prior to immobilization in a solid matrix, so as to properly estimate the amounts for the various components (*see* **Note 6**).

2. Add the appropriate amount of the tetraethylene glycol dimethacrylate (TEGDMA) to obtain a concentration of magnetic core of ~0.1 wt% (*see* **Note 7**). For example, to embed 0.01 g of magnetic nanoparticles with a 10 wt% of magnetic core (i.e., 10 wt% magnetic core and 90 wt% polymer coating), add 1.0 g of the monomer/initiator solution (1.0 mL assuming density ~1 g/mL), such as the final magnetic core concentration in the solution is 0.1 wt%.

3. Shake the solution vigorously. If necessary, use an ultrasonicator probe fitted with a tapered microtip (3/16 in.) to disperse the particles in the polymer.

4. Add AIBN to the previous solution to a concentration of 0.04 wt%.

5. Place 500 μL of the solution in glass test tubes with screw-cap.

6. Place the tube in an oil bath and increase the temperature to 70 °C.

7. Allow the reaction to proceed for 4 h. The monomer should polymerize, forming a solid matrix.

8. To release the polymer from the tube, tap the sides of the tube. Do this lightly and be careful not to break the sample or the glass tube.

9. Weigh the sample to account for any evaporation losses and recalculate particle concentration if needed.

**3.3 Suspending Nanoparticles in a Liquid Matrix**

1. Weigh ~10 mg of the magnetic nanoparticle sample. The weight percentage of the magnetic core in the sample must be measured prior to suspension in a liquid matrix, so as to properly estimate the amounts for the various components (*see* **Note 6**).

2. Add the approximate amount of solvent (organic for hydrophobic nanoparticles, water for hydrophilic nanoparticles) to obtain a concentration of magnetic cores of ~0.1 wt% (Fig. 1c, *see* **Note 7**).

3. Use a sonic bath to suspend the particles in the solvent.

4. Filter the solution using a syringe filter.

**3.4 Methods: Magnetic Measurements and Data Analysis**

For demonstration purposes this chapter describes the use of a solid sample prepared with oleic acid coated iron oxide nanoparticles embedded in a poly(styrene-divinyl benzene) matrix (PSDVB), which inhibits particle rotation, and nanoparticles suspended in 1-oactadecene to allow for nanoparticle rotation. The reader may adapt this protocol to characterize their specific nanoparticle of interest by taking into consideration the compatibility of the particle surface with the polymer or solvent selected.

In the following protocol, a Quantum Design MPMS3 Superconducting Quantum Interference Device (SQUID) magnetometer is used in Subheadings 3.5–3.8 to determine the saturation magnetization, remanence and coercivity, magnetic diameter, magnetocrystalline anisotropy constant, and blocking temperature solid samples. An Acreo DynoMag dynamic magnetic susceptometer is used in Subheading 3.9 to estimate the mechanism of magnetic relaxation of the nanoparticles in a liquid matrix. The reader is assumed to be familiar with the equipment and relevant software described associated with these instruments, or the instruments chosen to carry out similar characterization experiments.

### 3.5 Magnetization Vs. Magnetic Field (MH) at Constant Temperature

The MH curve allows one to verify the superparamagnetic behavior of the nanoparticles and determine their saturation magnetization. Furthermore, a superparamagnetic MH curve can be analyzed by fitting to the Langevin function and a lognormal size distribution to estimate the magnetic diameter of the nanoparticles [15, 16]. Finally, in cases where the nanoparticles are not superparamagnetic, the equilibrium magnetization curve can be used to determine the remanence and coercivity of the nanoparticles at a given temperature. It should be noted that to properly verify superparamagnetic behavior in a sample the nanoparticles should be prevented from rotating. This is because a collection of ferro/ferrimagnetic nanoparticles in liquid suspension will be able to physically rotate to align their magnetic dipoles in the direction of the magnetic field, resulting in an MH curve that lacks remanence and coercivity. This situation is called extrinsic superparamagnetism, whereas the case when the nanoparticles are fixed in a matrix and their MH curve lacks remanence and coercivity and follows the sigmoid function (typical S-shape) is called intrinsic superparamagnetism [16]. This is not to say that MH curves should not be obtained for liquid samples, as such measurements can still yield the magnetic diameter of the nanoparticles and can also be used to estimate the volume fraction of nanoparticles in the suspension. In a typical measurement, the magnetization (magnetic moment per unit of volume) of a sample is measured as a function of the applied magnetic field at constant temperature. The magnetic diameter distribution, saturation magnetization, coercive field and remanent magnetization can all be determined from MH curves.

1. Fix the magnetic nanoparticles in a solid matrix as explained in Subheadings 3.1 and 3.2. In this protocol we use iron oxide nanoparticles in a PSDVB matrix. The reader may adapt this protocol to use a polymer or solid matrix compatible with their nanoparticle system (see **Notes 1** and **2**).

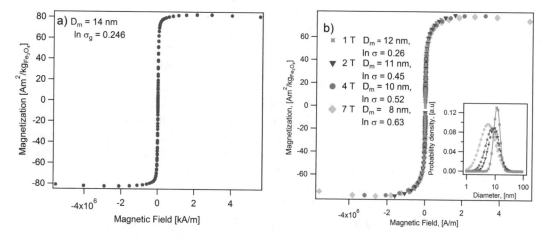

**Fig. 2 (a)** Magnetization curves of iron oxide nanoparticles in a poly(styrene-divinylbenzene) matrix measured at different saturation fields. Annotations indicate the magnetic diameter and the geometric deviation calculated using the Langevin-Chantrel model. *Inset* figure shows the magnetic diameter distribution. **(b)** Magnetization curve at 300 K for iron oxide nanoparticles in a poly(styrene-divinylbenzene) matrix. Annotations indicate the magnetic diameter (14 nm) and the geometric deviation (ln σ = 0.246)

2. Set the desired sample temperature using the equipment software (*see* **Note 8**). We use a Superconducting Quantum Interference Device (SQUID) magnetometer MPMS3 from Quantum Design. The reader is assumed to be familiar with the equipment and relevant software described in Subheading 2.4.

3. Measure the magnetization $M(H)$ of the sample with increasing magnetic field. Use of an applied magnetic field range of 7 to −7 T with a field ramp of 200 Oe/min is highly recommended (*see* **Note 9**). In Fig. 2a, the magnetization curve of iron nanoparticles is measured at different saturation fields.

4. The magnetic diameter and its distribution is a measure of the strength of the magnetic dipole within each nanoparticle. The volume median magnetic diameter ($D_m$) and geometric deviation ($\ln \sigma_g$) can be found by fitting the superparamagnetic equilibrium magnetization curve to the Langevin function, weighed using a lognormal size distribution $n_v(D_m)$ (*see* Eq. (5)), as suggested by Chantrel et al. [15, 16].

$$M(\alpha) = M_s \int_0^\infty n_v(D_m)L(\alpha)dD_m \qquad (3)$$

$$L(\alpha) = \coth(\alpha) - \frac{1}{\alpha}; \text{where } \alpha = \frac{\pi\mu_0 D_m^3 M_d H}{6k_B T} \qquad (4)$$

$$n_v(D_m) = \frac{1}{\sqrt{2\pi}D\ln\sigma_g}\exp\left(\frac{\ln(D_m/D_{pgv})}{2(\ln^2\sigma_g)}\right) \qquad (5)$$

In Eq. (4), $\alpha$ is the Langevin parameter (ratio of magnetic to thermal energy), $\mu_0$ is the permeability of free space, $k_B$ is Boltzmann's constant, and $T$ is the absolute temperature, and $M_d$ is the domain magnetization (446000 A/m or 86 $Am^2$/kg for iron oxide).

5. *See* Fig. 2a for an example of a MH curve for iron oxide nanoparticles embedded in a PSDVB matrix at a concentration of 0.057 $mg_{Fe_3O_4}$. Using Eq. (3), the volume median magnetic diameter is $D_m = 14$ nm with $\ln\sigma_g = 0.246$ at 300 K. The saturation magnetization is $\sim 83\,Am^2/kg_{Fe_3O_4}$, which is in the range of reported values for magnetite and maghemite (80 – 100$Am^2$/kg) [26].

6. For the same sample, the magnetic diameter, calculated using the Langevin-Chantrel model [15, 16], decreases as the saturation field increases (*see* Fig. 2b). This is because measurements carried in low field strength would cause only larger particles to respond to the magnetic field and thus narrow size distribution is obtained. A true representation of the magnetic diameter and its distribution can be obtained only if the measurements are run at fields much greater than the fields that saturate the particles, such that even smaller particles respond at large magnetic field strengths.

## 3.6 Low Field Magnetization

The extent of magnetic dipole-dipole interactions in a sample can be parameterized using the so-called interaction temperature parameter $T_0$. This parameter can be estimated from low field ($\pm$ 10-40 Oe, depending on the sample) MH curves for the sample, measured at various temperatures in a wide temperature range. The sample needs to be solid or embedded in a matrix to restrict particle rotation during the measurement (*see* Subheading 3.1 or 3.2). Under such conditions, the MH curves will be linear for superparamagnetic samples and the slope of the curves, which corresponds to the initial susceptibility of the sample, will be sensitive to the dipole-dipole interactions. The inverse of the initial susceptibility can then be plotted as a function of temperature and fitted to a Curie-Weiss model to obtain $T_0$ [27]. We note that $T_0$ can be a function of the state of aggregation of the iron oxide nanoparticles in a sample, or of the extent of dipole-dipole interactions in concentrated samples.

1. Embed the magnetic nanoparticles in a solid matrix as explained in Subheadings 3.1 and 3.2. In this protocol, we use iron oxide nanoparticles in a PSDVB matrix. The reader may adapt this protocol to use a polymer or solid matrix compatible with their nanoparticle system (*see* **Notes 1** and **2**).

2. Set the desired sample temperature using the equipment software (*see* **Note 8**). We use a Superconducting Quantum Interference Device (SQUID) magnetometer MPMS3 from Quantum

Design. The reader is assumed to be familiar with the equipment and relevant software described in Subheading 2.4.

3. Measure the magnetization of the sample as a function of increasing the applied magnetic field. Typically, the magnetization is measured for about ten magnetic field steps, uniformly distributed in the field range. The applied field may range from 0.004 to −0.004 T and a field ramp of 0.001 T/min is suitable.

4. Determine the initial susceptibility $\chi_0$ of the sample at different temperatures by calculating the slope of the $M(H)$ curve. Then plot $1/\chi_0$ versus $T$ and fit to the Curie-Weiss model [28]

$$\chi_0 = \frac{A}{(T - T_0)}. \tag{6}$$

where the data is expected to lie in a straight line whose intercept with the $1/\chi_0$ axis corresponds to the interaction temperature parameter $T_0$.

5. Figure 3a shows representative low field magnetization curves for iron oxide nanoparticles in a PSDVB matrix in a temperature range of 4–400 K. Note that for temperatures of 136 K and higher the low field MH curves are linear and cross the origin, indicating superpamagnetic behavior. For temperatures below 136 K the low field MH curves still appear linear but no longer cross the origin, indicating the sample has significant remanence and coercivity and is therefore no longer superparamagnetic.

6. Figure 3b shows the linear relation between the inverse of the initial susceptibility and temperature for the sample. Typically, this linear relation only holds for a limited temperature range, which usually starts much higher than the temperature for

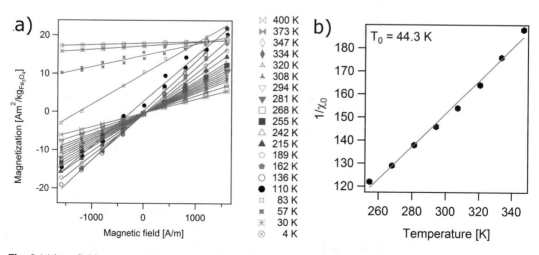

**Fig. 3** (**a**) Low field magnetization curves for iron oxide nanoparticles in a poly(styrene-divinylbenzene) matrix, at temperatures between 4 K and 400 K and (**b**) initial susceptibility data fitted to Curie-Weiss model

which the sample becomes superparamagnetic. For the sample in Fig. 3b the range starts at about ∼260 K, even though the sample appeared superpamagnetic at a temperature of 136 K in Fig. 3a. Using the model in Eq. (6), the interaction temperature parameter for this sample was determined to be $T_0 = 43.2$ K.

### 3.7 Zero-Field-Cooled/Field-Cooled (ZFC/FC) Magnetization Curves

Whether a given collection of magnetic nanoparticles displays superparamagnetic or ferro/ferrimagnetic behavior depends on the temperature at which the measurement is made. This is because superparamagnetism corresponds to a state where the energy barrier to dipole moment rotation in the crystal is much smaller than the thermal energy. As such, the temperature at which a collection of magnetic nanoparticles transitions from ferro/ferrimagnetic behavior to superparamagnetic behavior is an important property. This can be characterized through temperature-dependent magnetization $M(T)$ measurements. The most common way to do this is through so-called Zero-Field-Cooled and Field-Cooled (ZFC/FC) measurements. In such measurements a sample, usually solid or embedded in a solid matrix, begins at a high temperature and at zero field, such that thermal energy completely eliminates any magnetization in the sample. Then the sample is cooled to a low temperature in zero field. The ZFC portion of the plot is obtained by applying a magnetic field once the sample has equilibrated at the lowest temperature and then by measuring the sample's magnetization with increasing temperature. At the lowest temperature the dipoles in the nanoparticles will have the lowest amount of thermal energy, and as such will only align slightly with the applied field, resulting in a small magnetization value. As the temperature increases, the increasing thermal energy of the magnetic dipoles will free them from their initial states, resulting in increased alignment with the field and increasing sample magnetization. However, beyond a certain temperature, referred to as the blocking temperature, further increasing thermal energy will lead to a decrease in the extent of alignment of the magnetic dipoles with the applied field and hence to a decrease in the sample magnetization. The FC portion of the curve is obtained by then measuring magnetization as the sample is cooled back in the applied field to the initial low temperature. In an ideal sample the ZFC and FC curves will overlap at temperatures above the blocking temperature and diverge at temperature below the blocking temperature. For samples with significant dipole-dipole interactions or broad size distributions there will be a significant temperature range above the blocking temperature where the ZFC and FC curves do not overlap. Also, the shape of the FC curve below the blocking temperature can also be indicative of the extent of dipole-dipole interactions in a sample. Finally, the blocking temperature can be analyzed to obtain an estimate of the anisotropy constant of the nanoparticles in the sample, by using equations that consider the Néel and Volger-Fulcher models for the magnetic relaxation time.

1. Embed the magnetic nanoparticles in a solid matrix as explained in Subheadings 3.1 and 3.2. In this protocol, we use iron oxide nanoparticles in a PSDVB matrix. The reader may adapt this protocol to use a polymer or solid matrix compatible with their nanoparticle system (*see* **Notes 1** and **2**).

2. Heat the sample to the highest working temperature in the absence of an applied magnetic field (*see* **Note 10**). Hold this condition for at least 5 min. We use a SQUID magnetometer MPMS3 from Quantum Design. The reader is assumed to be familiar with the equipment and relevant software described in Subheading 2.4.

3. Set the sample temperature to the lowest value, depending on the actual sample and instrument limit (*see* **Note 11**). For iron oxide nanoparticles, we usually start at 10 K. Use cooling rates of 5–10 K/min while cooling down.

4. At the lowest temperature, apply a small magnetic field (10–100 Oe, *see* **Note 12**).

5. Measure the magnetization of the sample as the temperature increases from 10 to 400 K. Use a temperature sweep rate of 10 K/min and measure the magnetic moment at least every 2 K (*see* **Note 13**).

6. Decrease the temperature from 400 K to 10 K under a small magnetic field (10–100 Oe) and measure the magnetization at least every 2 K. We recommend using a temperature sweep rate of 10 K/min (*see* **Note 13**).

7. The temperature at which the ZFC magnetization curve exhibits a maximum is called the blocking temperature ($T_B$) (*see* Fig. 4). Many authors assume that at the blocking temperature the time scale of the measurements is comparable to the Néel relaxation time, given by Eq. (2) and therefore the anisotropy constant is determined by

$$K = \frac{k_B(T_B)}{V_m} \ln\left(\frac{\tau_{obs}}{\tau_0}\right) \qquad (7)$$

where $V_m$ is the magnetic volume, calculated using the magnetic diameter determined in Subheading 3.5, $\tau_{obs}$ is the observation time in seconds, $T_B$ is the blocking temperature, $\tau_0$ is the characteristic time, typically assumed to be $\sim 10^{-9} s$. The observation time used to calculate anisotropy constant is the ratio between blocking temperature and rate set to reach the temperature, i.e., the sweep rate during measurement (10 K/min for the procedure described above).

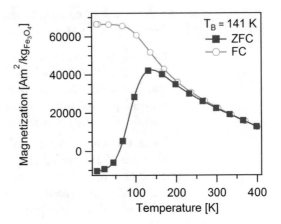

**Fig. 4** Zero field cooled (*closed symbols*) and field cooled (*open symbols*) magnetization curve for iron oxide nanoparticles in a poly(styrene-divinylbenzene) matrix, obtained at temperatures between 4 and 400 K using a 796 A/m (10 Oe) field

8. Recognizing the influence of magnetic dipole–dipole interactions in the relaxation time of the nanoparticles, shtrikman and wolhlfarth [27] proposed the Vogel-Fulcher law, which can be used to estimate the anisotropy constant in samples with significant interactions

$$K = \frac{k_B(T_B - T_0)}{V_m}\ln\left(\frac{\tau_{obs}}{\tau_0}\right) \tag{8}$$

where $T_0$ is the interaction temperature parameter. To estimate the anisotropy constant, one requires independent knowledge of $T_0$, $\tau_0$, and $\tau_{obs}$.

9. Polydispersity of the sample may be accounted for in Eqs. (7) and (8) by calculating $K$ according to

$$K = \frac{k_B(T_B - T_0)}{V_m}\ln\left(\frac{\tau_{obs}}{\tau_0}\right)\frac{1}{\exp(9/2\ln^2\sigma_g)} \tag{9}$$

10. Figure 4 shows a representative ZFC-FC magnetization curve for the same sample used in Subheading 3.5. The blocking temperature was $T_B = 137$ K, which is similar to the temperature for which the low field MH curves became linear and crossed the origin in Fig. 3a. Furthermore, note that the ZFC and FC curves do not overlap until temperatures above ~260 K. This is evidence of significant dipole-dipole interactions in the sample and also explains in part why the linear relationship between inverse initial susceptibility and temperature was observed at temperatures above ~260 K in Fig. 3b.

**Table 1**
**Anistropy constant for Iron Oxide nanoparticles ($D_m$ = 14 nm ) in a PSDVB matrix. Results from ZFC and DMS measurements**

|  |  | Néel model | | Vogel-Fulcher model | |
|---|---|---|---|---|---|
|  |  | $\ln\sigma_g = 0$ | $\ln\sigma_g = 0.246$ | $\ln\sigma_g = 0$ | $\ln\sigma_g = 0.246$ |
| ZFC | $K_{ZFC} [KJ/m^3]$ | 46.59 | 35.48 | 27.90 | 21.25 |
| DMS susceptibility | $\tau_o [s]$ | $3.18 \times 10^{-19}$ | $3.18 \times 10^{-19}$ | $1.09 \times 10^{-14}$ | $1.09 \times 10^{-14}$ |
|  | $K_{DMS} [KJ/m^3]$ | 61.15 | 46.58 | 32.28 | 24.59 |

The calculated values of the anisotropy constant $K$ using the maximum temperature of the ZFC curve, and the Néel or Vogel-Fulcher models for the relaxation time are summarized in Table 1. Because the Vogel-Fulcher model takes into account the particle-particle interaction parameter $T_0$, the value obtained using the Néel model is larger than that calculated using the Vogel-Fulcher model. However, both values are slightly larger than the magnetocrystalline anisotropy constant of bulk magnetite 13.5 kJ/m$^3$ [29]. When the polydispersity of the magnetic diameter is included in the analysis, the calculated values decrease and are comparable to the bulk value.

**3.8 Dynamic Magnetic Susceptibility (DMS) as a Function of Temperature**

The transition from ferro/ferrimagnetic behavior to superparamagnetic behavior can also be determined from measurements of the dynamic magnetization of the nanoparticles in response to an oscillating magnetic field as a function of temperature. The measurements are usually done in solid samples or samples in a solid matrix to inhibit the particle rotation (*see* Subheadings 3.1 and 3.2).

In these measurements, the so-called complex or dynamic magnetic susceptibility $\widehat{\chi}(T) = \chi' - i\chi''$ of the nanoparticles is determined as a function of temperature for various applied oscillating field frequencies. Here, $\chi'$ is referred to as the in-phase susceptibility and $\chi''$ is referred to as the out-of-phase susceptibility. This approach poses several advantages over ZFC/FC measurements to determine the anisotropy constant of the nanoparticles. First, the measurements are done in the absence of an applied constant field and the oscillating excitation field is of small magnitude, such that the Néel and Vogel-Fulcher models for the magnetic relaxation time are expected to be good descriptions of the behavior of the nanoparticles. Second, the observation time is precisely determined by the inverse of the frequency of the applied oscillating magnetic field and can be easily varied by changing the applied oscillating field frequency. This in turn allows one to determine the anisotropy constant without having to assume a value for $\tau_0$.

1. Fix the magnetic nanoparticles in a solid matrix as explained in Subheadings 3.1 and 3.2.

2. Heat the sample to the highest working temperature in the absence of an applied magnetic field (*see* **Note 10**). Hold this condition for at least 5 min. We use a SQUID magnetometer MPMS3 from Quantum Design in this protocol. The reader is assumed to be familiar with the equipment and relevant software described in Subheading 2.4.

3. Set the amplitude of the field in the range of 2–5 Oe.

4. Set the desired frequency of oscillation of the AC field. The same measurement must be repeated at multiple frequencies. The selected frequencies should be in a range that spans several orders of magnitude (0.1–1000 Hz).

5. Measure the in-phase and out-of-phase components of the magnetic susceptibility of the sample under a constant amplitude oscillating field as the temperature is decreased from 400 to 4 K. During the measurement, start with the highest temperature and decrease temperature at a rate of 6 K/min (*see* **Note 14**).

6. The in-phase susceptibility $\chi'(T)$ curves will display a peak at a temperature that decreases as the applied excitation field frequency decreases. It is assumed that at this peak of the $\chi'(T)$ curve the condition $\Omega\tau = 1$ applies. To use the Vogel-Fulcher model to interpret the temperature dependence of DMS data [30, 31], plot $\ln(1/\Omega)$ versus $1/T$ and compare to the equation

$$\ln\frac{1}{\Omega} = \ln\tau_0 + \left(\frac{KV_{\mathrm{m}}}{k_{\mathrm{B}}(T - T_0)}\right). \tag{10}$$

$V_{\mathrm{m}}$ is the magnetic volume. The graph should be linear, with the slope providing an estimate of $KV_{\mathrm{m}}$ and the infinite temperature intercept being a measure of $\tau_0$ (which should be in the range of $10^{-9} - 10^{-14}s$ [32], otherwise indicating significant interactions and casting doubt on the accuracy of the value of $K$).

7. To account for sample polydispersity, the expression in Eq. (10) is multiplied by the geometric deviation $\ln\sigma_{\mathrm{g}}$

$$\ln\frac{1}{\Omega} = \ln\tau_0 + \left(\frac{KV_{\mathrm{m}}}{k_{\mathrm{B}}(T - T_0)}\right)\exp\left(\frac{9}{2}\ln\sigma_{\mathrm{g}}^2\right) \tag{11}$$

In these equations, $T_0$ can be determined using the methods described in Subheading 3.6, or can be assumed to be $T_0 = 0$ K in the case of fitting to the Néel model.

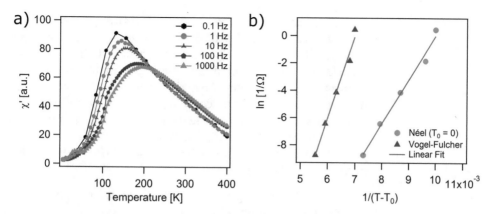

**Fig. 5** (**a**) In-phase component of the dynamic susceptibility with frequency for iron oxide nanoparticles in a poly(styrene-divinylbenzene) matrix. (**b**) Inverse applied field frequency as a function of the inverse temperature corresponding to the peak of in-phase component of dynamic susceptibility using the Néel and the Volger-Fuchler model

8. Figure 5a shows an example of the $\chi'(T)$ curve and Fig. 5b shows the corresponding plot of $\ln(1/\Omega)$ versus $1/(T - T_0)$ using the interaction temperature parameter determined with the Vogel-Fulcher model explained in Subheading 3.6 and assuming $T_0 = 0$ K for the Néel model. The sample was the same sample used in Subheadings 3.5 and 3.6, iron oxide nanoparticles embedded in a PSDVB matrix.

Table 1 summarizes the values of the anisotropy constant calculated using the temperature-dependent dynamic magnetic susceptibility measurements for magnetic nanoparticles with $D_m = 14$ nm, $\ln\sigma_g = 0.246$, and $V_m = 1370$ nm$^3$, calculated in **step 5** of Subheading 3.5 and the value of $T_0 = 43.2$ K determined from the low field MH measurements described in **step 6** of Subheading 3.6. The calculated values are larger than the magnetocrystalline anisotropy constant of bulk magnetite 13.5 kJ/m$^3$ [29], but decrease when the polydispersity of the magnetic diameter and the interaction temperature parameter are included in the analysis.

### 3.9 Dynamic Magnetic Susceptibility as a Function of Frequency

Dynamic magnetic susceptibility measurements can also be used to determine the mechanism of magnetic relaxation (Néel or Brownian) of a collection of nanoparticles, to estimate the hydrodynamic diameter distribution of particles with predominant Brownian relaxation suspended in a medium of known viscosity, and to determine the viscosity of a liquid with nanoparticles of known hydrodynamic diameter distribution. In these measurements, the dynamic magnetic susceptibility of a sample is measured in a small amplitude oscillating magnetic field at constant temperature and as a function of the frequency of the oscillating magnetic field. Ideally, the amplitude of the oscillating magnetic field remains constant in the whole frequency range of the measurement. However, if the amplitude of the oscillating magnetic field is small enough that the dynamic response is linear with the field amplitude, the measurement can still be completed even if the amplitude of the oscillating field decreases with frequency. Analysis of the frequency dependence of the dynamic magnetic susceptibility can be made for nanoparticles in suspension, for nanoparticles at various temperatures, for nanoparticles suspended in liquids of different viscosities, and for nanoparticles in a solid matrix to obtain information on the mechanism of magnetic relaxation.

1. Suspend the magnetic nanoparticles in a liquid matrix as explained in Subheading 3.3.

2. Set the desired sample temperature (*see* **Note 15**).

3. Set the amplitude of the oscillating magnetic field at a value in the range of 2–5 Oe.

4. Measure the dynamic magnetic susceptibility of the sample as a function of decreasing frequency of oscillation (*see* **Notes 16 and 17**).

5. Use the Debye model to interpret the measurements of DMS as a function of frequency and obtain information of the hydrodynamic diameter of nanoparticles with predominant Brownian relaxation mechanism. According to the model, when an alternating magnetic field of frequency $\Omega$ is applied to the sample, the in-phase and out-of-phase components of the dynamic magnetic susceptibility are given by

$$\chi' = \chi_\infty + \frac{\chi_0 - \chi_\infty}{1 + \Omega^2 \tau^2}; \ \chi'' = \frac{(\chi_0 - \chi_\infty)\Omega\tau}{1 + \Omega^2 \tau^2} \tag{12}$$

where $\chi_0$ is the low frequency susceptibility, $\chi_\infty$ is the high frequency susceptibility, and $\tau$ is the relaxation time. From Eq. (12), the real component decreases as the frequency increases, whereas the imaginary component has a maximum at $\Omega_{peak}\tau = 1$.

6. To account for polydispersity of nanoparticle hydrodynamic diameters, the susceptibility $\chi''$ in Eq. (12) can be weighed using a lognormal size distribution $n_v(D)$.

$$\chi' = \int_{D_{h,0}}^{D_{h,\infty}} n_v(D_h) \times \left[ \chi_\infty + \frac{\chi_0 - \chi_\infty}{1 + \Omega^2 \tau^2(D_h)} \right] dD_h \tag{13}$$

$$\chi'' = \int_{D_{h,0}}^{D_{h,\infty}} n_v(D_h) \times \left[ \frac{(\chi_0 - \chi_\infty) \times \Omega\tau(D_h)}{1 + \Omega^2 \tau^2(D_h)} \right] dD_h \tag{14}$$

$$n_v(D_h) = \frac{1}{\sqrt{2\pi}D_h \ln \sigma_g} \exp\left(\frac{\ln(D_h/D_{hgv})}{2(\ln^2 \sigma_g)}\right) \qquad (15)$$

where $D_{hgv}$ is the volume weighted hydrodynamic diameter, $\ln \sigma_g$ is the geometric deviation, and $\tau(D_h)$ is the volume weighted relaxation time corresponding to the volume weighted diameter.

7. Using the relation $\tau = 1/\Omega_{peak}$, obtain the effective relaxation time for a collection of monodisperse nanoparticles. For particles that relax by the Brownian mechanism, $\tau = \tau_B$, and the hydrodynamic diameter of the nanoparticles can be calculated using the peak frequency and the relation

$$D_h = \sqrt{\frac{2k_B T}{\pi \eta \Omega_{peak}}} \qquad (16)$$

8. Similarly, for nanoparticles with predominant Néel relaxation the peak frequency could in principle be used to determine the magnetic diameter of the nanoparticles or the anisotropy constant. However, in those cases, the peak frequency usually occurs at frequencies that are outside the range of commercially available equipment.

9. To verify if the observed peak corresponds to a Brownian peak, measurements at different temperatures or solvents with different viscosities can be used to observe a shift in the peak frequency due to changes in solvent viscosity, whereas the frequency corresponding to the Néel peak would remain constant [20].

10. Plot the DMS spectra of the magnetic nanoparticle samples. Figure 6 shows examples of DMS spectra for two magnetic nanoparticle samples (cobalt ferrite and iron oxide) in 1-octadecene ($\eta = 0.0412 \; Pa \cdot s$) at 298 K. The difference in the shape of the curves is attributed to the fact that cobalt ferrite nanoparticles possess a much larger anisotropy constant than the iron oxide nanoparticles, and therefore the cobalt ferrite nanoparticles have predominant Brownian relaxation whereas the iron oxide nanoparticles have predominant Néel relaxation. For the iron oxide nanoparticles the peak frequency corresponding to the inverse of the Néel relaxation time would appear at a frequency that is much higher than the maximum frequency that the instrument can apply.

11. Figure 6a corresponds to the DMS spectra of iron oxide nanoparticles with $D_h = 17 \; nm$ ($\ln \sigma_g = 0.065$) and $D_m = 14 \; nm$ ($\ln \sigma_g = 0.246$). Assuming the anisotropy constant for the iron oxide nanoparticles corresponds to the bulk value $K = 13 kJ/mol$ [33], the Néel relaxation time of the particles is

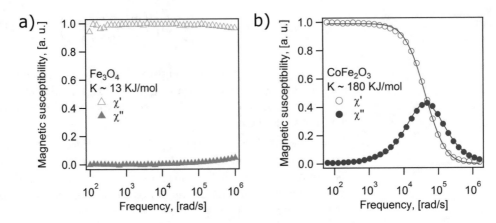

**Fig. 6.** DMS spectra for liquid samples of (**a**) iron oxide and (**b**) cobalt ferrite nanoparticles in 1-octadecene at 298 K

estimated to be $\tau_N \approx 9.4 \times 10^{-8}$s using Eq. (2), whereas using the hydrodynamic diameter the Brownian relaxation time is estimated to be $\tau_B \approx 7.73 \times 10^{-6}$s using Eq. (1). Since $\tau_N \ll \tau_B$, the nanoparticles respond primarily through the Néel relaxation mechanism. The corresponding peak frequencies for these two characteristic relaxation times would be $2.06 \times 10^4$ rad/s for the Brownian mechanism and $1.7 \times 10^6$ rad/s for the Néel relaxation mechanism. The fact that there is no peak at $2.06 \times 10^4$ rad/s in Fig. 6a indicates that the Brownian mechanism is not dominant for these nanoparticles. Instead, the Néel relaxation mechanism appears to dominate, but the peak cannot be observed as it is expected to lie over one order of magnitude above the frequency range of the instrument. In fact, it appears that the low-frequency tail of the peak is evident at the highest frequencies in the plot of Fig. 6a, although it is difficult to tell if this is instrument artifact introduced at the extreme frequency range for the instrument.

12. The DMS spectrum of Fig. 6b may be interpreted using similar arguments. In this case the nanoparticles are cobalt ferrite, with a bulk anisotropy constant of $K \sim 180$ kJ/mol [34]. According to dynamic light scattering measurements of these nanoparticles, the hydrodynamic diameter is $D_h = 20$ nm ($\ln\sigma_g = 0.051$). Using Eqs. (1) and (2), the expected Néel relaxation time would be $\tau_N \approx 10^{39}$s, whereas the expected Brownian relaxation time would be $\tau_B \approx 1.26 \times 10^{-5}$s. Since $\tau_B \ll \tau_N$, the nanoparticles respond primarily through the Brownian relaxation mechanism. The corresponding peak frequencies are $10^{-40}$rad/s for the Néel mechanism and $7.94 \times 10^3$ rad/s for the Brownian mechanism. The experimental peak frequency Fig. 6b is $\Omega_{peak} = 4.78 \times 10^4$ rad/s ($7.61 \times 10^3$ Hz). The

close agreement between the experimental and Brownian peak frequencies indicates that this sample responds predominantly through Brownian relaxation.

# 4  Notes

1. Any compatible solid matrix can be used to restrict nanoparticle rotation, for example paraffin wax, docosane, or higher hydrocarbons. A solid polymer matrix of polystyrene–divinyl benzene is used for this protocol due to its high thermal resistance, allowing measurements at up to 400 K.

2. Any compatible solid matrix can be used to restrict nanoparticle rotation, for example silica, agar, etc. A solid polymer matrix of TEGDMA is used for this protocol due to its higher thermal resistance, allowing measurements at up to 400 K.

3. The membrane used should be selected based on its compatibility with the nanoparticle surface chemistry. The most commonly used filters are nylon filters for hydrophilic solutions, i.e., nanoparticles suspended in water-based solvents, and PTFE filters for hydrophobic solutions, i.e., nanoparticles suspended in organic solvents. The size of the filter is selected based on the particle size. Typically, 0.2 μm filters are used for nanoparticles with a size range between 10 and 100 nm. After filtration, the solution may look diluted. This is because some particles and/or particle aggregates are retained by the filter. One should quantify the magnetic concentration of the solution after filtration using a suitable method (*see* **Note 6**). Some particles tend to aggregate or become unstable in solution. We recommend not to filter such solutions since the particles are trapped by the filter.

4. SQUID magnetometers are designed to be extremely sensitive ($10^{-8}$ emu), whereas the most commonly used vibrating sample magnetometer with an inductive pick up coil are less sensitive ($10^{-6}$ emu) but can make faster measurements.

5. Work with all chemicals inside a fume hood. The stock solutions of monomers and initiator must have a 6.66:1 volume ratio of styrene to DVB, and 3.3 mg of AIBN per 1 mL of styrene. We scale up the solution volume to 3 mL to accurately weigh the AIBN. The monomer/initiator solution can be stored at 4 °C for 1 month.

6. This is the concentration of inorganic magnetic core, i.e., not counting the ligands or polymers on the particle surface. To study magnetic properties, determining the inorganic core content is important. A few commonly used techniques include quantification using UV spectrometric assays [35], inductively

coupled plasma mass spectrometry (ICP-MS) [36], electron paramagnetic resonance [37], and thermal gravimetric analysis [38]. Of these, we prefer EPR, ICP-MS, and UV spectrophotometric assays because they can determine the amount of iron in a sample accurately. For UV spectrophotometric and ICP-MS assays care should be taken while digesting samples in concentrated $HNO_3$ or HCl during sample preparation.

7. Although the concentration of magnetic core can be increased, we have found that for concentrations of 1 wt% (mg of nanoparticles per mg of polymer) and higher, the calculated value of magnetic properties such as anisotropy constant and the characteristic time are apparent and represent the effective property of the collection of nanoparticles and not an intrinsic property [33].

8. Typically, to demonstrate superparamagnetic behavior in a sample fixed in a solid matrix the field-dependent magnetization curves are recorded at temperatures above the blocking temperature. At temperatures below the blocking temperature, the magnetization will reveal hysteresis loops and the sample coercivity can be measured.

9. Typically, the magnetization is measured for about 100 magnetic field points, logarithmically spaced. This allows one to acquire sufficient data at all logarithmic decades of magnetization, distributed in the field range. The field range available is dependent on the equipment used and must be enough to saturate the ferrimagnets.

10. The low and high temperatures will depend on the actual sample, i.e., transition temperatures of the sample, and equipment temperature range. Polymer matrices have a higher melting temperature (PSDVB >400 K) than long hydrocarbon chains (docosane 315 K, paraffin 326 K). We recommend using the widest possible range for the sample and instrument.

11. We recommend working with the lowest temperature first and increasing the temperature in 30 K increments. The temperature range of the experiment depends on the transition temperatures of the sample and equipment temperature range.

12. The blocking temperature becomes a function of the magnitude of this field. At very low field values the blocking temperature is constant, but above 100 Oe it certainly decreases with increasing applied field [39]. The analysis for the anisotropy constant is only valid for the range in which the blocking temperature does not vary with field strength.

13. For better resolution, we recommend measuring the magnetic moment continuously as the field is increased. Also, it is recommended to use small temperature increments to increase the analysis resolution. The blocking temperature will also be a function of the sweep rate for the temperature. This is evident

from Eq. (7) if one realizes that the anisotropy constant does not vary, but the observation time changes with temperature sweep rate [29].

14. We prefer to sweep temperature at fixed frequency, warming the sample after each sweep before starting a new sweep for another frequency. We find this mode of measurement is fast using a Quantum Design MPMS-3, because the instrument can measure the ac susceptibility while it sweeps the field. However, one could also step temperature and measure each frequency at fixed temperature during the cool-down step. We find this mode of measurement to be more effective on a Quantum Design MPMS-XL.

15. Since the frequency of the Brownian peak depends on the viscosity of the carrier fluid, the peak frequency can be shifted to the frequency range of the instrument by changing the temperature of the measurement or the viscosity of the carrier liquid. During calculations use the appropriate sample temperature and solvent viscosity.

16. One can also measure the dynamic magnetic susceptibility of the sample as the frequency of oscillation is increased. Both the methods should yield similar results. It is recommended to use low concentrations of particles in solution since the relaxation times given in Eqs. (1) and (2) only apply for infinitely dilute systems with no particle-particle interactions.

17. The oscillation frequency of some commercially available instruments is in the range of a few Hz to kHz to 100 s of kHz, corresponding to over four orders of magnitude in range. For better resolution measure the DMS for at least ten frequencies per decade, using logarithmically spaced oscillation frequencies.

## Acknowledgments

This work was supported by the US National Science Foundation (CBET-143993) and by the US National Institutes of Health (1R01AR068424-01).

## References

1. Pankhurst QA et al (2009) Progress in applications of magnetic nanoparticles in biomedicine. J Phys D Appl Phys **42** Pag 224001 (15 pp)

2. Roca AG et al (2009) Progress in the preparation of magnetic nanoparticles for applications in biomedicine. J Phys D Appl Phys 42 Pag 224002 (11 pp)

3. Alvarez-Berrios MP et al (2014) Magnetic fluid hyperthermia enhances cytotoxicity of bortezomib in sensitive and resistant cancer cell lines. Int J Nanomedicine 9:145–153

4. Kim DH, Nikles DE, Brazel CS (2010) Synthesis and characterization of multifunctional chitosan- $MnFe_2O_4$ nanoparticles for

magnetic hyperthermia and drug delivery. Materials 3(7):4051–4065

5. Bonini M, Berti D, Baglioni P (2013) Nanostructures for magnetically triggered release of drugs and biomolecules. Curr Opin Colloid Interface Sci 18(5):459–467

6. Hu SH et al (2008) Core/single-crystal-shell nanospheres for controlled drug release via a magnetically triggered rupturing mechanism. Adv Mater 20(14):2690–2696

7. Kami D et al (2011) Application of magnetic nanoparticles to gene delivery. Int J Mol Sci 12 (6):3705–3722

8. Prijic S et al (2010) Increased cellular uptake of biocompatible superparamagnetic iron oxide nanoparticles into malignant cells by an external magnetic field. J Membr Biol 236 (1):167–179

9. Na HB, Song IC, Hyeon T (2009) Inorganic nanoparticles for MRI contrast agents. Adv Mater 21(21):2133–2148

10. Qiao RR, Yang CH, Gao MY (2009) Superparamagnetic iron oxide nanoparticles: from preparations to in vivo MRI applications. J Mater Chem 19(35):6274–6293

11. Dhavalikar R et al (2015) Ferrohydrodynamic modeling of magnetic nanoparticle harmonic spectra for magnetic particle imaging. J Appl Phys 118(17)

12. Goodwill PW et al (2012) X-Space MPI: magnetic nanoparticles for safe medical imaging. Adv Mater 24(28):3870–3877

13. Krishnan KM (2010) Biomedical nanomagnetics: a spin through possibilities in imaging, diagnostics, and therapy. IEEE Trans Magnet 46(7):2523–2558

14. Neel L (1953) Thermoremanent magnetization of fine powders. Rev Mod Phys 25 (1):293–296

15. Chantrell RW, Popplewell J, Charles SW (1978) Measurements of particle size distribution parameters in ferrofluids. IEEE Trans Magnet 14(5):975–977

16. Rosensweig RE (2013) Ferrohydrodynamics. Courier Corporation. Chapter 2, Pag 33–73

17. Jiles D (2015) Introduction to magnetism and magnetic materials. CRC Press, Boca Raton, FL

18. Fannin PC, Perov PA, Charles SW (1999) Complex susceptibility measurements of magnetic fluids over the frequency range 50 MHz to 18 GHz. J Phys D Appl Phys 32 (14):1583–1586

19. Shliomis M, Stepanov V (1994) Relaxation phenomena in condensed matter. Advances in chemical physics series. Wiley, New York, p. 1

20. Calero-DdelC VL, Santiago-Quinonez DI, Rinaldi C (2011) Quantitative nanoscale viscosity measurements using magnetic nanoparticles and SQUID AC susceptibility measurements. Soft Matter 7(9):4497–4503

21. Dobrotă C-I, Stancu A (2013) What does a first-order reversal curve diagram really mean? A study case: array of ferromagnetic nanowires. J Appl Phys 113(4):043928

22. Carvallo C, Muxworthy AR, Dunlop DJ (2006) First-order reversal curve (FORC) diagrams of magnetic mixtures: micromagnetic models and measurements. Phys Earth Planet In 154(3–4):308–322

23. Bruvera I et al (2015) Determination of the blocking temperature of magnetic nanoparticles: The good, the bad, and the ugly. J Appl Phys 118(18):184304

24. Fabian K (2003) Some additional parameters to estimate domain state from isothermal magnetization measurements. Earth Planet Sci Lett 213(3–4):337–345

25. Caron L et al (2009) On the determination of the magnetic entropy change in materials with first-order transitions. J Magn Magn Mater 321(21):3559–3566

26. Cornell RM, Schwertmann U (2003) The iron oxides: structure, properties, reactions, occurrences and uses. John Wiley, New York

27. Shtrikman S, Wohlfarth EP (1981) The theory of the Vogel-Fulcher law of spin-glasses. Phys Lett A 85(8-9):467–470

28. Bradbury A et al (1984) Magnetic size determination for interacting fine particle systems. IEEE Trans Magnet 20(5):1846–1848

29. Goya GF et al (2003) Static and dynamic magnetic properties of spherical magnetite nanoparticles. J Appl Phys 94(5):3520–3528

30. Taketomi S (1998) Spin-glass-like complex susceptibility of frozen magnetic fluids. Phys Rev E 57(3):3073–3087

31. Zhang JL, Boyd C, Luo WL (1996) Two mechanisms and a scaling relation for dynamics in ferrofluids. Phys Rev Lett 77(2):390–393

32. Monson TC et al (2013) Large enhancements of magnetic anisotropy in oxide-free iron nanoparticles. J Magn Magn Mater 331:156–161

33. del Castillo VLCD, Rinaldi C (2010) Effect of sample concentration on the determination of the anisotropy constant of magnetic nanoparticles. IEEE Trans Magnet 46(3):852–859

34. Tung LD et al (2003) Magnetic properties of ultrafine cobalt ferrite particles. J Appl Phys 93 (10):7486–7488

35. ASTM (2000) Standard test method for iron in trace quantities using the 1,10-phenanthroline

method. In: ASTM standard E394-09, West Conshohocken, PA

36. Boutry S et al (2009) How to quantify iron in an aqueous or biological matrix: a technical note. Contrast Media Mol Imaging 4(6):299–304

37. Danhier P et al (2012) Electron paramagnetic resonance as a sensitive tool to assess the iron oxide content in cells for MRI cell labeling studies. Contrast Media Mol Imaging 7(3):302–307

38. Mahdavi M et al (2013) Synthesis, surface modification and characterisation of biocompatible magnetic iron oxide nanoparticles for biomedical applications. Molecules 18 (7):7533

39. Goya, G.F. and M. Morales.(2004) Field dependence of blocking temperature in magnetite nanoparticles. J Metast Nanocryst Mater Trans Tech Publ

# Chapter 5

# Preparation of Magnetic Nanoparticles for Biomedical Applications

## Xiaolian Sun and Shouheng Sun

## Abstract

Magnetic nanoparticles have obtained great attention in the field of biomedicine in recent years owing to their excellent biocompatibility, unique magnetic properties, and ease of functionalization. Potential applications for functionalized magnetic nanoparticles span biomedical imaging, treatment via magnetic hyperthermia, drug delivery, and biosensing. This chapter provides detailed procedures for the synthesis, PEGylation, and bioconjugation of monodispersed $Fe_3O_4$ nanoparticles, hollow $Fe_3O_4$ nanoparticles, porous hollow $Fe_3O_4$ nanoparticles, and dumbbell-like Au-$Fe_3O_4$ nanoparticles. We hope this article can help readers design and reproducibly prepare high-quality magnetic nanoparticles with their specific goals in mind.

Key words Organic phase synthesis, Iron oxide nanoparticles, PEGylation, EDC/NHS coupling, IgG

## 1 Introduction

Magnetic nanoparticles (NPs) are an emerging class of magnetic probes for biomedical applications. They can be made to show superparamagnetism with high susceptibility at biologically relevant temperatures [1–4]. In the presence of an external magnetic field (either permanent or alternating), their magnetization direction can be aligned along the field direction. Upon removal of the field, their magnetization directions are randomized due to temperature-induced thermal relaxation and their overall magnetic moment is reduced to zero, minimizing magnetic interactions among the NPs and, as a result, stabilizing the NP dispersion. Once introduced into biological systems, these stable and strongly superparamagnetic NPs can serve as desirable probes for biological imaging, drug delivery, and therapeutic applications [5].

Iron oxide NPs, especially magnetite $Fe_3O_4$ NPs, have been studied extensively for biomedical applications due to their favorable biocompatibility and high magnetic moments [6–8]. These $Fe_3O_4$ NPs are commonly prepared in aqueous solution and have

Sarah Hurst Petrosko and Emily S. Day (eds.), *Biomedical Nanotechnology: Methods and Protocols*, Methods in Molecular Biology, vol. 1570, DOI 10.1007/978-1-4939-6840-4_5, © Springer Science+Business Media LLC 2017

found applications in clinical imaging. For example, Feridex, a cluster of iron oxide NPs with sizes of 120–180 nm, has already been approved by the FDA for clinical liver imaging. Despite the progress made thus far in aqueous phase syntheses, these conventional approaches have certain limits with regards to controlling NP dimensions and structures and their resulting magnetic performance. Furthermore, recent studies indicate that to maximize medical detection sensitivity and accuracy, new multi-modality probes should be developed. This requires $Fe_3O_4$ NPs to couple with other NPs with different functionalities [1, 8], which has been challenging to achieve via aqueous phase methods. This limitation has motivated the vigorous search for an alternative approach that can be used to synthesize $Fe_3O_4$-based NPs via organic phase reactions.

Compared with aqueous phase syntheses, ones in the organic phase have the following advantages: (1) an organic solvent can be either polar or nonpolar, which allows a wide variety of surfactants to be chosen for NP stabilization; (2) depending on the solvent used, the synthetic reaction can be controlled at temperatures higher than 100 °C or lower than 0 °C, outside the range that aqueous solution-based methods allow (0–100 °C); the nucleation and growth of $Fe_3O_4$ NPs with tighter dimensions and a more controlled structure can be facilitated in the organic phase; (3) an organic solvent is chemically inert under the synthetic conditions, making it possible to couple $Fe_3O_4$ NPs with other functional components to achieve multifunctionality within the composite NP structure. Here, we summarize a common method of preparing monodisperse magnetic NPs in an organic phase reaction at temperatures between 200 and 300 °C [9, 10]. By controlling the reaction parameters including the types and concentrations of the metal precursor/surfactant, reaction temperature, and reaction time, we have been able to finely tune NP sizes, shapes, and compositions, as well as composite multifunctionality.

Magnetic NPs prepared from a nonpolar hydrocarbon solvent or a weakly polar ether solvent can be stabilized using a long chain bipolar surfactant. Oleic acid and oleylamine are two common surfactants applied for NP surface passivation via –COO- and/or $-NH_2$-bonding to the NP surface and for NP dispersion stabilization via the presence of a "thick" hydrocarbon coating. Such NPs can be dispersed easily in a nonpolar solvent, such as hexane, or a polar solvent, such as methylene chloride or chloroform, but they are not dispersible in water or aqueous biological solutions. Before applying these NPs for biological uses, they must be modified such that they exhibit not only dispersion stability, but also the desired bio-circulation, bio-distribution, and bio-elimination properties. Efforts have been made to understand the interactions between such dispersible NPs and biological systems; indeed, this is an extremely active research area in nanomedicine nowadays [11, 12].

Here, we highlight two examples that have been demonstrated to be reliable in our labs for NP functionalization via surfactant exchange or surfactant addition reactions. We further present an example in which the modified NPs were coupled to IgG antibodies via common EDC/NHS chemistry.

# 2  Materials

## 2.1  Organic Phase Synthesis

### 2.1.1  Chemicals

Prepare and store all reagents at room temperature (unless indicated otherwise). Diligently follow all waste disposal regulations when disposing of waste materials.

1. $Fe(acac)_3$ (acac = aceylacetonate), (99.9% trace metals basis, Sigma-Aldrich).

2. Phenyl ether (>99%, Sigma-Aldrich).

3. 1,2-Hexadecanediol (>98%, Sigma-Aldrich).

4. Oleic acid (OA, technical grade, 90%, Sigma-Aldrich).

5. Oleylamine (OAm, technical grade, >70%, Sigma-Aldrich).

6. Iron pentacarbonyl ($Fe(CO)_5$, >99.9% trace metals basis, Sigma-Aldrich, stored at 4 °C under Ar protection).

7. 1-Octadecene (ODE, technical grade, 90%, Sigma-Aldrich).

8. Trimethyl amine N-oxide (98%, Sigma-Aldrich).

9. Benzyl ether (98%, Sigma-Aldrich).

10. Hydrogen tetrachloroaurate (III) hydrate ($HAuCl_4 \cdot 3H_2O$) (99.999% trace metals basis, Strem chemicals).

11. 1,2,3,4-Tetrahydronaphthalene (Tetralin, anhydrous, 99%, Sigma-Aldrich).

12. tert-Butylamine-borane complex (TBAB, 97%, Sigma-Aldrich).

13. Hexane (reagent grade).

14. Ethanol (reagent grade, >99.9%).

15. Isopropanol (reagent grade, >99.5%).

### 2.1.2  Synthetic Setup

1. Glasswares (Ace Glass, Inc).

2. Digital temperature controller (Dyna-Sense®, VWR).

3. Magnetic stirrer (IKA) with a maximum stirring speed up to 1500 rpm (1680 × $g$).

## 2.2  NP Surface Modification

1. Chloroform (reagent grade).

2. $N,N$-Dimethylformamide (anhydrous, 99.8%).

3. Dopamine hydrochloride (reagent grade, >98%).

4. NHS-PEG-COOH   (NHS  =   N-hydroxysuccinimide, PEG = polyethylene glycol, molecular weight of 3 K, Nanocs, stored at −20 °C).

5. Phospholipid-PEG-COOH (molecular weight of 2 K, Avanti Polar Lipids, stored at −20 °C).

6. Dialysis tubing cellulose membrane (with a typical molecular weight cutoff of 5 K, Sigma-Aldrich, stored at 4 °C).

7. Solution A: Weigh and transfer 6 mg of NHS-PEG-COOH (2 μmol) to a 10 mL glass vial and add 2 mL of chloroform to dissolve the molecule.

8. Solution B: Weigh and transfer 0.38 mg of dopamine (2 μmol) to another glass vial. Dissolve the dopamine in 1 mL of $N,N$-dimethylformamide (DMF). Add 0.5 mg of $Na_2CO_3$ powder.

9. Solution E. Weigh and transfer 3 mg of phospholipid-PEG-COOH into a glass vial. Add 1 mL of chloroform to dissolve the powder.

## 2.3   NP Conjugation with IgG

Prepare all aqueous solutions using ultrapure water (prepared by purifying deionized water via a Millipore ultra-pure system to obtain a resistivity of ~18 MΩ cm at 25 °C).

1. 1-Ethyl-3-[3-dimethylaminopropyl]carbodiimide hydrochloride (EDC, reagent grade, >98%, stored at −20 °C in a desiccated container).

2. $N$-hydroxysulfosuccinimide   (Sulfo-NHS,   reagent   grade, >98%, stored at −20 °C).

3. IgG (reagent grade, > 95%, stored at −20 °C).

4. Activation   buffer:   2-($N$-morpholino)ethanesulfonic   acid (MES) buffer (0.1 M, pH 6).
     Dissolve  2.13  g  MES  (2-($N$-morpholino)ethanesulfonic acid, Sigma-Aldrich, reagent grade) in 80 mL ultrapure water. Adjust the pH to 6.0 with 10 M NaOH and make up to 100 mL total volume. Filter through a 0.2 μm filter.

5. Coupling buffer: 1× Phosphate Buffered Saline (PBS).
     Dissolve 8 g NaCl, 0.2 g KCl, 1.44 g $Na_2HPO_4$, and 0.24 g $KH_2PO_4$ in 800 mL of ultrapure water. Adjust the pH to 7.4 with HCl. Add $H_2O$ to 1 L. Filter through a 0.2 μm filter.

6. Washing buffer: 1× Phosphate Buffered Saline + 0.05% Tween 20 (PBST).
     Dissolve 8 g NaCl, 0.2 g KCl, 1.44 g $Na_2HPO_4$, and 0.24 g $KH_2PO_4$ and 2 mL of Tween-20 in 800 mL of ultrapure water. Adjust the pH to 7.4 with HCl. Add $H_2O$ to 1 L. Filter through a 0.2 μm filter.

**2.4 Sample Characterization**

1. Philips EM 420 (120 kV) transmission electron microscopy (TEM) instrument. The microscope is operated under an accelerating voltage of 120 kV.

2. Bruker AXS D8-Advanced diffractometer with Cu Kα radiation ($\lambda = 1.5418$ Å).

3. Lakeshore*7404 high sensitivity vibrating sample magnetometer (VSM).

4. Malvern Zeta Sizer Nano S-90 dynamic light scattering (DLS).

# 3  Methods

**3.1 Set Up the Synthetic Apparatus (See Note 1).**

Figure 1 shows a schematic setup used in the high-temperature organic phase synthesis of magnetic NPs. The synthetic reaction is carried out in a flask under the protection of an inert gas (Ar or $N_2$). The flask is connected with a temperature probe and temperature

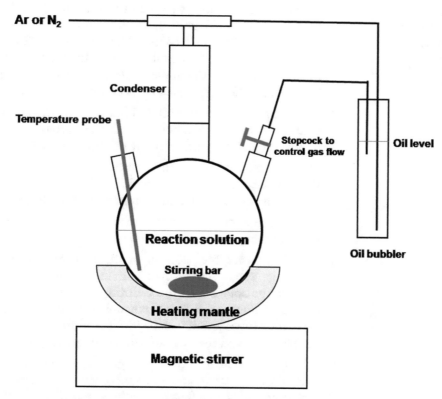

**Fig. 1** Schematic illustration of a common synthetic setup used to prepare magnetic NPs. The setup should be placed in a chemical safety hood with good ventilation. The stopcock illustrated in the figure is used to control the gas flow initially from the reaction system to remove air and moisture. Before heating to high temperature or before adding a volatile chemical into the reaction solution, the stopcock should be closed so that the protection gas can by-pass the reaction system and the reaction system is kept in a positive pressure (determined by the depth of the oil level in the oil bubbler)

control and an air-cooled condenser (note that a water cooling condenser should not be used here due to the high temperature applied to the reaction solution). The outlet of the reaction flask should be sealed with a rubber septum (or other proper sealing method) and oil bubbler. The inert gas is allowed to flush through the reaction system by leaving the stopcock open, which can help to remove air and moisture from the reaction solution. Before a high temperature is applied to the reaction system or before a volatile precursor is added to the reaction solution, the stopcock should be closed so that the inert gas can by-pass the reaction system and the reaction can proceed under a blanket of inert gas protection. The reaction solution is stirred with a Teflon (or glass)-coated magnetic stirring bar driven by a stand magnetic stirrer. The reaction flask is heated using a hemispherical heating mantle (80 W–115 V), and the heating power should be carefully controlled so that the reaction temperature does not over-shoot the preset value by more than 10 °C.

## 3.2 Synthesis of Magnetic NPs (See Note 2)

### 3.2.1 Synthesis of 4 nm Fe₃O₄ NPs [13]

*3.2.1 Synthesis of 4 nm $Fe_3O_4$ NPs [13]*

1. At room temperature, weigh and transfer 0.706 g Fe(acac)₃ (2 mmol) and 2.58 g 1,2-hexadecanediol (10 mmol) into a reaction flask. Pipette 2.12 mL OA (6 mmol), 2.82 mL OAm (6 mmol), and 20 mL phenyl ether into the flask and then seal the flask outlet with the rubber septum. Set the stirring speed at 900 rpm (1008 × $g$) to mix the solution (*see* **Note 3**).

2. Under a gentle Ar flow with 2–3 bubbles per second, heat the mixture to 120 °C and hold it at this temperature for 1 h to remove air and moisture from the reaction system (*see* **Note 4**).

3. Under a blanket of Ar (by closing the stopcock shown in Fig. 1), heat the solution to 265 °C at a heating rate of 10 °C/min and keep it at this temperature for 2 h (*see* **Note 5**).

4. Remove the system from heat and let the reaction mixture cool to room temperature (*see* **Note 6**).

5. Equally transfer the solution in the flask into four 15 mL centrifuge tubes. Use 4 mL hexane to rinse the flask and transfer the solution to the centrifuge tubes equally. Add to each centrifuge tube 8 mL ethanol and vortex. Centrifuge these four tubes at 8500 rpm (9520 × $g$) for 10 min at room temperature, and then discard the supernatant (*see* **Note 7**).

6. Add 5 mL hexane to each centrifuge tube and vortex the tubes to re-disperse the samples. Add 8 mL of ethanol subsequently to each tube and vortex. Precipitate the product by centrifugation at 8500 rpm (9520 × $g$) for 10 min at room temperature, and then discard the supernatant. Repeat this step once more (*see* **Note 8**).

7. Disperse the product in each tube in 5 mL hexane, and then combine and store the NP dispersion in a vial and seal the vial to prevent hexane evaporation (*see* **Note 9**).

8. Analyze/characterize the sample using TEM, x-ray diffraction (XRD), and VSM to confirm that $Fe_3O_4$ nanoparticles have been prepared before further use (*see* Subheading 3.5).

*3.2.2  Synthesis of Fe NPs [14]*

1. At room temperature, pipette 0.3 mL OAm (0.9 mmol) and 20 mL ODE into a flask and then flush the reaction system with Ar (2–3 bubbles per second). Set the stirring speed at 900 rpm (1008 × *g*) to mix the solution (*see* **Note3**).

2. Heat the mixture to 120 °C for 30 min to remove air and moisture (*see* **Note 4**).

3. Heat the solution to 180 °C at a heating rate of 5 °C/min. As soon as the temperature reaches 180 °C, close the stopcock so that the reaction is now protected under the blanket of Ar. Inject 0.7 mL of $Fe(CO)_5$ into the reaction solution using a syringe and keep the reaction at this temperature for 30 min (*see* **Note 10**).

4. Turn off the heat, remove the heating mantle, and let the reaction mixture cool to room temperature (*see* **Note 11**).

5. Transfer the reaction solution into four equal 15 mL centrifuge tubes, rinse the stir bar with 16 mL hexane, and combine the washing solutions with the reaction solution in four tubes. Add 8 mL of isopropanol to each of the tubes. Vortex them and centrifuge them at 8500 rpm (9520 × *g*) for 10 min at room temperature. Discard the supernatant (*see* **Note 12**).

6. Add 5 mL hexane to each centrifuge tube, and vortex each of them to re-disperse the precipitates. Add 8 mL of ethanol subsequently to each tube and vortex them. Centrifuge the tubes at 8500 rpm (9520 × *g*) for 10 min at room temperature. Discard the supernatants. Repeat this process another time (*see* **Note 8**).

7. Add 5 mL of hexane in each tube to disperse the precipitates, and then store the products in a well-sealed vial under $N_2$. Use the products as soon as possible (*see* **Note 13**).

*3.2.3  Synthesis of Hollow Fe₃O₄ NPs [15]*

1. At room temperature, weigh and transfer 30 mg of trimethyl amine N-oxide to the flask and then pipette 2 mL of methanol into the flask to dissolve it. After the white power is completely dissolved, add 20 mL of ODE to the flask. Set the stirring speed to 900 rpm (1008 × *g*) and mix the solution (*see* **Note 14**).

2. Under a gentle argon flow with 2–3 bubbles per second, heat the mixture to 130 °C for 1 h to remove air, moisture, and other low boiling solvents (*see* **Note 15**).

3. Add (via a syringe) 2 mL of the hexane dispersion of Fe NPs prepared above to the reaction mixture, and keep the mixture

heated under the gentle Ar flow at 130 °C for 2 h to remove hexane (*see* **Note 16**).

4. Heat the solution to 250 °C at a heating rate of 2 °C/min and keep heating at this temperature for 1 h (*see* **Note 17**).

5. Follow **steps 4–8** in Subheading 3.2.1 (*see* **Note 18**) to obtain a hexane dispersion of hollow $Fe_3O_4$ NPs.

*3.2.4 Synthesis of Porous Hollow $Fe_3O_4$ NPs [15]*

1. At room temperature, pipette 0.16 mL of OA (0.5 mmol), 0.17 mL of OAm (0.5 mmol), and 20 mL of benzyl ether into the flask. Pipette 2 mL of the hexane dispersion of hollow $Fe_3O_4$ NPs prepared above into the mixture and then seal the inlet with a rubber septum. Set the stirring speed to 900 rpm ($1008 \times g$) (*see* **Note 3**).

2. Let Ar flow gently (2–3 bubbles per second) through the reaction system. Heat the mixture to 130 °C for 1 h to remove hexane, air, and moisture (*see* **Note 4**).

3. Heat the solution to 260 °C at a heating rate of 5 °C/min and keep heating at this temperature for 30 min (*see* **Note 19**).

4. Repeat **steps 4–8** in Subheading 3.2.1.

*3.2.5 Synthesis of 4 nm Au NPs [16]*

1. Weigh and transfer 0.2 g of $HAuCl_4 \cdot 3H_2O$ (0.5 mmol) into a 25 mL glass vial. Pipette 10 mL OAm and 10 mL tetralin into the vial to dissolve the Au salt (*see* **Note 20**).

2. Weigh and transfer 87 mg of TBAB (1 mmol) into a 5 mL glass vial. Pipette 2 mL of OAm and 2 mL tetralin into the vial. Sonicate for 10 min to dissolve the TBAB (*see* **Note 21**).

3. Inject the TBAB solution into the $HAuCl_4$ solution under vigorous stirring and let the reaction proceed for 1 h (*see* **Note 22**).

4. Follow **steps 5–8** in Subheading 3.2.1 to obtain a hexane dispersion of Au NPs.

*3.2.6 Synthesis of Dumbbell-like Au-$Fe_3O_4$ NPs [17]*

1. Calculate the concentration of the Au NP solution: Weigh an empty tube ($W_1$), transfer 100 μL of the Au NP dispersion into the tube, remove hexane under a nitrogen flow, and weigh the tube again ($W_2$) to obtain the Au NP weight ($W_2–W_1$) and concentration ($W_2–W_1$)/0.1 (in mg/mL) (*see* **Note 23**).

2. At room temperature, pipette 1 mL of OA (3 mmol), 1.1 mL of OAm (3 mmol), 2.5/($W_2–W_1$) mL of Au NPs (25 mg), and 20 mL ODE into the flask. Set the stirring speed to 900 rpm ($1008 \times g$) and mix the solution (*see* **Note 3**).

3. Let Ar flow gently (2–3 bubbles per second) through the reaction system. Heat the mixture to 120 °C for 30 min to remove air, hexane, and moisture (*see* **Note 4**).

4. Close the stopcock of the flask outlet and inject 0.15 mL of $Fe(CO)_5$. Heat the solution to 310 °C at a heating rate of 25 °C/min. Keep it refluxing for 30 min (*see* **Notes 10** and **24**).

5. Follow **steps 5–8** in Subheading 3.2.1 to obtain a hexane dispersion of dumbbell-like NPs.

### 3.3 NP Surface Modification with PEG

#### 3.3.1 Concentration Calculation

1. Transfer 0.5 mL of a hexane dispersion of NPs (including $Fe_3O_4$ NPs, Fe NPs, hollow $Fe_3O_4$ NPs, porous hollow $Fe_3O_4$ NPs, and dumbbell-like Au-$Fe_3O_4$ NPs) to a 2 mL centrifuge tube, add 1 mL of ethanol and vortex. Centrifuge at 8500 rpm ($9520 \times g$) for 10 min at room temperature to separate the product. Re-disperse the product into 0.5 mL of chloroform (*see* **Note 25**).

2. Calculate the NP concentration: Weigh an empty tube ($W_1$), transfer 100 μL of solution into the tube, dry it completely under a nitrogen flow, and weigh the tube again ($W_2$) to obtain the NP weight ($W_2$–$W_1$) and concentration ($W_2$–$W_1$)/0.1 (in mg/mL) (*see* **Note 26**).

#### 3.3.2 Ligand Exchange via Dopamine-PEG-COOH [15]

1. Add solution A to solution B dropwise to make solution C and keep the mixture stirring at room temperature for 3 h (*see* **Note 27**).

2. Transfer 1 mg of NPs into a glass vial. Add chloroform to a total volume of 1 mL (*see* **Note 28**). This is solution D.

3. Add solution D to solution C and shake for 24 h (*see* **Note 29**).

4. Add 2 mL of hexane to the above solution. Use a magnetic bar to collect the precipitate. If the supernatant is clear and colorless, then discard it. If the supernatant shows a deep brownish color, indicating the presence of NPs, then transfer the supernatant to a 15 mL centrifuge tube, and centrifuge at 8000 rpm ($8960 \times g$) for 1 min to collect the precipitate. Dry the precipitate under a gentle $N_2$ flow for 1 h. Dissolve the product in 1 mL of ultrapure water (*see* **Note 30**).

5. Transfer the aqueous solution to a dialysis tube and dialyze it for 2 days to remove unbound polymer and excess dopamine. Filter the purified samples through the 0.2 μm filter to remove any large aggregates and store the aqueous dispersion at 4 °C (*see* **Note 31**).

6. Use TEM and DLS to ensure that the NP modification process was successful (*see* Subheading 3.5).

#### 3.3.3 Ligand Addition via Phospholipid-PEG-COOH [18]

1. Add solution E into solution D dropwise under vigorous stirring. Seal the vial and keep stirring for 2 h at room temperature (*see* **Note 32**).

2. Dry the product under a gentle nitrogen flow. Add 1 mL of ultrapure water to dissolve the product (*see* **Note 33**).

3. Follow **steps** 7 and **8** in Subheading 3.3.2 to obtain PEG-modified NPs.

### 3.4 Conjugation of the PEG-NPs with IgG (See Note 34)

1. Prepare a fresh EDC/NHS-sulfo solution in MES buffer; concentrations of both EDC and NHS-sulfo are 30 mg/mL (*see* **Note 35**).

2. Mix 250 μL of NPs (~0.6 mg/mL) with 250 μL of EDC/NHS solution prepared in **step 1** in Subheading 3.4. Incubate the mixture at room temperature for 1 h under mild shaking (*see* **Note 36**).

3. Add 1 mL of PBST and vortex the solution for 1 min. Purify the solution by centrifugation at $8500 \times g$ for 10 min and remove the supernatant. Repeat this step two more times (*see* **Note 37**).

4. Add 200 μL 1× PBS to the pellet container and sonicate it in a water bath sonicator for 30 s to fully disperse the NPs. Add 200 μL of IgG (1 mg/mL in 1× PBS) and incubate the solution for 2–4 h at room temperature under mild shaking (*see* **Note 38**).

5. Add 1 mL of PBST and vortex the solution thoroughly. Centrifuge at $5000 \times g$ for 10 min to remove the supernatant. Repeat this step two more times (*see* **Note 39**).

6. Add 100 μL of PBS to re-disperse the NPs and store the dispersion at 4 °C for further use.

### 3.5 NP Characterization

1. Prepare samples for imaging by dropping one drop of NP solution on a copper grid coated with a thin layer of carbon and slowly letting it dry. NPs synthesized in the organic phase are usually dissolved in hexane and can be dried within 5 min under ambient conditions. Water-soluble NPs deposited on the TEM grid need to be dried overnight under ambient conditions (*see* **Note 40**).

2. Collect XRD patterns of the NPs on a Bruker AXS D8-Advanced diffractometer with Cu Kα radiation ($\lambda = 1.5418$ Å). For the sample preparation, drops of the NP hexane dispersion are carefully dried on a glass glide to form a thin film on the substrate (*see* **Note 41**).

3. Perform magnetic measurements on a Lakeshore 7404 high sensitivity vibrating sample magnetometer (VSM) with fields up to 1.5 T. Obtain powder samples by drying the colloidal dispersion under $N_2$. Wrap 2–5 mg of powder with Teflon tape and measure at room temperature (*see* **Note 42**).

4. Use a Malvern Zeta Sizer Nano S-90 dynamic light scattering (DLS) to measure the hydrodynamic size of the NPs. To prepare the sample, clean and dry a quartz cuvette, and fill the cuvette with a diluted hexane or water dispersion of the NPs. The cuvette is placed in the sample holder for the measurement (*see* **Note 43**).

# 4  Notes

1. Set up the reaction system with dry glassware. Calibrate the thermocouple with a mercury thermometer to make sure that the temperature reading from the controller is the actual reaction temperature.

2. We provide the detailed protocols for the synthesis of 4 nm $Fe_3O_4$, hollow $Fe_3O_4$, porous hollow $Fe_3O_4$, and dumbbell-like Au-$Fe_3O_4$ NPs. These protocols can be applied to prepare a wide variety of magnetic NPs with controlled dimensions and magnetic properties in airless and moisture-less conditions.

3. The weights or volumes of the reagents must be adjusted according to the certificates of analysis from the suppliers. For example, the exact volume of OA should be $(6 \times 10^{-3} \text{ mol} \times 282.46 \text{ g/mol})/(0.9 \times 0.887 \text{ g/mL})$, in which 0.9 is the purity of the reagent. Some of the chemical reagents may cause burns and irritation, make sure to use them with caution and wear personal protective equipment (PPE, gloves). The stirring speed should be controlled to avoid solution splashing but fast enough to ensure uniform mixing/reaction.

4. The stopcock attached to one outlet of the reaction flask should be left open in the initial heating process to remove air, moisture, and low-boiling solvents.

5. The stopcock should be closed before a volatile chemical is added or before the reaction solution is heated up to a higher temperature. Otherwise, the solvent and other volatile chemicals could evaporate quickly from the open stopcock due to the continued Ar or $N_2$ flow through the reaction system. When the protection gas (Ar or $N_2$) by-passes the reaction system, the reaction proceeds under a blanket of inert gas at a slightly positive pressure (depending on the oil level in the oil bubbler) to prevent air/moisture from diffusing back into the reaction system. The air condenser is needed to ensure the cooling/return of the boiled solvent back to the reaction solution. If the reaction temperature stops short of the boiling point of the solvent used, open the stopcock for 2–3 min to allow the evaporation of the low-boiling components produced during the initial reaction.

6. Once the reaction is over, the heating power should be turned off and the heating mantle should be removed to allow the reaction flask to cool to room temperature. *Caution: the heating mantle and the reaction flask will be very hot, protective gloves (thermo-resistance) should be used.* The cooled solution can be kept under Ar or $N_2$ for days without affecting the quality of the NPs.

7. In the separation process, do not fill the centrifuge tubes too full, or the solution may spill out during the centrifugation process. Add more ethanol and centrifuge for a longer time if the initial centrifugation process does not give a solid precipitate. Discard the supernatant slowly and carefully to ensure the NP powder is not washed away with the supernatant. To prevent sample loss during the decanting process, it is recommended that a permanent magnet is attached to the centrifuge tube so that the magnetic NP product is attracted to the magnet and better separated from the supernatant.

8. The purification process washes away high boiling point solvents and excess surfactants. If the NPs become less dispersible in hexane after this process, then add one or two drops of oleic acid/oleylamine.

9. A hexane dispersion of NPs sealed properly in a glass vial can be stored at room temperature for up to a few months without noticeable morphological changes or changes to their chemical/physical properties. Do not use a plastic vial to store the hexane dispersion as it can interact with the solvent or surfactant, reducing the shelf life of the NP dispersion.

10. $Fe(CO)_5$ is toxic and sensitive to oxygen and moisture. Handle it with care in a well-ventilated hood or an Ar glove box. The fresh $Fe(CO)_5$ should be clear and straw-colored. Discard it when the liquid becomes dark brown with precipitate (metallic Fe). $Fe(CO)_5$ is highly volatile, and it is better to use a syringe to transfer it. Before adding $Fe(CO)_5$, the stopcork should be closed so that the reaction system is protected by a blanket of Ar. When heated above 150 °C, $Fe(CO)_5$ evaporates and decomposes, and a light yellow fog can fill the reaction flask. As the reaction proceeds, the $Fe(CO)_5$ vapor should disappear and the space above the reaction solution should appear clear.

11. After it is cooled to room temperature, the magnetic product may stick to the magnetic stir bar, and the reaction solution may appear transparent. Separate the Fe NPs from the reaction solution with minimum exposure to air.

12. Collect the product attached to the stir bar only. The product dispersed in the solution may not be of good quality. Once exposed to air, the Fe NPs will be oxidized.

13. Nitrogen protection can only slow down the Fe oxidation rate. It is advised to use the Fe NPs when they are freshly prepared.

14. Trimethyl amine N-oxide is difficult to dissolve in ODE directly. Dissolve it in methanol first and then transfer it into ODE. After adding ODE, the solvent becomes turbid immediately.

15. The stopcock is open to allow the removal of air, moisture, and methanol. Once the methanol is evaporated, the trimethyl

amine N-oxide is precipitated out as a thin layer of white crystals around the side-wall of the flask.

16. The temperature is above the boiling temperature of hexane. Do the injection with the tip of the needle inside the solvent to reduce the loss of NPs caused by bumping. As the reaction proceeds, the white crystals will disappear gradually.

17. The heating rate is extremely important for this step. Faster heating may cause uneven oxidation around each NP, resulting in the formation of cracks around the oxide shell.

18. To precipitate the NP product from ODE solution, isopropanol should be chosen as a non-solvent (ethanol is not miscible with ODE). Ethanol is used as the non-solvent to precipitate NPs from their hexane dispersion.

19. The heating rate is extremely important for this step. A heating rate faster than 5 °C/min is a must for synthesizing porous hollow NPs. Too fast a heating rate might lead to NP cracking.

20. $HAuCl_4 \cdot 3H_2O$ is acidic and corrosive. Handle it with a glass spoon. Tetralin has an aromatic smell and should be handled in a good ventilated hood.

21. TBAB is difficult to dissolve in either OAm or tetralin at room temperature. It may be necessary to heat the mixture to 35 °C to dissolve the TBAB. Let the solution cool to room temperature before further use.

22. The TBAB should be injected quickly so that monodisperse Au NPs can be synthesized. Upon TBAB injection, the solution changes to a purple color immediately, indicating that Au particle nucleation has occurred. Stir the solution for 1 h to make sure that all of the TBAB has been consumed.

23. The hexane can be removed within 5 min. The estimated weight is not the true weight of the Au NPs, rather it is the weight of the Au NPs plus the surfactant.

24. The Au NPs tend to aggregate at higher temperatures. The Fe atoms produced via decomposition of $Fe(CO)_5$ can help stabilize the Au NPs from aggregation. $Fe(CO)_5$ starts to decompose at 120 °C and fully decomposes by 180 °C. It is important to inject $Fe(CO)_5$ quickly and use a fast heating rate so that Fe atoms form and nucleate on the Au NPs.

25. Chloroform is toxic and highly volatile. Handle it inside a well-ventilated fume hood. Carefully seal the sample vial and use the sample as soon as possible to avoid concentration changes due to evaporation.

26. Chloroform can be removed within 1 min. The estimated weight includes the inorganic NP cores plus the surfactants surrounding the NP cores.

27. The NHS ester is reactive and the solution must be immediately used once prepared. This ester reacts with a primary amine under physiological to slightly alkaline conditions to yield stable amides. $Na_2CO_3$ helps to adjust the acidic environment of solution B. The $Na_2CO_3$ will not be completely dissolved. It is normal to see some white crystal powder the whole time. Dopamine is not soluble in chloroform. After solution A is added to solution B, it is possible to see some new precipitates. If so, carefully adjust the DMF to chloroform ratio to make sure both the NHS-PEG-COOH and the dopamine are dissolved. Dopamine is easily oxidized at high temperature. Run the reaction at room temperature in the dark.

28. Make sure the NPs are well dispersed.

29. Mechanical shaking is better than magnetic stirring since magnetic NPs tend to stick to the magnetic bar. It is likely to see NP agglomeration during the first 2 h of shaking since the hydrophobic NPs prepared from the organic phase reaction are not dispersible in DMF. However, the NPs should finally be dispersed due to the successful ligand exchange.

30. If there is no precipitation after centrifugation, add more hexane. If 5 mL of hexane has been added and precipitates still are not seen, carry out **steps 1–6** again, but decrease the iron oxide amount to 0.5 mg. Pour the supernatant out to discard the displaced hydrophobic ligands. Ensure the product is dry under a gentle $N_2$ flow, otherwise turbid micelles will likely form when water is added.

31. The molecular weight cutoff of the dialysis tubing should be bigger than the average molecular weight of the polymer. Replace the solution with fresh water after the first 2 h of dialysis and then every 8 h in the following 2 days thereafter.

32. Since chloroform is volatile, it is normal to see a decrease of the solution volume after 2 h of stirring. Carefully seal the vial to reduce the rate of solvent evaporation.

33. Chloroform is non-miscible with water. Thus, dry the product first before dispersing it in water (even a trace amount of chloroform can cause the formation of micelles).

34. Here, we just give an example of how to couple IgG to the PEG-$Fe_3O_4$ NPs via common EDC/NHS chemistry. Lysine residues are the primary target sites for EDC conjugation. A protein with a high number of lysine groups on the outer surface usually has higher conjugation efficiency. The purity of the protein is also of great concern. Any other molecules containing primary amines will compete with the protein in the conjugation reaction.

35. EDC should be stored at $-20\,^{\circ}\text{C}$ in a desiccated container and allowed to equilibrate to room temperature right before use. EDC is extremely sensitive to moisture. If it clumps, do not use it. Activation using EDC and sulfo-NHS is most efficient between pH 4.5 and 7.2. The activation buffer should not contain any primary amine or other groups that compete with the activation reaction. Therefore, it is often preferred to use MES buffer at pH 6 for the activation. EDC/NHS can hydrolyze rapidly in an aqueous solution and must be prepared fresh just before conjugation.

36. After dialysis, the volume of the sample typically increases 1.5 times. It is estimated to be 0.6 mg iron oxide per mL (1 mg NPs/1.5 mL water). It is necessary to know the accurate concentration here as EDC/NHS is in large excess. It is important that the NPs are well dispersed.

37. Tween-20 can help to stabilize the NPs and remove nonspecifically bound entities. Adjust the centrifuge speed and time carefully to ensure the formation of a loose NP precipitate—too firm/packed a precipitate can be difficult to redisperse.

38. The concentration of the conjugated protein may vary depending on the NP size and concentration. The protein added can be estimated to be 10x to 1000x in excess of the amount of full surface coverage.

39. Adjust the centrifuge speed and time carefully to ensure a loose NP precipitate.

40. TEM is used to monitor the quality of the NPs with respect to their shape, size, and uniformity (Fig. 2a–d). The sample must be thoroughly purified before TEM characterization. Upon the high energy beam irradiation, organic ligands tend to decompose into carbonaceous deposits, contaminating the EM vacuum column and deteriorating the EM imaging quality. After PEGylation, the NPs should maintain their morphology with no obvious aggregation (Fig. 2e).

41. XRD is used to characterize the structure of a group of NPs.

42. VSM is used to measure the magnetic properties of the NP product and monitor NP magnetic stability. Further oxidation of $Fe_3O_4$ or Fe should give lower magnetization values. Larger magnetic NPs with high magnetic moment are needed to enhance $T_2$ relaxation while smaller NPs are good to enhance $T_1$ relaxation, which is important to enhance MRI sensitivity.

43. DLS is used to check the hydrodynamic size of the NPs in the dispersed state (Fig. 2f). The hydrodynamic size includes the size of the inorganic core plus the thickness of the surfactant coating. After PEGylation, the hydrodynamic size of the NPs

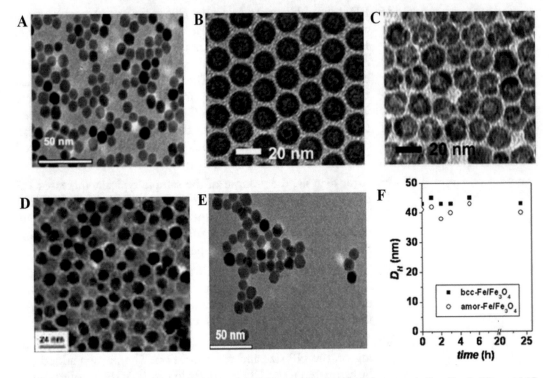

**Fig. 2** (**a–d**) TEM of as-synthesized (**a**) $Fe_3O_4$ NPs, (**b**) hollow $Fe_3O_4$ NPs, (**c**) porous hollow $Fe_3O_4$ NPs, and (**d**) dumbbell-like Au-$Fe_3O_4$ NPs. (**e**) TEM of $Fe_3O_4$ NPs after ligand exchange with dopamine-PEG6000. (**f**) A representative hydrodynamic size change of the PEGylated Fe/$Fe_3O_4$ core/shell NPs incubated in PBS (pH = 7) at 37 °C. No obvious size increase indicates a good stability after surface coating. (**a**, **e**) Reprinted with permission from [19]. Copyright 2007 Wiley. (**b**, **c**) Reprinted with permission from [15]. Copyright 2009 American Chemical Society. (**d**) Reprinted with permission from [17]. Copyright 2005 American Chemical Society. (**f**) Reprinted with permission from [20]. Copyright 2011 American Chemical Society

will increase tens of nanometers depending on the PEG used. After protein conjugation, the hydrodynamic size will increase tens of nanometers again depending on the type of protein used. It is also suggested to refer to a specific protein immune-dot blot assay or cell binding assay to confirm the protein functionality after the conjugation.

# References

1. Huang J, Zhong X, Wang L, Yang L, Mao H (2012) Improving the magnetic resonance imaging contrast and detection methods with engineered magnetic nanoaprticles. Theranostics 2:86–102
2. Gallo J, Long NJ, Aboagye EO (2013) Magnetic nanoparticles as contrast agents in the diagnosis and treatment of cancer. Chem Soc Rev 42:7816–7833
3. Lee H, Shin T, Cheon J, Weissleder R (2015) Recent developments in magnetic diagnostic systems. Chem Rev 115:10690–10724
4. Ho D, Sun X, Sun S (2011) Monodisperse magnetic nanoparticles for theranostic applications. Acc Chem Res 44:875–882
5. Frey NA, Peng S, Cheng K, Sun S (2009) Magentic nanoparticles: synthesis, functionalization, and applications in bioimaging and

magnetic energy storage. Chem Soc Rev 38:2532–2542

6. Lee N, Hyeon T (2012) Designed synthesis of uniformly sized iron oxide nanoparticles for efficient magnetic resonance imaging contrast agents. Chem Soc Rev 41:2575–2589

7. Ling D, Lee N, Hyeon T (2015) Chemical synthesis and assembly of uniformly sized iron oxide nanoparticles for medical applications. ACS 48:1276–1285

8. Lee N, Yoo D, Ling D, Cho M, Hyeon T, Cheon J (2015) Iron oxide based nanoparticles for multimodal imaging and magnetoresponsive therapy. Chem Rev 115:10637–10689

9. Laurent S, Forge D, Port M, Roch A, Robic C, Elst LV, Muller RN (2008) Magnetic iron oxide nanoparticles: synthesis, stabilization, vectorization, physicochemical characterizations, and biological applications. Chem Rev 108:2064–2110

10. Gupta AK, Gupta M (2005) Synthesis and surface engineering of iron oxide nanoparticles for biomedical applications. Biomaterials. 26: 3995–4021

11. Palui G, Aldeek F, Wang W, Mattoussi H (2015) Strategies for interfacing inorganic nanocrystals with biological systems based on polymer-coating. Chem Soc Rev 44: 193–227

12. Jin X, Liu G, Henry SE, Ai H, Chen X (2011) Surface-engineered magnetic nanoparticle platforms for cancer imaging and therapy. Acc Chem Res 44:883–892

13. Sun S, Zeng H (2002) Size-controlled synthesis of magnetite nanoparticles. J Am Chem Soc 124:8204–8205

14. Peng S, Wang C, Xie J, Sun S (2006) Synthesis and stabilization of monodisperse Fe nanoparticles. J Am Chem Soc 128:10676–10677

15. Cheng K, Peng S, Xu C, Sun S (2009) Porous hollow $Fe_3O_4$ nanoparticles for targeted delivery and controlled release of cisplatin. J Am Chem Soc 131:10637–10644

16. Peng S, Lee Y, Wang C, Yin H, Dai S, Sun S (2008) A facile synthesis of monodisperes Au nanoparticles and their catalysis of CO oxidation. Nano Res 1:229–234

17. Yu H, Chen M, Rice PM, Wang SX, White RL, Sun S (2005) Dumbbell-like bifunctional Au-$Fe_3O_4$ nanoparticles. Nano Lett 5: 379–382

18. Lee N, Choi Y, Lee Y, Park M, Moon WK, Choi SH, Hyeon T (2012) Water-dispersible ferrimagnetic iron oxide nanocubes with extremely high r2 relaxivity for highly sensitive in vivo MRI of tumors. Nano Lett 12:3127–3131

19. Xie J, Xu C, Kohler N, Hou Y, Sun S (2007) Controlled PEGylation of monodisperse Fe3O4 nanoparticles for reduced non-specific uptake by macrophage cells. Adv Mater 19:3163–3166

20. Lacroix L, Huls N, Ho D, Sun X, Cheng K, Sun S (2011) Stable single-crystalline body centered cubic Fe nanoparticles. Nano Lett 11:1641–1645

# Chapter 6

## Brain-Penetrating Nanoparticles for Analysis of the Brain Microenvironment

### Elizabeth Nance

### Abstract

The past decade has witnessed explosive growth in the development of nanoparticle-based therapies for the treatment of neurological disorders and diseases. The systemic delivery of therapeutic carriers to the central nervous system (CNS) is hindered by both the blood–brain barrier (BBB) and the porous and electrostatically charged brain extracellular matrix (ECM), which acts as a steric and adhesive barrier. Therapeutic delivery to the brain is influenced by changes in the brain microenvironment, which can occur as a function of physiology, biology, pathology, and developmental age. Brain-penetrating nanoparticles (BPNs) are an optimal platform not only for therapeutic delivery to the brain, but also for evaluating changes in the brain microenvironment. BPNs possess both the capability to readily move within their local environment to survey their surroundings and the ability to reach the diffuse disease cells often associated with CNS disorders. To achieve effective delivery of BPNs to specific locations within the brain requires careful control over the nanoparticle's transport properties. Here, we describe the process of conjugating a dense layer of poly(ethylene glycol) (PEG) to the surface of nonbiodegradable nanoparticles to achieve brain-penetrating capabilities.

**Key words** Brain-penetrating nanoparticle, PEGylation, Polystyrene, Carboxyl-amine reaction, PEG density

## 1 Introduction

Nanoparticle delivery to specific locations within the brain requires the ability to control the transport properties of the nanoparticle platform. As neurological injury or disease is not generally limited to one location or one cell type, a therapeutic or diagnostic platform must have the ability to move from the point of access in the brain to diffuse disease-related cells. This is especially important in diseases like cancer or mediated by neuroinflammation, where access across an impaired blood–brain barrier (BBB) is heterogeneous and variable. Importantly, delivery platforms that can readily move within the brain, regardless of how they access the brain microenvironment, can reach these diffuse disease cells to maximize therapeutic effect [1]. Nanoparticles with no electrostatic,

Sarah Hurst Petrosko and Emily S. Day (eds.), *Biomedical Nanotechnology: Methods and Protocols*, Methods in Molecular Biology, vol. 1570, DOI 10.1007/978-1-4939-6840-4_6, © Springer Science+Business Media LLC 2017

hydrophobic or hydrogen bonding interactions within the brain are classified as brain-penetrating [2]. These brain-penetrating nanoparticles (BPNs) can also probe the steric limitations of nanoparticle transport within the brain microenvironment, and be used to characterize microrheological properties of the brain extracellular space. BPNs are therefore an optimal platform for evaluating changes in the brain microenvironment as a function of normal development, developmental age, disease etiology, or disease progression, and can then be a more effective platform for therapeutic intervention.

For a nanoparticle to efficiently penetrate within the brain microenvironment, it must avoid any steric limitation it would experience as it moves between cells, vessels, and extracellular matrix (ECM) components. The particle must also avoid any adhesive interactions, via hydrogen bonding, hydrophobic interactions, or electrostatic interactions between the particle surface and a cell membrane, protein, or ECM component. Steric limitations are primarily influenced by the brain's volume fraction and tortuosity. The volume fraction of the brain is a ratio of the volume of the extracellular space to the total brain volume. This volume fraction changes with sleep, position (supine or upright), stress, and chronic inflammation. Tortuosity in the brain microenvironment is defined by a molecule's hindrance to diffusion within the brain. To maximize diffusion within the brain and accurately assess the steric limitations that occur due to changes in volume fraction and tortuosity, it is important to have a bio-inert particle that will avoid adhesive interactions. Here, we describe the process of conjugating a dense layer of PEG to the surface of nonbiodegradable polymer nanoparticles to create a bio-inert particle that can achieve brain-penetrating capabilities to survey the brain microenvironment in normal and diseased states (Fig. 1). PEG is a hydrophilic, FDA approved polymer that, when coated on the surface of nanoparticles, can provide a bio-inert nanoparticle platform that minimizes interactions with the environment. Although we only focus on polystyrene particles, this protocol can also apply to quantum dots with similar starting surface chemistry and surface functionality density [2].

This chapter will only cover nonbiodegradable nanoparticle platforms. Biodegradable polymeric nanoparticles and DNA nanoparticles with high PEG density have also been reported in the literature, demonstrating that improvement in diffusive capability in the brain can improve therapeutic outcome [3–5]. DNA nanoparticles with high PEG density show increased penetration and a larger area of transfection compared to nonpenetrating DNA platforms [6, 7]. The PEG density on the surface of a biodegradable polymeric particle is equally important for diffusion within the brain; yet the surfactant used for emulsifying the particles will also play a role and should be further evaluated for brain-penetrating capabilities. There is evidence that commonly used surfactants,

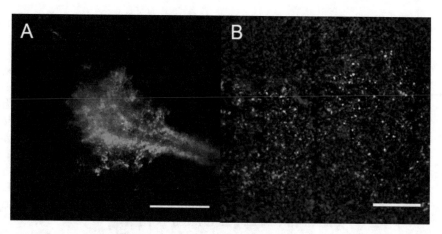

**Fig. 1** BPN distribution in the rodent brain. (**a**) PS-PEG (BPN, *green*) penetrate in vivo in living mice away from the site of injection more readily compared to nonpenetrating PS-COOH (*red*). Scale bar: 50 μm. From Sci Transl Med. 2012 Aug 29;4(149): 149ra119. Reprinted with permission from AAAS. Copyright 2012 AAAS. (**b**) PS-PEG (BPN, *red*) distribute more evenly through a 9 L gliosarcoma following direct injection intracranially into the tumor, compared to PS-COOH (*green*). Cell nuclei are stained with DAPI (*blue*). *Yellow* in both panels represent overlay of the two particle platforms. Scale bar: 100 μm. Reprinted in part with permission from ACS Nano. 2014 Oct 28; 8(10): 10655–10664. Copyright 2014 American Chemical Society

including polysorbate 80 [8, 9], poly(vinyl alcohol) [4, 10], and pluronics [11, 12], do not provide similar diffusive capability in biological environments. The effect of these surfactants in combination with PEG on the diffusion of biodegradable nanoparticles in the brain should be further characterized and understood before classifying a particle as brain-penetrating. PEG provides steric stabilization, shielding of hydrophobic domains on the particle surface, and a near-neutral surface charge. However, not all nanoparticle platforms will require PEG to enable brain penetration. Lipophilic and hydrophilic particles, like dendrimers, with near-neutral surface chemistry, can also readily move within the brain to reach disease associated cells [13, 14].

The end goal of this protocol is to obtain nanoparticles coated with a high density of PEG. PEGylation has been used to create stealth nanoparticles [15], increase circulation time of nanoparticles, and aid in penetration of nanoparticles across mucosal barriers [16]. Unlike many protocols that focus on specific size ranges (40–100 nm [17] or larger than 200 nm [18]), or require bottom-up fabrication and layer-by-layer techniques like PRINT [19], this protocol focuses on utilizing a single step, stable carboxyl-amine chemistry to achieve dense PEG coatings for particles ranging from 20 nm up to micron-sized particles. In addition, materials other than PEG, including zwitterionic compounds [20], surfactants [10], and polymers [21], can provide stealth-like qualities, increased circulation time, and improved penetration of nanoparticles within a tissue environment. Many of these materials have emerged to address the limitation that repeat administration of PEG conjugated nanoparticles has shown to induce a PEG-specific immune response [22]. The concept of nanoparticle penetration within the brain microenvironment is relatively new, where most of the focus on drug delivery to the brain has been on developing nanoparticles that can bypass or overcome the BBB. However, it cannot be assumed that coatings that provide stealth properties also provide brain-penetrating properties, and thus, many of these coatings that could create bio-inert nanoparticles, like zwitterions or surfactants, have not thoroughly been tested to determine if they maintain brain-penetrating properties. This is an area for future exploration that may greatly expand the nanoparticle platforms that can be utilized as BPNs.

There are multiple methods for calculating PEG density [19]. PEG chains have a Flory radius defined as $R_f \sim \alpha N^{3/5}$, where $N$ is the degree of polymerization and $\alpha$ is the effective monomer length. For example, an unconstrained 5000 Da PEG chain has a diameter of 5.4 nm and occupies a surface area of 22.7 nm$^2$ (assuming an unconstrained random walk). Although not covered in detail in this chapter, PEG surface densities, provided by the ratio of number of PEG molecule chains to 100 nm$^2$ of nanoparticle surface are, for these nonbiodegradable polystyrene (PS) nanoparticles can be calculated using NMR analysis, using an adapted method published previously [23]. The average PEG surface density (chains/100 nm$^2$) on the surface of the NPs is then determined by taking into account the total quantity of PEG detected by NMR and the total NP surface area. The surface area of PS NPs is calculated by assuming that the NPs are made of individual particles of diameter equal to that measured by laser Doppler anemometry, with smooth surfaces, and a density of 1.055 g/mL, as provided by the manufacturer. In general, this protocol will produce nonbiodegradable nanoparticles with PEG densities greater than nine PEG per 100 nm$^2$, which allows penetration in the brain [2]. The packing of PEG can be determined by calculating the ratio of PEG chains per 100 nm$^2$ to the number of unconstrained PEG molecules that occupy a surface area of 100 nm$^2$ on a nanoparticle. The required PEG density necessary for other nanomaterials to achieve rapid brain penetration is dependent on the PEG molecular weight (MW), the PEG structure (linear vs branched), and the material properties of the particle core, including its composition, porosity, size, and MW.

## 2    Materials

Prepare all solutions using deionized water or ultrapure water and analytical-grade reagents. Prepare and store all reagents at room temperature or as otherwise noted below. Follow all local and university waste-disposal regulations when disposing of waste materials from this protocol.

**2.1 Reaction Solutions**

1. Reaction buffer: 200 mM borate buffer, pH 8.2. Add 200 mL of ultrapure water to a 500 mL glass pyrex bottle. Weigh 3.71 g of boric acid and transfer to the bottle. Add ultrapure water to a total volume of 300 mL. Mix well and adjust pH to 8.2 using 1 N NaOH. Store at room temperature (*see* **Note 1**).

2. Working suspension of fluorescent PS-COOH nanoparticles: 4× dilution (*see* **Note 2**). Fluorescent PS-COOH nanoparticles may be purchased from a variety of vendors (*see* **Note 3**). In a water bath sonicator, place the stock bottle in a floating tube holder and sonicate the stock polystyrene nanoparticle solution for 10 min. In a 1.5 mL siliconized centrifuge tube, add 50 μL of fluorescent polystyrene nanoparticles from the stock bottle to 150 μL of ultrapure water (*see* **Note 4**). Particles should be sonicated in a water bath sonicator for 5 min after dilution. Stock particle solutions sometimes contain trace amounts of sodium azide (*see* **Note 5**).

**2.2 Reagents**

Allow all reagents to thaw for 20 min prior to starting the reaction.

1. Methoxy-PEG-NH$_2$: Remove methoxy-PEG-NH$_2$ (mPEG-NH$_2$, PEG MW: 5000 Da) from the −20 °C freezer. Do not open until ready to weigh out desired quantities. Keep mPEG-NH$_2$ stored in a jar containing indicating desiccant.

2. *N*-Hydroxysulfosuccinimide (Sulfo-NHS): Remove Sulfo-NHS sodium salt from the 4 °C refrigerator. Do not open until ready to weigh out desired quantities. Keep NHS in a jar containing indicating desiccant.

3. 1-Ethyl-3-(3-dimethylaminopropyl) carbodiimide (EDC): Remove EDC from the −20 °C freezer. Do not open until ready to weigh out desired quantities. Keep EDC stored in a jar containing indicating desiccant.

**2.3 Characterization Solutions**

1. Size and zeta-potential measuring solutions: 10 mM NaCl, pH 7.4. Add 200 mL of ultrapure water to a 500 mL glass pyrex bottle. Weight 175.32 mg of NaCl and transfer to the bottle. Add ultrapure water to a total volume of 300 mL. Mix well and use 1 N HCl to adjust pH to 7.4. Store at room temperature.

2. In situ measuring solution: artificial cerebrospinal fluid (ACSF), pH 7.4. Add 200 mL of deionized water to a 500 mL glass pyrex bottle. Add 2.07 g NaCl (119 mM), 660 mg NaHCO$_3$ (26.2 mM), 56 mg KCl (2.5 mM), 36 mg NaH$_2$PO$_4$ (1 mM), 37 mg MgCl$_2$ (1.3 mM), and 540 mg glucose (10 mM). Mix the solution thoroughly then bring the final volume to 300 mL with deionized water. Gas the

solution with 5% $CO_2$/95% $O_2$ for 10–15 min. Then add 83 mg $CaCl_2$ (2.5 mM). Filter using a 0.22 μm filter apparatus. Store at 4 °C (*see* **Notes 6** and **7**).

## 3 Methods

Each carboxyl-modified fluorescent PS bead batch has a charge (milliequivalents, mEq), percent solid, and density (particles/mL). These numbers need to be obtained from the Certificate of Analysis (CoA) provided by the manufacturer prior to starting the protocol. There are four stages of the protocol: (1) calculation to determine the amount of reagents needed, (2) the PEGylation reaction, (3) collection of the nanoparticles, and (4) characterization of the nanoparticles' physicochemical properties.

### 3.1 Calculation of Number of COOH Per Bead

The reaction protocol is based on calculations that associate the mole ratio of number of COOH per bead to the number of moles of each reagent, which are often added in excess. Therefore, the number of COOH per bead needs to be calculated first. These calculations will be different for each size polystyrene particle and for each lot number of a specific particle type. A summary of sample calculations is shown in Table 1.

1. Number of beads per gram: Calculate density provided in CoA divided by the percent solid provided in CoA.

2. Number of milliequivalents per bead: Divide the charge provided in CoA by the number of beads per gram calculated in **step 1**.

3. Number of COOH groups per bead: Multiply the number of mEq per bead calculated in **step 2** by Avogadro's Number ($6.022 \times 10^{23}$) and divide by 1000 (*see* **Note 8**).

**Table 1**
**Sample calculation to determine number of COOH per bead**

| Starting particles | 100 nm PS-COOH red | Notes |
|---|---|---|
| % solid | 0.02 | In CoA, based on nanoparticle size |
| Density (particles/mL) | $2.73 \times 10^{13}$ | In CoA, based on lot number |
| Charge (mEq) | 0.309 | In CoA, based on lot number |
| | **Value** | **Calculation** |
| Number of beads per gram | $1.37 \times 10^{15}$ | $2.73 \times 10^{13}/0.02$ |
| Number of millieq per bead | $2.26 \times 10^{-16}$ | $1.37 \times 10^{15}/0.309$ |
| Number of COOH per bead | 1,36,322 | $(2.26 \times 10^{-16}) \times 6.022 \times 10^{23}$ |

**3.2 Calculation of Number of Moles of PEG Needed**

The following calculations are based on a starting volume of 50 μL of 100 nm PS-COOH nanoparticles. A summary of sample calculations is shown in Table 2.

1. The number of beads in the working nanoparticle solution is calculated by multiplying the volume of stock beads (50 μL) by the density of beads provided in the CoA, divided by 1000.

2. The total number of COOH in the working solution is the number of beads added in **step 1** of Subheading 3.2 multiplied by the number of COOH groups per bead calculated in **step 3** of Subheading 3.1.

3. The number of moles of PEG needed is calculated by dividing the total number of COOH in the working solution by Avogadro's Number (*see* **Notes 9** and **10**).

**3.3 PEGylation Reaction**

The following reaction protocol is based on a starting volume of 50 μL of 100 nm PS-COOH nanoparticles. A sample calculation is provided in Table 3. The reaction schematic for PEGylation of PS nanoparticles is shown in Fig. 2.

Allow all reagents to thaw for 20 min prior to use.

**Table 2**
**Sample calculation for number of moles of PEG needed for the conjugated of mPEG-NH$_2$ to PS-COOH**

| Starting particles | 100 nm PS-COOH | Notes |
|---|---|---|
| % solid | 0.02 | In CoA, based on nanoparticle size |
| Density (particles/mL) | $2.73 \times 10^{13}$ | In CoA, based on lot number |
| Charge (mEq) | 0.309 | In CoA, based on lot number |
| | **Value** | **Calculations** |
| Number of beads per gram | $1.37 \times 10^{15}$ | $2.73 \times 10^{13}/0.02$ |
| Number of millieq per bead | $2.26 \times 10^{-16}$ | $1.37 \times 10^{15}/0.309$ |
| Number of COOH per bead | 1,36,322 | $(2.26 \times 10^{-16}) \times 6.023 \times 10^{23}$ |
| Volume of beads added (μL) | 50 | |
| Dilution factor | 4 | *See* **Note 2** |
| Number of beads added | $1.37 \times 10^{12}$ | $50 \times 2.73 \times 10^{13}/1000$ |
| Total number of COOH | $1.86 \times 10^{17}$ | $(1.37 \times 10^{12}) \times 136322$ |
| Number of mole of PEG needed | $3.09 \times 10^{-7}$ | $1.86 \times 10^{17}/6.023 \times 10^{23}$ |
| Molar mass of PEG (Da) | 5000 | *See* **Note 9** |
| Excess of PEG | 4 | *See* **Note 10** |
| Mass of PEG needed (mg) | 6.18 | $(3.09 \times 10^{-7}) \times 5000 \times 4 \times 1000$ |

**Table 3**
**Sample calculation table for PEGylation reaction**

|  | 100 nm PS-COOH | Calculations/notes |
|---|---|---|
| Volume of beads added (µL) | 50 | |
| Dilution factor | 4 | *See* **Note 2** |
| Number of beads added | $1.37 \times 10^{12}$ | $50 \times 2.73 \times 10^{13}/1000$ |
| Total number of COOH | $1.86 \times 10^{17}$ | $(1.37 \times 10^{12}) \times 136322$ |
| Number of mole of PEG needed | $3.09 \times 10^{-7}$ | $1.86 \times 10^{17}/6.023 \times 10^{23}$ |
| Molar mass of PEG (Da) | 5000 | *See* **Note 9** |
| Excess of PEG | 4 | *See* **Note 10** |
| Mass of PEG needed (mg) | 1.55 | $(3.09 \times 10^{-7}) \times 5000 \times 1000$ |
| Mass of PEG to add (mg) | 6.18 | $1.55 \times 4$ |
| Mass of NHS to add (mg) | 2.68 | $(3.09 \times 10^{-7}) \times 4 \times 217.13 \times 10 \times 1000$ |
| Amount of Borate buffer to add (µL) | 800 | *See* **Note 11** |
| Mass of EDC to add (mg) | 2.37 | $(6.18/5000) \times 191.7 \times 10$ |

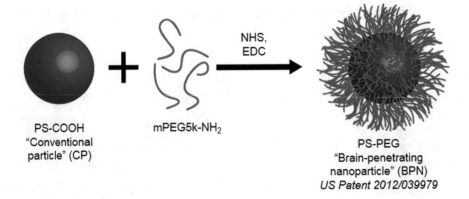

PS-COOH
"Conventional
particle" (CP)

mPEG5k-NH$_2$

NHS,
EDC

PS-PEG
"Brain-penetrating
nanoparticle" (BPN)
*US Patent 2012/039979*

**Fig. 2** Conjugation of methoxy PEG (5000 Da)-NH$_2$ to carboxy-modified polystyrene nanoparticles, using NHS-and EDC-assisted chemistry

1. In a water bath sonicator, sonicate the diluted working nano-particle solution prepared as described in Subheading 2.1, **item 2** for 5 min.

2. Add 1.55 mg mPEG-NH$_2$ to the 1.5 mL centrifuge tube containing the diluted particle solution. Vortex ~30 s to 1 min to mix (*see* **Notes 9** and **10**).

3. The next steps should be done quickly. Add 2.68 mg Sulfo-NHS to the 1.5 mL centrifuge tube containing the particle solution. Vortex briefly (~1–5 s) to mix solution.

4. Add 800 µL 200 mM borate buffer, pH 8.2, to the 1.5 mL centrifuge tube containing the particle solution. Vortex briefly (~5 s) to mix. The amount of borate is based on a 4× dilution from the starting dilute PS volume (200 µL in this example) (*see* **Note 11**).

5. Add 2.37 mg 1-Ethyl-3-(3-dimethylaminopropyl) carbodiimide (EDC) to the centrifuge tube containing the particle solution. Vortex ~30 s to 1 min to mix thoroughly.

6. Wrap 1.5 mL tube in aluminum foil and place on a rotary incubator for 4 h at 25 °C (*see* **Note 12**).

**3.4 Nanoparticle Collection**

Particle collection methods are based on the size of the particle and are therefore provided below based on the size of the particles used.

*3.4.1 For Sub-100 nm Particles*

1. At the end of the incubation time, add 500 µL of the PEGylated particle solution to an Amicon Ultra 0.5 mL 100 kDa molecular weight cut-off (MWCO) spin filter tube.

2. Place in a microcentrifuge at 4 °C.

3. Centrifuge particles for 12 min at 14,000 × $g$ (*see* **Note 13**).

4. Pour off supernatant and add 500 µL ultrapure water to the top of the filter. Pipet up and down to resuspend particles. Particles should resuspend easily if well-PEGylated (*see* **Note 14**).

5. Repeat **steps 3** and **4**. This wash step should be repeated twice. More washes can risk loss of particles or cause the filter membrane to break.

6. After the last wash step, invert the filter membrane upside down in a new tube. Spin at 1000 × $g$ for 2 min to collect the particles from the membrane. The volume of particles will be 20–50 µL.

7. If multiple tubes were used, combine into one 1.5 mL centrifuge tube. Bring the total volume to 100 µL in ultrapure water and store at 4 °C in the dark until use.

*3.4.2 For 100 nm and Larger Particles*

The centrifugation speed and time is dependent on the particle diameter and is outlined for different size particles that are greater than 100 nm in Table 4. All other steps in the collection process are the same.

1. At the end of the incubation time, remove the aluminum foil and place the 1.5 mL tube in a microcentrifuge at 4 °C. These particles, and particles larger than 100 nm, do not use filter membranes.

2. Centrifuge the particles for 25 min at 21,000 × $g$.

3. Pipet off the supernatant and add 200 µL of ultrapure water to the 1.5 mL tube. Pipet up and down to resuspend the particles. Particles should resuspend easily if well PEGylated (*see* **Note 14**).

**Table 4**
**Suggested collection times and speeds for nonfiltration centrifugation for 100 nm PS particles or larger**

| Particle diameter (nm) | Centrifugation speed ($\times g$) | Centrifugation time (min) | Number of wash steps |
|---|---|---|---|
| 100 | 21,000 | 25 | 2 |
| 200 | 18,000 | 15 | 2 |
| 500 | 15,000 | 15 | 2 |
| 1000 or larger | 10,000 | 10 | 2 |

4. Bring final volume in tube to 1 mL using ultrapure water. Repeat **step 2**. This wash step (**steps 2** and **3**) should be repeated twice. More washes can risk loss of particles.

5. After the last wash step, pipet off the supernatant. Bring the total volume to 100 μL in ultrapure water and store at 4 °C in the dark until use (*see* **Note 15**).

*3.5 Brain-Penetrating Nanoparticle Characterization*

The net surface charge (ξ-potential), polydispersity index (PDI), and hydrodynamic diameter should be measured for COOH- and PEG-coated fluorescent nanoparticles (NPs) of all sizes. COOH- and PEG-coated fluorescent NPs of all sizes can be measured by laser Doppler anemometry for net surface charge (ξ-potential) and dynamic light scattering (DLS) for PDI, and hydrodynamic diameter. The Malvern Zetasizer NanoZS is the most commonly employed instrument for these measurements.

1. In the instrument settings, polystyrene should be chosen for the material (for other materials, *see* **Note 16**).

2. In the instrument settings, ACSF needs to be added as a dispersant based on the concentrations provided in Subheading 2.3, **item 2**.

3. Dilute ACSF from 119 mM NaCl to 10 mM NaCl to obtain accurate ζ-potential measurements.

4. Dilute nanoparticle samples 1000-fold in ACSF at pH 7.0 to run the size and zeta potential measurements.

5. Perform size measurements and PDI at 25 °C at a scattering angle of 90°. A minimum of three separate measurements should be taken with a minimum of ten runs per measurement. Typical size, zeta potential, and PDI for particles referenced in this protocol are provided in Table 5.

**Table 5**
**Representative physicochemical properties of PS nanoparticles after PEGylation. Mean diameter in ACSF at pH 7.0 was measured with dynamic light scattering. $\zeta$-potential and PDI were measured in ACSF at pH 7.0. Adapted From Sci Transl Med. 2012 Aug 29;4(149): 149ra119. Reprinted with permission from AAAS. Copyright 2012 AAAS**

| Starting mean diameter of PS-COOH ± SEM (nm) | Mean diameter with PEG ± SEM (nm) | Mean Zeta potential with ± SEM (mV) | PDI |
|---|---|---|---|
| 57 ± 2 | 69 ± 2 | −2.8 ± 0.4 | 0.05 |
| 94 ± 3 | 106 ± 4 | −4.4 ± 0.2 | 0.03 |
| 185 ± 1 | 198 ± 6 | −7.8 ± 0.6 | 0.03 |

# 4    Notes

1. Borate buffer is stable for 3–6 weeks when stored at room temperature. The solution should be clear and not contain any precipitates. If the solution becomes cloudy or has precipitates, a new solution should be made.

2. Dilution of the stock PS-COOH particles is based on the nanoparticle size. Each size has a different starting concentration, with smaller particles having a higher concentration than larger particles. Therefore, smaller particles should be diluted more. Particles with starting concentrations greater than $10^{14}$ particles/mL (i.e., 20 and 40 nm) should be diluted six-fold. Particles with starting concentrations of $10^{12}$–$10^{13}$ particles/mL (i.e., 100 and 200 nm) should be diluted four-fold. Particles with starting concentrations less than $10^{11}$ particles/mL should be diluted two-fold.

3. Fluorescent particles are necessary for visualization using confocal imaging. COOH surface functionalization is useful for creating a stable, nondegradable amine bond, using carboxyl-amine chemistry. However, other surface functionalities on the nanoparticle surface can be utilized depending on the desired degradation or removal of PEG coating on the surface of the particle.

4. The dilution of the stock particles based on the particle size and concentration, as described in **Note 2**, can be scaled to any starting volume of stock particle solution.

5. Stock solutions of PS particles may contain trace amounts of sodium azide. If this is a concern, the particles can be purified by adding up to 1 mL of ultrapure water to the solution, centrifuging the particles based on their size to form a pellet, then removing the supernatant as described in Subheading 3.4.

This process should be repeated at least three times to ensure adequate removal of the sodium azide.

6. Nanoparticles in the brain microenvironment are exposed to approximately a 300 mM ion concentration in both in vivo and ex vivo CSF; however, at this high ion concentration, the accuracy of $\xi$-potential measurements can be negatively impacted. Therefore dilution of ACSF down to a 10 mM NaCl concentration can provide more accurate and stable zeta potential measurements.

7. ACSF is stable for 3–4 weeks when stored at 4 °C. The solution should be clear and not contain any precipitates. If the solution becomes cloudy or has precipitates, a new solution should be made.

8. The number of COOH per bead can be replaced with any surface functionality, provided the surface functionality can be directly associated with the Charge (mEq) provided in the CoA.

9. The PEG MW used in this protocol to achieve brain penetration is 5000 Da. Linear PEG chains with MWs from 1000 to 20,000 Da have shown the ability to penetrate through biological mediums like mucus and tissue [24]; however, thus far, 5000 Da PEG has led to the highest effective diffusivities for nanoparticles in the brain microenvironment. Branched PEGs or PEG MWs above 10 kDa have not been extensively explored and are an avenue for future investigation, in addition to alternative surface coatings discussed in subheading 1.

10. PEG excess is dependent on the starting nanoparticle size. Generally speaking, given the high density of COOH groups on sub-100 nm particles, and the surface curvature, a one-to-one mole ratio of PEG to COOH groups is used. For particles 100 to sub-500 nm, a four-fold excess of PEG to COOH groups can be used. For larger particles, greater than 500 nm, a two- to four-fold or higher excess of PEG to COOH groups can be used. If too much PEG is used, it will be immediately obvious once the borate buffer is added. The solution will become gel-like or very viscous. This indicates potential cross-linking or entanglement of the PEG, which will lead to poor reaction outcomes.

11. Borate should be added at four times the diluted PS-COOH particle volume. However, when determining the starting volume, it is important to remember the max volume of microcentrifuge tubes is either 1.5 or 2.0 mL. This protocol can scale to volumes large enough to justify 15 mL conical tubes; however, preparing multiple small batches in 1.5–2.0 mL tubes is more reproducible, and the batches can be combined after the final wash step.

12. Particles can be incubated for 4–24 h at room temperature; however, there is no benefit to incubating longer than 4 h.

13. If all the particle solution has not filtered through the membrane, add five more minutes of centrifugation time. Continue to add 5 min until all particle solution has filtered through the membrane. It is better to add time instead of increasing the speed.

14. PS-PEG particles that do not readily resuspend after centrifugation suggest that the PEG conjugation was not efficient. After the first collection, add 200 µL of ultrapure water to the pellet and pipet up and down over the pellet. Vortex the solution. After the pellet is resuspended in this volume, bring the total volume up to 800 µL. If the pellet is not dissolving, then vortex the solution longer. Do not sonicate the solution, as the sonication could disrupt the PEG layer.

15. Particles can be resuspended to the starting volume of stock PS particles used, or can be resuspended to a two-fold dilution.

16. Depending on the instrument being used for the size and zeta potential measurements, quantum dot and other materials can be listed in the material setting. If the material is not listed, it should be added. The dielectric constant and refractive index of the material must be known to add it to the list.

## Acknowledgements

The author would like to acknowledge J. Hanes and the Center for Nanomedicine at Johns Hopkins University.

## References

1. Patel T, Zhou J, Piepmeier JM, Saltzman WM (2012) Polymeric nanoparticles for drug delivery to the central nervous system. Adv Drug Deliv Rev 64(7):701–705. doi:10.1016/j.addr.2011.12.006

2. Nance EA, Woodworth GF, Sailor KA, Shih TY, Xu Q, Swaminathan G, Xiang D, Eberhart C, Hanes J (2012) A dense poly(ethylene glycol) coating improves penetration of large polymeric nanoparticles within brain tissue. Sci Transl Med 4(149):149ra119. doi:10.1126/scitranslmed.3003594

3. Nance E, Timbie K, Miller GW, Song J, Louttit C, Klibanov AL, Shih TY, Swaminathan G, Tamargo RJ, Woodworth GF, Hanes J, Price RJ (2014) Non-invasive delivery of stealth, brain-penetrating nanoparticles across the blood-brain barrier using MRI-guided focused ultrasound. J Control Release 189:123–132. doi:10.1016/j.jconrel.2014.06.031

4. Strohbehn G, Coman D, Han L, Ragheb RR, Fahmy TM, Huttner AJ, Hyder F, Piepmeier JM, Saltzman WM, Zhou J (2015) Imaging the delivery of brain-penetrating PLGA nanoparticles in the brain using magnetic resonance. J Neurooncol 121(3):441–449. doi:10.1007/s11060-014-1658-0

5. Nance E, Zhang C, Shih TY, Xu Q, Schuster BS, Hanes J (2014) Brain-penetrating nanoparticles improve paclitaxel efficacy in malignant glioma following local administration. ACS Nano 8(10):10655–10664. doi:10.1021/nn504210g

6. Mead BP, Mastorakos P, Suk JS, Klibanov AL, Hanes J, Price RJ (2016) Targeted gene transfer to the brain via the delivery of brain-penetrating DNA nanoparticles with focused

ultrasound. J Control Release 223:109–117. doi:10.1016/j.jconrel.2015.12.034

7. Burke CW, Suk JS, Kim AJ, Hsiang YH, Klibanov AL, Hanes J, Price RJ (2012) Markedly enhanced skeletal muscle transfection achieved by the ultrasound-targeted delivery of non-viral gene nanocarriers with microbubbles. J Control Release 162(2):414–421. doi:10.1016/j.jconrel.2012.07.005

8. Wohlfart S, Gelperina S, Kreuter J (2012) Transport of drugs across the blood-brain barrier by nanoparticles. J Control Release 161(2): 264–273. doi:10.1016/j.jconrel.2011.08.017

9. Kreuter J (2014) Drug delivery to the central nervous system by polymeric nanoparticles: what do we know? Adv Drug Deliv Rev 71:2–14. doi:10.1016/j.addr.2013.08.008

10. Zhou J, Patel TR, Sirianni RW, Strohbehn G, Zheng MQ, Duong N, Schafbauer T, Huttner AJ, Huang Y, Carson RE, Zhang Y, Sullivan DJ Jr, Piepmeier JM, Saltzman WM (2013) Highly penetrative, drug-loaded nanocarriers improve treatment of glioblastoma. Proc Natl Acad Sci USA 110(29):11751–11756. doi:10.1073/pnas.1304504110

11. Yang M, Lai SK, Wang YY, Zhong W, Happe C, Zhang M, Fu J, Hanes J (2011) Biodegradable nanoparticles composed entirely of safe materials that rapidly penetrate human mucus. Angew Chem Int Ed Engl 50(11):2597–2600. doi:10.1002/anie.201006849

12. Tang BC, Dawson M, Lai SK, Wang YY, Suk JS, Yang M, Zeitlin P, Boyle MP, Fu J, Hanes J (2009) Biodegradable polymer nanoparticles that rapidly penetrate the human mucus barrier. Proc Natl Acad Sci USA 106 (46):19268–19273. doi:10.1073/pnas.0905998106

13. Nance E, Porambo M, Zhang F, Mishra MK, Buelow M, Getzenberg R, Johnston M, Kannan RM, Fatemi A, Kannan S (2015) Systemic dendrimer-drug treatment of ischemia-induced neonatal white matter injury. J Control Release 214:112–120. doi:10.1016/j.jconrel.2015.07.009

14. Kannan S, Dai H, Navath RS, Balakrishnan B, Jyoti A, Janisse J, Romero R, Kannan RM (2012) Dendrimer-based postnatal therapy for neuroinflammation and cerebral palsy in a rabbit model. Sci Transl Med 4(130):130ra146. doi:10.1126/scitranslmed.3003162

15. Jenkins SI, Weinberg D, Al-Shakli AF, Fernandes AR, Yiu HH, Telling ND, Roach P, Chari DM (2016) 'Stealth' nanoparticles evade neural immune cells but also evade major brain cell populations: implications for PEG-based neurotherapeutics. J Control Release 224:136–145. doi:10.1016/j.jconrel.2016.01.013

16. Suk JS, Xu Q, Kim N, Hanes J, Ensign LM (2015) PEGylation as a strategy for improving nanoparticle-based drug and gene delivery, Adv Drug Deliv Rev. doi:10.1016/j.addr.2015.09.012

17. Popielarski SR, Pun SH, Davis ME (2005) A nanoparticle-based model delivery system to guide the rational design of gene delivery to the liver. 1. Synthesis and characterization. Bioconjug Chem 16(5):1063–1070. doi:10.1021/bc050113d

18. Wang YY, Lai SK, Suk JS, Pace A, Cone R, Hanes J (2008) Addressing the PEG mucoadhesivity paradox to engineer nanoparticles that "slip" through the human mucus barrier. Angew Chem Int Ed Engl 47(50): 9726–9729. doi:10.1002/anie.200803526

19. Perry JL, Reuter KG, Kai MP, Herlihy KP, Jones SW, Luft JC, Napier M, Bear JE, DeSimone JM (2012) PEGylated PRINT nanoparticles: the impact of PEG density on protein binding, macrophage association, biodistribution, and pharmacokinetics. Nano Lett 12 (10):5304–5310. doi:10.1021/nl302638g

20. Pombo Garcia K, Zarschler K, Barbaro L, Barreto JA, O'Malley W, Spiccia L, Stephan H, Graham B (2014) Zwitterionic-coated "stealth" nanoparticles for biomedical applications: recent advances in countering biomolecular corona formation and uptake by the mononuclear phagocyte system. Small 10 (13):2516–2529. doi:10.1002/smll.201303540

21. Amoozgar Z, Yeo Y (2012) Recent advances in stealth coating of nanoparticle drug delivery systems. Wiley Interdiscip Rev Nanomed Nanobiotechnol 4(2):219–233. doi:10.1002/wnan.1157

22. Yang Q, Lai SK (2015) Anti-PEG immunity: emergence, characteristics, and unaddressed questions. Wiley Interdiscip Rev Nanomed Nanobiotechnol 7(5):655–677. doi:10.1002/wnan.1339

23. Rizzo V, Pinciroli V (2005) Quantitative NMR in synthetic and combinatorial chemistry. J Pharm Biomed Anal 38(5):851–857. doi:10.1016/j.jpba.2005.01.045

24. Suk JS, Xu Q, Kim N, Hanes J, Ensign LM (2016) PEGylation as a strategy for improving nanoparticle-based drug and gene delivery. Adv Drug Deliv Rev 99(Pt A):28–51. doi:10.1016/j.addr.2015.09.012

# Chapter 7

# Volumetric Bar-Chart Chips for Biosensing

## Yujun Song, Ying Li, and Lidong Qin

## Abstract

The volumetric bar-chart chip (V-Chip) is a microfluidics-based, point-of-care (POC) device for the multiplexed and quantitative measurement of biomarkers. Volumetric readouts, based on the measurement of oxygen generated by a reaction between catalase and hydrogen peroxide, allow instant visual quantitation of target biomarkers and provide visualized bar charts without any assistance from instruments and without the need for data processing or graphics plotting. V-Chip shows potential capabilities in POC and personalized diagnostics; for instance, it can be utilized for making high-throughput, multiplexed, and quantitative measurements. Further, this system is highly portable and can be performed at low cost. The development of the V-Chip thus marks a POC milestone and opens up the possibility of instrument-free personalized diagnostics. Here, we describe the protocols for the fabrication of V-Chip and the use of silica nanoparticles as the probe carrier for the V-Chip-based enzyme-linked immunosorbent assay (ELISA) for the detection of biomarkers.

Key words Point-of-care, Multiplex, Quantitation, ELISA, Biomarkers

## 1 Introduction

Current protein-based biomarker methods mostly use enzyme-linked immunosorbent assay (ELISA), which serves as the clinical gold standard [1, 2]. In traditional ELISA methods, colorimetric, fluorescent, electrochemical, or magnetic signals are introduced to transduce the binding event of a protein to a specific recognition molecule into a readout signal [3–7]. As advanced instrumentation is required for quantitative detection, most of these methods are not ideal for point-of-care (POC) applications, due to their high cost and/or complicated operation [8–10].

We have created a new volumetric biosensing platform that allows quantitative, multiplexed, and instrument-free protein measurement based on Slipchip technology, which performs multiplexed microfluidic reactions without pumps or valves [1, 11–19]. Instead of using chemiluminescence or fluorescence readouts, the V-Chip presents an on-chip visualized bar chart, based on the volumetric measurement of oxygen generation on-chip, integrated

Sarah Hurst Petrosko and Emily S. Day (eds.), *Biomedical Nanotechnology: Methods and Protocols*, Methods in Molecular Biology, vol. 1570, DOI 10.1007/978-1-4939-6840-4_7, © Springer Science+Business Media LLC 2017

with an ELISA reaction. The design employs catalase as the ELISA probe, which is conjugated to silica nanoparticles functionalized with ELISA detection antibodies. The silica nanoparticles used here increase the amount of catalase probes per antibody, thus efficiently improving the detection sensitivity. Catalase reacts with hydrogen peroxide to release oxygen. The reaction is very sensitive and activated within seconds after the two agents are brought in contact with each other [20–22]. The generated oxygen accumulates within the limited volume of the microfluidics channels and causes an increase in pressure, which pushes preloaded inked bars. The advancement of each individual inked bar independently indicates the amount of catalase that reacted in that ELISA well, which correlates with the concentration of the corresponding ELISA target. This approach allows for the measurement of target protein biomarkers in both a quantitative and multiplexed manner.

## 2  Materials

### 2.1  Photolithography Components

1. All devices are designed as computer graphics using Auto-CAD software and then printed out as transparency photomasks by CAD/Art Services Inc. (Bandon, OR) with resolution at 10 μm.
2. Glass slides ($75 \times 50 \times 1$ mm, Corning, NY).
3. Hexamethyldisilazane (HMDS, Sigma-Aldrich).
4. SPR 220-7 photoresist (MicroChem Corp., MA), stored at room temperature in the dark.
5. SPR developer solution (MF-CD26, MicroChem Corp., MA), stored at room temperature.

### 2.2  Glass Etching and Drilling Components

1. Chloride sealing tape.
2. Glass etchant solution: 1:0.5:0.75 mol/L $HF/NH_4F/HNO_3$. Prepare 3 L in a plastic bottle and store at room temperature (*see* **Note 1**).
3. Diamond drill tips (0.031 in. in diameter, Harvey Tool, MA).
4. Cameron 164 drill press (Cameron Micro Drill Presses, CA).

### 2.3  Surface Hydrophobic Modification Components

1. Piranha solution: 70% $H_2SO_4$, 30% $H_2O_2$, stored at room temperature in the dark for up to one week (*see* **Note 2**).
2. Tridecafluoro-1, 1, 2, 2-tetrahydrooctyl-1-trichlorosilane (Gelest, Inc., PA), stored at room temperature.

### 2.4  Device Assembly Components

1. Fluorinert liquid FC-70 (3 M, MN), stored at room temperature.

**2.5 Capture Antibody Immobilization Components**

1. Polyclonal antibody (Abcam, MA) to increase sensitivity, stored at −80 °C.

2. Toluene.

3. 10% (3-glycidoxypropyl) trimethoxysilane (3-GPS) in toluene (Sigma-Aldrich).

4. 2% Bovine serum albumin (BSA) in PBS.

5. 5% BSA in PBS.

6. Washing solution: 2% BSA and 0.05% Tween-20 in PBS buffer.

**2.6 Detection Antibody Conjugation**

1. Monoclonal antibody (Abcam, MA) to increase the specificity, stored at −80 °C.

2. Epoxy-functionalized $SiO_2$ nanoparticles (SkySpring Nanomaterials, Inc., TX).

3. Catalase from bovine liver (Sigma-Aldrich).

4. Phosphate-buffered Saline (PBS, pH 7.4).

**2.7 CEA Serum Samples Collection and Storage**

1. Here, the patient sera were collected under the protocol IRB0412-0066 approved by the institutional review board of the Methodist Hospital Research Institute (*see* **Note 3**), stored at −80 °C, volumes of 20 μL.

**2.8 On-Chip Assay**

1. Red ink.

# 3 Methods

Carry out all procedures at room temperature unless otherwise specified.

**3.1 Fabrication of V-Chip**

The V-chip device was fabricated using standard lithography and wet-etching processes, as shown in Fig. 1.

*3.1.1 Photolithography*

1. Spin-coat glass slides with a 10-μm-thick layer of SPR220-7 photoresist (*see* **Note 4**) [23, 24].

2. Bake the glass slides at 75 °C for 3 min and then 115 °C for 5 min to promote resistance to adhesion.

3. Align the photoresist-coated glass slides with a photomask containing the designed wells and channels after cooling the slides to room temperature (*see* **Note 5**).

4. Expose the photomask and glass slides with ultraviolet light (i-line, 365 nm) for 50 s.

5. Remove the photomask from the glass slide and immerse the glass slides in SPR developer solution (MF-CD26) for 3 min.

6. Rinse the slides with Millipore water thoroughly (*see* **Note 6**) and dry the slides with nitrogen gas.

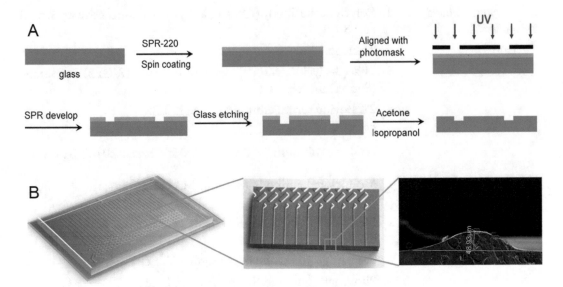

**Fig. 1** (**a**) Schematic representation of the procedures of photolithography, glass etching, and removal of photoresist. (**b**) Channel height measurement. Polydimethylsiloxane elastomer was used to replicate the channels on the glass slide for the height measurement. The polydimethylsiloxane elastomer was cured and removed from the glass, cut into small pieces, and coated with a thin layer of gold with a sputter coater for measurement via scanning electron microscopy (SEM). The SEM image on the *right* shows the channel height; 50 μm was used for all the V-Chip devices

*3.1.2 Glass Etching and Preparation of Loading Holes*

1. Cover the back of the glass slides with polyvinyl chloride sealing tape to protect it from etching.

2. Immerse the taped glass slides carefully into the glass etchant in a plastic container to etch the photolithography pattern [1].

3. Keep the plastic container in a 35 °C constant-temperature water bath to control etching speed.

4. Remove the glass slides from the etchant after 45 min, which will generate wells and channels with the depth of approximately 50 μm (*see* **Note 7**).

5. Peel off the tape thoroughly, rinse the slides with acetone and isopropanol to remove the photoresist.

*3.1.3 Preparation of Loading Holes*

1. Manually prepare access holes with a diamond drill, 0.031 in. in diameter (*see* **Note 8**).

*3.1.4 Surface Hydrophobic Modification*

To avoid liquid or gas leakage after assembling the two glass slides with FC-70, it is necessary to modify the surface hydrophobicity of the slides.

1. Immerse the glass slides in piranha solution for 1 h to clean them and then rinse the slides with Millipore water.

2. Silanize the glass slides with tridecafluoro-1, 1, 2, 2-tetrahydrooctyl-1-trichlorosilane by putting them in a vacuum desiccator after oxygen plasma treatment.

3. Bake the glass plates at 120 °C for 30 min, then rinse them with isopropanol and Millipore water, and dry them with nitrogen gas.

*3.1.5   Assembly of V-Chip*

1. Add 5 μL of Fluorinert liquid FC-70 to the top device plate with the patterns facing up and then assemble the top plate with the bottom plate.

2. Slide the two plates against each other repeatedly to distribute the FC-70 oil, which seals the two glass slides together and prevents solution leakage.

*3.1.6   Operation of V-Chip*

1. Align the two plates to make the relevant wells partially overlap and form a continuous "N"-shaped fluidic path in the horizontal direction (Fig. 2).

2. Insert a pipette tip containing the sample or reagent into the left inlet of the fluidic path in the assembled device and load the sample/reagents into the wells by pushing the pipette manually.

3. Slide the top plate obliquely to make the wells connect in a "Z" shape in the vertical direction, which will initiate the reaction between catalase and $H_2O_2$.

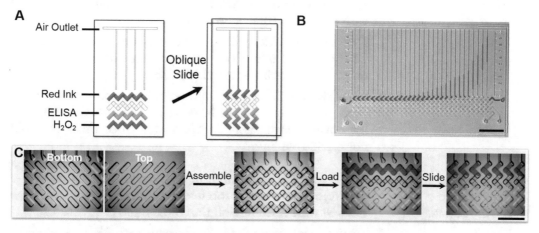

**Fig. 2 (a)** Schematic view of a typical V-Chip. On the *left* is a view of an assembled V-Chip with the flow path at the horizontal position. Ink and $H_2O_2$ can be preloaded and the ELISA assay can be performed in the designated lanes. An oblique slide breaks the flow path and forms the structure on the *right*, causing catalase and $H_2O_2$ to react and push the inked bars. **(b)** Progressive ink advancement in 30-channel V-Chips generated by the 3-h diffusion of catalase from the drilled holes on the *right* to the ELISA lane at room temperature. Scale bar, 1 cm. **(c)** Zoomed microscopic images of as-fabricated bottom and top plates, device assembly, reagent loading, and assay operation in a 50-plexed V-Chip. Scale bar, 2.5 mm

### 3.2 Antibody Immobilization and Conjugation

#### 3.2.1 Procedures for the Synthesis of Detecting Antibody-SiO$_2$-Catalase Complexes (Catalase Probes)

1. Add 20 μL detecting antibodies (1 mg/mL) to 1 mL epoxy-SiO$_2$ nanoparticles suspension (2 mg/mL in PBS) and incubate with shaking at 800 rpm for 1 h. Antibodies will be conjugated to the nanoparticle based on the reaction between the epoxy and amine groups.

2. Add 144 μL catalase (2 mg/mL in PBS) to the above solution and incubate with shaking at 800 rpm for 30 min. Catalase will be conjugated to the nanoparticle based on the reaction between the epoxy and amine group. Then add 100 μL of 12 M ammonium sulfate solution to the mixture to speed up the reaction. Incubate the solutions with shaking at 800 rpm for 3 h.

3. Wash the mixture three times with 2% BSA (in PBS) with a 10-min incubation each time. Then, separate the nanoparticles by centrifugation at 5000 rpm (1000 × g).

4. Suspend the antibody-SiO$_2$-catalase particles in 1 mL PBS as the stock solution (see **Note 9**).

#### 3.2.2 Capture Antibody Immobilization

1. Apply drops of piranha solution to the ELISA wells (the second layer of the bottom plate) and keep for 1 h to clean the surface, then rinse the wells with Millipore water, and dry them with nitrogen gas.

2. Add 4 μL of 10% 3-GPS (in toluene) into each well using a pipette.

3. Keep the solution in each well for 30 min, and then rinse the wells with fresh toluene to remove extra 3-GPS.

4. Dry the glass slides with nitrogen gas and bake the slides at 120 °C for 30 min. Epoxy groups were covalently modified onto the surface in each well (Fig. 3a).

5. Carefully add the capture antibodies into each well and incubate at 4 °C overnight (see **Note 10**). Then wash the wells with 5% bovine serum albumin (BSA) for three times to avoid nonspecific binding.

#### 3.2.3 Parallel ELISA for Carcinoembryonic Antigen (CEA) Detection

1. Add 1 μL of each serum sample to each ELISA well with a pipette, and incubate at room temperature for 1 h. Then, wash the well with Washing Solution three times (see **Note 11**).

2. Further wash the ELISA fluidic path with 2% BSA twice to block the ELISA wells after assembling the top plate and bottom plate and align the wells to form a "N"-shaped path (see **Note 12**).

3. Add the catalase probe from the inlet and incubate for 1 h. Wash the wells four times with Washing Solution and once with PBS buffer.

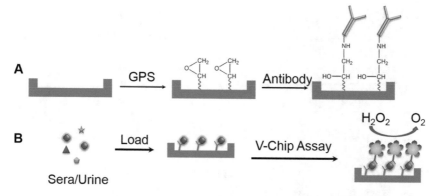

**Fig. 3** (**a**) The ELISA wells are modified with epoxy groups and then reacted with capture antibodies. The capture antibodies are covalently immobilized on the glass surface by the reaction between their amino groups and the epoxy groups. GPS is (3-glycidoxypropyl) trimethoxysilane. (**b**) Serum, urine, or PBS-based samples are loaded in the ELISA wells and the biomarkers react with the capture antibodies. Following this, the catalase probes with detecting antibodies are bound, generating the ELISA "sandwich." This complex is then reacted with hydrogen peroxide to generate oxygen and push the inked bars

4. Load 10 μL red ink and 10 μL 0.45 M $H_2O_2$ into the respective fluidic paths from the left inlets (*see* **Note 13**).

5. Slide the top plate obliquely to make the wells connect in a "Z" shape in the vertical direction and initiate the reaction between catalase and $H_2O_2$ to generate oxygen (Fig. 3b), which pushes the inked bars.

6. Record the frontiers of ink advancement with a camera. The distance of the ink bars, which is used for quantitation, can be read out based on the ruler beside the small reading channels (Fig. 4).

*3.2.4 Multiplexed V-Chip ELISA*

1. Coat the relevant capture antibody for each marker in the well following the protocol in Subheading 3.2.2 (*see* **Note 14**).

2. Load 10 μL of the cell lysate into the ELISA wells through the left inlet and this loading step was repeated four times (*see* **Note 15**).

3. Incubate the device at room temperature for 1 h and wash the wells with Washing Solution three times (*see* **Note 16**).

4. Add the catalase probe through the inlet and incubate for 1 h. Then wash the wells four times using Washing Solution and once with PBS buffer.

5. Load 10 μL red ink and 10 μL 0.45 M $H_2O_2$ into the respective fluidic paths from the left inlets.

6. Slide the top plate obliquely to make the wells connect in a "Z" shape in the vertical direction and initiate the reaction between catalase and $H_2O_2$ to generate oxygen, which pushes the inked bars.

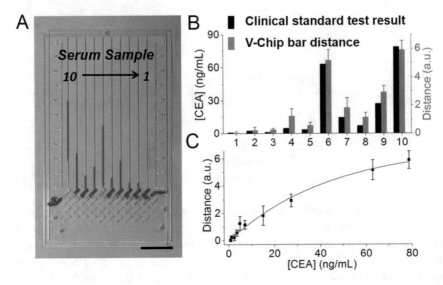

**Fig. 4** V-Chip CEA assays on patient sera. (**a**) A typical CEA readout of ten patient sera analyzed on the V-Chip. The readout of the distance of the bar-chart is obtained from the rulers besides the reading channel. Scale bar, 1 cm. (**b**) Bar graphs of the CEA concentration measured by the ADVIA Centaur Instrument (*black*) and the V-Chip distance (*red*), side-by-side. The error bars represent the s.d. of three V-Chip measurements. (**c**) Plot of V-Chip bar distances against CEA concentration. The error bars represent the standard deviation of three measurements

7. Record the frontiers of ink advancement with a camera. The distance of the ink bars, which is used for quantitation, can be read out based on the ruler besides the small reading channels (Fig. 5).

# 4    Notes

1. Hydrogen fluoride (HF) is highly corrosive and toxic, and it can irritate to the skin, eyes, and mucous membranes. Inhalation may cause respiratory irritation or hemorrhage. One should be well-trained before using HF and wear appropriate personal protective equipment (PPE) at all times.

2. It is essential to clean the coverslips with piranha solution. Note that piranha solution is very energetic and potentially explosive. Handle with great care and wear proper PPE.

3. Sera and antibodies should be stored at −20 °C for long-term storage (>3 months). Avoid repeated freezing and thawing before use in the assay.

4. Dispense 1 mL photoresist on the glass slide, and spin at 500 rpm for 5 s, then 1500 rpm for 25 s.

5. To prevent rapid cooling, the hot plate should be turned off and allowed to cool to room temperature.

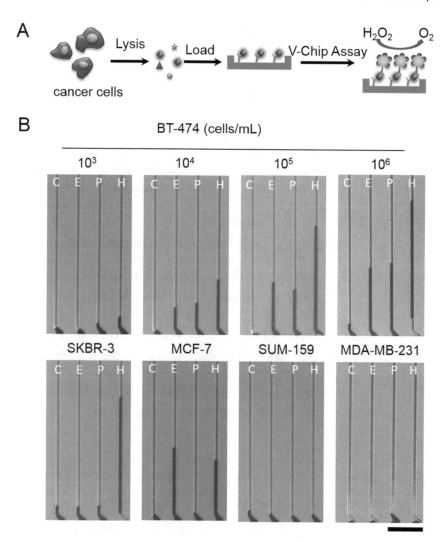

**Fig. 5** (**a**) Cancer cell lysates are loaded into the ELISA wells and the biomarkers are bound to the capture antibodies. The catalase probes with detecting antibodies are then loaded to form the sandwich structures. Afterward, the complex is reacted with hydrogen peroxide to generate oxygen and move the inked bars. (**b**) The *top four ink advancement images* show the results of assays on BT-474 lysates with increasing numbers of cells; the *bottom four images* show the results of the assays on SKBR-3, MCF-7, SUM-159, and MDA-MB-231 cell lysates with a concentration of $1 \times 10^6$ cells/mL. *C, E, P*, and *H* represent control, ER, PR and HER2, respectively. Scale bar, 0.5 cm

6. The glass slides should be thoroughly washed to remove the developer solution.

7. During the etching process, the glass slides are fixed in the bottom of the plastic container by using tape to avoid floating.

8. As glass dust is very bad for lungs, drilling under water can keep the dust from entering the air.

9. The catalase molecules were conjugated to $SiO_2$ nanoparticles and formed antibody-$SiO_2$-catalase complexes. In the complex, the mole ratio between the antibodies and catalase was approximately 1:9 based on the calculation of the concentration and the molecular weight. These $SiO_2$ nanoparticle-based complexes can increase catalase loading.

10. The glass slides are kept in a moisture environment at 4 °C to avoid the antibody solution drying.

11. To test a specific biomarker for serum samples from different patients, the samples are loaded into each well before the assembly of the top plate and bottom plate. This is also suitable for urine samples.

12. Liquid FC-70 is uniformly spread by sliding the device plate with a blank glass slide. The excess FC-70 was removed by spinning the device at 2000 rpm for 20 s.

13. The glass surface in the fluidic path can be pretreated using 1 M NaOH to produce hydrophilic surface.

14. To test multiple biomarkers in the same sample.

15. The repeated loading of cell lysates is essential to generate uniform binding of biomarkers in each ELISA well.

16. The device should be kept at humid environment to avoid the solution drying near the drilling holes.

## References

1. Song Y, Zhang Y, Bernard PE, Reuben JM, Ueno NT, Arlinghaus RB, Zu Y, Qin L (2012) Multiplexed volumetric bar-chart chip for point-of-care diagnostics. Nat Commun 3:12831–12839

2. Duan B, Hockaday LA, Kang KH, Butcher JT (2013) 3D bioprinting of heterogeneous aortic valve conduits with alginate/gelatin hydrogels. J Biomed Mater Res A 101:1255–1264

3. Gervais L, de Rooij N, Delamarche E (2011) Microfluidic chips for point-of-care immuno-diagnostics. Adv Mater 23:H151–H176

4. Chin CD, Laksanasopin T, Cheung YK, Steinmiller D, Linder V, Parsa H, Wang J, Moore H, Rouse R, Umviligihozo G, Karita E, Mwambarangwe L, Braunstein SL, van de Wijgert J, Sahabo R, Justman JE, El-Sadr W, Sia SK (2011) Microfluidics-based diagnostics of infectious diseases in the developing world. Nat Med 17:1015–1019

5. Fan R, Vermesh O, Srivastava A, Yen BK, Qin L, Ahmad H, Kwong GA, Liu CC, Gould J, Hood L, Heath JR (2008) Integrated barcode chips for rapid, multiplexed analysis of proteins in microliter quantities of blood. Nat Biotechnol 26:1373–1378

6. Stern E, Vacic A, Rajan NK, Criscione JM, Park J, Ilic BR, Mooney DJ, Reed MA, Fahmy TM (2010) Label-free biomarker detection from whole blood. Nat Nanotechnol 5:138–142

7. Xiang Y, Lu Y (2011) Using personal glucose meters and functional DNA sensors to quantify a variety of analytical targets. Nat Chem 3:697–703

8. Song YJ, Huang YY, Liu XW, Zhang XJ, Ferrari M, Qin LD (2014) Point-of-care technologies for molecular diagnostics using a drop of blood. Trends Biotechnol 32:132–139

9. Addae-Mensah KA, Cheung YK, Fekete V, Rendely MS, Sia SK (2010) Actuation of elastomeric microvalves in point-of-care settings using handheld, battery-powered instrumentation. Lab Chip 10:1618–1622

10. Qin LD, Vermesh O, Shi QH, Heath JR (2009) Self-powered microfluidic chips for

multiplexed protein assays from whole blood. Lab Chip 9:2016–2020

11. Du WB, Li L, Nichols KP, Ismagilov RF (2009) SlipChip. Lab Chip 9:2286–2292

12. Belder D (2010) Screening in one sweep using the slipchip. Angew Chem Int Ed Engl 49:6484–6486

13. Li L, Du WB, Ismagilov RF (2010) User-loaded SlipChip for equipment-free multiplexed nanoliter-scale experiments. J Am Chem Soc 132:106–111

14. Shen F, Davydova EK, Du WB, Kreutz JE, Piepenburg O, Ismagilov RF (2011) Digital isothermal quantification of nucleic acids via simultaneous chemical initiation of recombinase polymerase amplification reactions on SlipChip. Anal Chem 83:3533–3540

15. Shen F, Du WB, Davydova EK, Karymov MA, Pandey J, Ismagilov RF (2010) Nanoliter multiplex PCR arrays on a SlipChip. Anal Chem 82:4606–4612

16. Shen F, Sun B, Kreutz JE, Davydova EK, Du WB, Reddy PL, Joseph LJ, Ismagilov RF (2011) Multiplexed quantification of nucleic acids with large dynamic range using multivolume digital RT-PCR on a rotational SlipChip tested with HIV and hepatitis C viral load. J Am Chem Soc 133:17705–17712

17. Li Y, Xuan J, Song Y, Qi W, He B, Wang P, Qin L (2016) Nanoporous glass integrated in volumetric bar-chart chip for point-of-care diagnostics of non-small cell lung cancer. ACS Nano 10:1640–1647

18. Li Y, Xuan J, Xia T, Han X, Song Y, Cao Z, Jiang X, Guo Y, Wang P, Qin L (2015) Competitive volumetric bar-chart chip with real-time internal control for point-of-care diagnostics. Anal Chem 87:3771–3777

19. Li Y, Xuan J, Song Y, Wang P, Qin L (2015) A microfluidic platform with digital readout and ultra-low detection limit for quantitative point-of-care diagnostics. Lab Chip 15:3300–3306

20. George P (1947) Reaction between catalase and hydrogen peroxide. Nature 160:41–43

21. Song Y, Wang Y, Qin L (2013) A multistage volumetric bar chart chip for visualized quantification of DNA. J Am Chem Soc 135:16785–16788

22. Song Y, Xia X, Wu X, Wang P, Qin L (2014) Integration of platinum nanoparticles with a volumetric bar-chart chip for biomarker assays. Angew Chem Int Ed Engl 53:12451–12455

23. Eichelsdoerfer DJ, Liao X, Cabezas MD, Morris W, Radha B, Brown KA, Giam LR, Braunschweig AB, Mirkin CA (2013) Large-area molecular patterning with polymer pen lithography. Nat Protoc 8:2548–2560

24. Heyries KA, Tropini C, VanInsberghe M, Doolin C, Petriv OI, Singhal A, Leung K, Hughesman CB, Hansen CL (2011) Megapixel digital PCR. Nat Methods 8:649–651

# Chapter 8

## qFlow Cytometry-Based Receptoromic Screening: A High-Throughput Quantification Approach Informing Biomarker Selection and Nanosensor Development

### Si Chen, Jared Weddell, Pavan Gupta, Grace Conard, James Parkin, and Princess I. Imoukhuede

## Abstract

Nanosensor-based detection of biomarkers can improve medical diagnosis; however, a critical factor in nanosensor development is deciding which biomarker to target, as most diseases present several biomarkers. Biomarker-targeting decisions can be informed via an understanding of biomarker expression. Currently, immunohistochemistry (IHC) is the accepted standard for profiling biomarker expression. While IHC provides a relative mapping of biomarker expression, it does not provide cell-by-cell readouts of biomarker expression or absolute biomarker quantification. Flow cytometry overcomes both these IHC challenges by offering biomarker expression on a cell-by-cell basis, and when combined with calibration standards, providing quantitation of biomarker concentrations: this is known as qFlow cytometry. Here, we outline the key components for applying qFlow cytometry to detect biomarkers within the angiogenic vascular endothelial growth factor receptor family. The key aspects of the qFlow cytometry methodology include: antibody specificity testing, immunofluorescent cell labeling, saturation analysis, fluorescent microsphere calibration, and quantitative analysis of both ensemble and cell-by-cell data. Together, these methods enable high-throughput quantification of biomarker expression.

**Key words** Quantitative flow cytometry, qFlow cytometry, Immuno-labeling, Systems biology, Vascular Endothelial Growth Factor (VEGF), Platelet-Derived Growth Factor (PDGF), Angiogenesis, Background subtraction, Mixture modeling, Heterogeneity

## 1 Introduction

### 1.1 Nanosensor-Based Detection of Membrane Angiogenic Receptors Can Inform Drug Development

Nanosensors are playing an increasingly important role in biomedicine [1–3] with exciting new applications to cardiovascular diseases and cancer. This is due to the fact that nanosensors enable highly sensitive, early-stage disease detection, which is linked with better clinical outcomes [4]. For example, the use of RGD-peptide targeted Copper-64 ($^{64}$Cu)-quantum dots (QDs) as nanosensors to provide contrast for Cerenkov luminescence imaging has enabled

Sarah Hurst Petrosko and Emily S. Day (eds.), *Biomedical Nanotechnology: Methods and Protocols*, Methods in Molecular Biology, vol. 1570, DOI 10.1007/978-1-4939-6840-4_8, © Springer Science+Business Media LLC 2017

atherosclerotic plaque detection in a rodent model [5]. Similarly, $\alpha_v\beta_3$-targeted Cu-nanoparticles have been coupled with photoacoustic imaging to visualize angiogenesis in a rodent model [6], and single-chain cysteine-tagged recombinant vascular endothelial growth factor (VEGF)-121 molecules have been used as nanosensors with near-infrared fluorescence imaging (NIRF) to assess $VEGF_{121}$ uptake in tumor-activated host vasculature in a mouse model [7]. These examples demonstrate the immense potential of nanosensors in disease management, and highlight a common nanosensor feature: they can target specific biomarkers important to disease progression.

As new nanosensors are developed to guide therapy selection and disease management [8], nanosensor development requires considering the biomarker localization. Most diseases present several overexpressed plasma membrane and intracellular biomarkers, typically genes, proteins, or other biomolecules [9–11]. Therefore, determining the biomarker target is an important step in nanosensor development. The simplest biomarker targets are plasma membrane proteins: their extracellular residues render them highly accessible. Conversely, intracellular biomarkers require manipulating membrane permeability or cell trafficking to enable nanosensor binding. Thus, plasma membrane proteins are promising disease biomarkers.

Nanosensor development also requires considering biomarker abundance; biomarkers expressed at low levels require high-affinity nanosensors, whereas lower-affinity nanosensors may sufficiently target biomarkers expressed at high levels. To measure plasma membrane protein abundance on the cell scale, many studies examine mRNA and total protein expression using quantitative real-time polymerase chain reaction (qRTPCR) and Western blot, respectively [12–16]. However, identifying plasma membrane protein abundance from these assays requires correlations between DNA or mRNA and translated protein that are inconsistent [17], protein-specific [18], or require trafficking insights [19]. To measure plasma membrane protein abundance on the tissue scale, immunohistochemistry (IHC) is the employed standard [20]. However, IHC only provides a relative protein expression mapping within an area, including the membrane, and does not provide an absolute protein quantification or cell-by-cell protein expression analysis. Therefore, new high-resolution techniques are necessary for accurate plasma membrane protein quantification.

**1.2   Rationale for Quantitative Flow Cytometric Assays**

Traditional flow cytometry directly profiles membrane proteins using an affinity probe conjugated to a fluorophore, providing biomarker expression readout on the plasma membrane. Advantages of traditional flow cytometry are its amenability to live-cell analysis and its inherent multidimension data obtained. In particular, both fluorescence intensity and light scattering data are obtained on a cell-by-cell basis, providing cell-subpopulation information. Furthermore, biomarker expression can be dynamically observed in response to experimental parameters, such as temperature or drug administration [21]. However, a major disadvantage of traditional flow cytometry is its nonquantitative nature. Traditional flow cytometry provides a fluorescence signal correlating with protein abundance: higher fluorescence intensity indicates higher protein expression. To translate fluorescence intensity to protein abundance, housekeeping proteins (positive control) are used to provide comparative insight into target protein expression levels. This comparative estimation allows trends and differences in samples to be identified; however, these trends and differences in samples can be erroneous if the positive control is not fully established. For example, housekeeping proteins may exhibit unexpected shifts due to internal factors, such as sample conditions, or external factors, such as variation across flow cytometry instruments [22–26]. To avoid these erroneous measurements, recent advances have made traditional flow cytometry quantitative by including fluorophore calibration standards [27], a technique termed qFlow cytometry [27–29].

qFlow cytometry advances nonquantitative traditional flow cytometry by converting the arbitrary flow cytometry signal to absolute protein concentration. Absolute protein quantification overcomes the shortcomings inherent to positive control comparisons used in traditional flow cytometry. For example, we have observed changing protein concentrations across slightly over-confluent to under-confluent cell cultures (data not shown), as have others [30]. Again, when such changes happen in both a housekeeping protein and a target protein, the relative differences may not be detected or falsely translated using traditional flow cytometry. Since qFlow cytometry reports absolute protein concentrations, it can detect such differences, alerting the researcher to possible problems in their experimental protocol. Thus, qFlow cytometry allows for experimental standardization, allowing researchers to understand experimental variation and easily compare data across labs.

qFlow cytometry offers a promising approach to advance computational modeling, which is widely used to accelerate scientific discovery and optimize therapeutic approaches by delineating the complex behaviors inherent to biological systems. For example, Weddell and Imoukhuede found that anti-VEGF efficacy depends on endothelial VEGFR1 plasma membrane concentration, with high VEGFR1 concentrations resulting in ineffective anti-VEGF treatment, using a whole-body computational model [31]. Likewise, our lab found that small increases in plasma membrane receptor concentrations (<1000 receptors/cell) double nuclear-based receptor signaling [32] using an endocytosis computational model. However, such computational models require parameterization with physiological data, including protein concentrations [33], to accurately represent the biological system. Accurate biomarker concentrations are therefore necessary to develop computational models [21, 31, 33]. qFlow cytometry renders this much needed accuracy to biomarker quantification, ensuring optimal parameterization and physiologically relevant computational models.

Here, we describe the method to successfully quantify membrane-localized biomarkers on a cell-by-cell basis. We discuss antibody specificity, establishing saturation conditions, immuno-fluorescent labeling strategies, live-cell versus fixed cell methods, and cell-by-cell analysis considerations (e.g., background subtraction, quantification, and statistically analyzing protein heterogeneity). We describe this method in the context of quantifying the angiogenesis-related membrane proteins vascular endothelial growth factor receptor 1 (VEGFR1), VEGFR2, VEGFR3, and neuropilin 1 (NRP1) on human umbilical vein endothelial cells (HUVECs), and platelet-derived growth factor receptor alpha (PDGFRα) and PDGFRβ on adult human dermal fibroblasts (HDFs). However, the method presented here can be adapted and applied to any cell type and biomarker.

## 2 Materials

### 2.1 Cell Culturing

1. Human umbilical vein endothelial cells (HUVECs).

2. EGM™-2 BulletKit endothelial cell growth medium.

3. Adult human dermal fibroblasts (HDFs).

4. FGM™-2 BulletKit fibroblast growth medium.

5. Phosphate-buffered saline (PBS), pH 7.4.

6. Sterile Disposable Bottle Top Filters with PES Membrane. The filter membrane pore size should be 0.20 μm.

7. TrypLE™ Express.

### 2.2 Cell Harvest and Membrane Receptor Staining for qFlow Cytometry

1. PBS, pH 7.4.

2. Cellstripper™ (see **Note 1**).

3. 5 mL polystyrene round-bottom tubes.

4. Stain buffer: PBS supplemented with 0.2% bovine serum albumin (BSA) and 0.05% $NaN_3$, pH 7.4.

5. Phycoerythrin (PE)-conjugated antibodies specific for human VEGFR1, VEGFR2, VEGFR3, NRP1, PDGFRα, and PDGFRβ (see **Note 2**).

**2.3  Quantitative Flow Cytometry**

1. LSR Fortessa (BD Biosciences) or equivalent flow cytometer.

2. SYTOX™ Blue Dead Cell Stain, for flow cytometry (*see* **Note 3**).

3. QuantiBRITE™ PE beads (this protocol uses beads from BD Biosciences, but equivalent beads from other manufacturers could be used).

4. Stain buffer from Subheading 2.2, **item 4**.

**2.4  Data Analysis Software**

1. FlowJo (TreeStar) software is used for analyzing and exporting flow cytometry data.

2. Excel software is used for performing calculations on exported flow cytometry data.

3. The R programming language software is used for mixture modeling.

4. MATLAB software is used for statistically describing unipopulation data (*see* **Note 4**).

# 3  Methods

Readers are assumed to have knowledge of aseptic cell culture technique and have access to the necessary equipment for growing cells: a biosafety cabinet in which the cell flasks can be opened and culture media can be changed, an incubator in which the cells can be kept at an appropriate temperature for growth, a microscope for cell observation, and pipettes/pipette tips.

**3.1  Cell Culturing**

1. Culture HUVECs in EGM™-2 BulletKit medium per standard cell culture protocols [21, 34–37].

2. Culture HDFs in FGM™-2 BulletKit medium per standard cell culture protocols [21]. Plasma membrane receptor concentrations may differ depending on the serum level in the medium (*see* **Note 5**).

3. Remove and discard culture media upon cell passaging.

4. Briefly rinse the cell layer with 5–10 mL PBS to remove all traces of serum that inhibit the action of trypsin.

5. Add 3.0–5.0 mL of TrypLE™ Express to flask and incubate at 37 °C in a 5% $CO_2$ humidified incubator for 5 min.

6. Remove flask from the incubator. Gently tap on the side of flasks and monitor cell release from the bottom of the flask using a microscope.

7. Add 3.0–5.0 mL of complete growth media and aspirate cells by gently pipetting.

8. Add appropriate aliquots of the cell suspension to new culture flasks. A subculture ratio of 1:3–1:5 is recommended, split every 4–5 days for HUVECs and 9–12 days for HDFs.

9. Incubate cell cultures at 37 °C/5% $CO_2$.

*3.2 Cell Harvest from T-175 Cell Culture Flasks*

1. Harvest cells when they grow to 75–85% confluent (*see* **Note 6**).

2. Remove culture media from cells with an aspirating pipette. In this and all subsequent steps, be careful not to scrape the cell layer with the tip of the pipette.

3. Gently add 10 mL PBS to the cells and let sit for 5–10 s. Remove PBS from the cells.

4. Add 10 mL Cellstripper™ to the cell culture flask.

5. Incubate at 37 °C in a 5% $CO_2$ humidified incubator for 5 min.

6. Remove flask from the incubator. Gently tap on the side of flasks and monitor cell release from the bottom of the flask using a microscope. If the cells are not releasing, subject flasks to abrupt mechanical force to dislodge cell adherence (*see* **Note 7**).

7. Collect the released cells in a 50 mL conical tube, add 10 mL of stain buffer, and keep on ice.

8. Perform a cell count using a hemocytometer or automated cell counter.

9. Centrifuge the cell suspension from **step 6** at $500 \times g$ for 5 min at 4 °C to pellet the cells. Remove the supernatant, being careful not to remove any cells from the pellet.

10. Resuspend the cell pellet in stain buffer to a final concentration of $4 \times 10^6$ cells/mL based on the cell count determined in **step 8**, and keep the cells on ice.

*3.3 Cell Surface Staining with PE-Conjugated Monoclonal Antibodies*

1. Prepare and label the 5 mL polystyrene round-bottom tubes. The number of tubes used in one experiment depends on how many samples you have.

2. In a biosafety cabinet, transfer a 25 μL aliquot of cell suspension ($1 \times 10^5$ cells) to each 5 mL polystyrene round-bottom tube.

3. Add PE-conjugated monoclonal antibodies to each tube (*see* **Note 8**). For non-labeled cell samples, do not add antibodies. We recommend having 2–4 replicates for each antibody.

4. Incubate cells with added antibodies for 40 min on ice in the dark.

5. Add 4 mL stain buffer to each tube.

6. Centrifuge at $500 \times g$ for 4 min at 4 °C to form a cell pellet, and then remove the supernatant.

7. Repeat washing as described in **steps 5** and **6**.

8. Resuspend cells in 250 μL of stain buffer, and keep on ice.

**3.4 Data Acquisition Using Flow Cytometry**

We describe data acquisition using a LSR Fortessa (BD) Flow cytometer with BD FACSDIVA software. Other flow cytometers and software should also work if the correct lasers and filters are included in the system. When using a LSR Fortessa (BD) Flow cytometer, use the Pacific Blue channel to measure fluorescence intensity of SYTOX Blue Stain, and use the PE channel to measure PE fluorescence intensity. If the reader is using a different flow cytometer, SYTOX Blue can be excited by a violet laser and its fluorescence intensity can be detected with a 450/50 band pass filter; PE can be excited by a yellow-green laser and detected with a 582/15 band pass filter.

1. Reconstitute one tube of QuantiBRITE™ PE, which contains a lyophilized pellet of beads, with 500 μL of stain buffer and vortex briefly. Each QuantiBRITE™ PE tube can be reused up to 2–3 times within a month.

2. Place the reconstituted PE beads from **step 1** at the inlet to the flow cytometer and begin analysis following proper protocols for your instrument.

3. Adjust the voltage for PE channel or equivalent to ensure all four bead populations are distinctively displayed on a PE histogram (*see* **Note 9**). Collect 10,000 events above the threshold. The geometric mean of each bead population will be used to determine the calibration curve for PE, as described in Subheading 3.5, **step 2, 3**.

4. Do not adjust the voltage for PE channels or the speed of the flow after acquiring these events.

5. Add 5 μg/mL SYTOX™ Blue Dead Cell Stain to a sample tube (from Subheading 3.3, **step 8**) and vortex briefly immediately prior to placement in a flow cytometer (*see* **Note 10**).

6. Within the flow cytometer software, display a forward scatter area (FSC-A) versus side scatter area (SSC-A) dot plot for cell samples. Adjust voltages for FSC-A and SSC-A to ensure gating on single-cell populations can be achieved (*see* **Note 11**).

7. Display the histogram of the cell samples for Pacific Blue channel or equivalent. Adjust voltages for Pacific Blue channel to ensure that two cell populations are distinctively displayed. Gate the population on the left side (expressing lower SYTOX™ Blue) and collect at least 10,000 gated events in each sample.

**3.5 qFlow Cytometric Analysis: Ensemble-Averaged Plasma Membrane Receptor Concentrations**

FlowJo (TreeStar) software and Excel are used for data analysis.

1. Within FlowJo software, plot FSC-A versus SSC-A for the QuantiBRITE™ PE calibration beads and gate the single-bead population. Representative gating is shown in Fig. 1a.

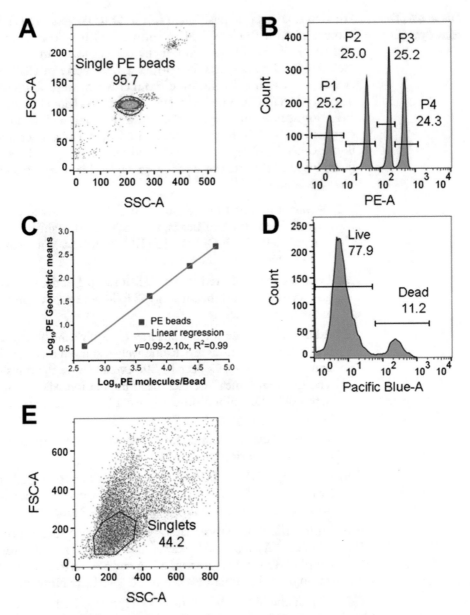

**Fig. 1** Gating on the HDFs for qFlow cytometric analysis using FlowJo. (**a**) Gate single-bead population on a FSC-A vs. SSC-A dot plot. (**b**) Gate each population of beads conjugated with different numbers of PE on a histogram of PE-A. (**c**) PE calibration curve (*red*) obtained by fitting $log_{10}$ PE geometric mean against $log_{10}$ PE molecules/bead. (**d**) Gate live-cell and dead-cell population on a histogram of Pacific Blue-A. (**e**) Gate single-cell population from live cells on a FSC-A vs. SSC-A dot plot

2. Plot a histogram of PE of the gated single-bead population. Gate the four distinctive peaks respectively. Representative gating is shown in Fig. 1b.

3. Using Excel, plot a linear regression of $\log_{10}$ PE molecules per bead against $\log_{10}$ PE geometric mean using the equation $y = mx + b$, where m and b represent the slope and intercept of the linear regression respectively. A representative PE calibration curve is shown in Fig. 1c.

4. Using FlowJo, plot histograms of Pacific Blue channel for cell samples stained with SYTOX™ Blue Dead Cell Stain and gate the live-cell populations. Representative gating is shown in Fig. 1d.

5. Plot FSC-A versus SSC-A for gated live cells and gate the single-cell populations. Representative gating is shown in Fig. 1e.

6. Plot a histogram of PE of the gated live single-cell population for each sample. The PE geometric means of both live-cell samples labeled with PE-conjugated antibodies ($PE_{labeled}$ geometric mean) and unlabeled live-cell samples ($PE_{unlabeled}$ geometric mean) are then quantified.

7. Using Excel, determine the number of PE-conjugated antibodies per cell for both PE-stained samples and non-labeled live-cell samples using the equations:

$$(\text{number of PE-antibodies/cell})_{\text{labeled}} = 10^{\frac{1}{m}\left(\log_{10}PE_{\text{labeled}} \text{ geometric mean}-b\right)}$$

$$(\text{number of PE-antibodies/cell})_{\text{unlabeled}} = 10^{\frac{1}{m}\left(\log_{10}PE_{\text{unlabeled}} \text{ geometric mean}-b\right)}$$

8. Determine the number of receptors per cell using the equation:

number of receptors/cell = (number of PE − antibodies/cell)$_{\text{labeled}}$ − (number of PE − antibodies/cell)$_{\text{unlabeled}}$.

9. Express ensemble-averaged plasma membrane receptor concentration as mean of number of receptors/cell ± standard error from replicates of each antibody.

### 3.6 qFlow Cytometric Analysis: Cell-by-Cell Analysis

1. Export cell-by-cell PE fluorescence intensity from both labeled and unlabeled live-cell samples as a CSV file from FlowJo.

2. Open the CSV file exported in **step 1** with Excel. Sum PE fluorescence intensity of all cells labeled with PE-conjugated antibody, $\sum PE_{\text{labeled}}$, and divide that sum by the number of labeled cells, $n_{\text{labeled}}$.

3. Sum PE fluorescence intensity of all unlabeled cells, $\sum PE_{\text{unlabeled}}$, and divide the sum by the number of unlabeled cells, $n_{\text{unlabeled}}$.

4. Calculate the actual PE fluorescence intensity by subtracting the background signal, $PE_{\text{real}}$, using the equation (*see* **Note 12**):

$$PE_{real} = PE_{labeled} \cdot \left( 1 - \frac{\left( \sum PE_{labeled} \right) / n_{labeled}}{\left( \sum PE_{unlabeled} \right) / n_{unlabeled}} \right)$$

5. Loop through the cell-by-cell $PE_{real}$ data and calculate the number of receptors/cell for each cell using the equation described in Subheading 3.5, **step 8**.

6. For each sample, construct a histogram $H$ where the number of receptors/cell is contained in equally spaced bins. The histogram is defined by bin centers $s$, the mean number of receptors/cell defined by each bin, and frequency $w$, the fraction of total cells contained in each bin. After constructing the histogram, eliminate outliers [31], store the bincenters as vector $s$, and store the frequency as vector $w$.

**3.7  qFlow Cytometric Analysis: Mixture Modeling for Statistically Describing Subpopulations**

R programming software is used for this data analysis.

1. Import the number of receptors/cell, with outliers removed as described in Subheading 3.6, **step 6**, as vector $v_1$ into the R programming language software. Note this will need to be done individually for each sample.

2. Take the natural logarithm of $v_1$ using the "log" command and store into a second vector $v_2$.

3. Use the "normalmixEM" command in the "mixtools" package to fit vector $v_2$ to a logarithm mixture model with two subpopulations. Store the mixture model fit as a new variable $L_{fit,2}$. The logarithm mixture model for any number of subpopulations $n$ is defined by:

$$L(v_2) = \sum_{i=1}^{n} p_i l(\mu_i, \sigma_i)$$

where $L$ is the lognormal mixture and $l$ defines the lognormal subpopulation with index $i$, mean $\mu_i$, standard deviation $\sigma_i$, and density $p_i$.

4. Repeat **step 3**, this time fitting to a logarithm mixture model with three subpopulations, storing as a new variable $L_{fit,3}$.

5. Create the 2 subpopulation mixture model:

$$L_2(s) = p_1 l(s|\mu_1, \sigma_1) + p_2 l(s|\mu_2, \sigma_2)$$

where $s$ defines the bincenter positions (number of receptors/cell), as described in Subheading 3.6, **step 6**, and the densities, means, and standard deviations are given by $L_{fit,2}$.

6. Create the 3 subpopulation mixture model:

$$L_3(s) = p_1 l(s|\mu_1, \sigma_1) + p_2 l(s|\mu_2, \sigma_2) + p_3 l(s|\mu_3, \sigma_3)$$

where the densities, means, and standard deviations are given by $L_{fit, 3}$.

7. Calculate the sum of squared error between the 2 subpopulation mixture model $SSE_2$ and the number of receptors/cell data by:

$$SSE_2 = \sum_{i=1}^{nbins} (L_2(s_i) - w_i)^2$$

where $i$ is the bin index with bincenter $s_i$, *nbins* is the number of bins, $L_2(s_i)$ is the 2 subpopulation mixture model value at bincenter $s_i$, and $w_i$ is the frequency of cells contained within $i$ as described in Subheading 3.6, **step 6**.

8. Calculate the sum of squared error between the 3 subpopulation mixture model $SSE_3$ and the number of receptors/cell data by:

$$SSE_3 = \sum_{i=1}^{nbins} (L_3(s_i) - w_i)^2$$

where $L_3(s_i)$ is the 3 subpopulation mixture model value at bincenter $s_i$.

9. If $SSE_2$ (**step 7**) is less than $SSE_3$ (**step 8**), define the best fit mixture model $L_B(s_i)$ as $L_2(s_i)$. Otherwise, define $L_B(s_i)$ as $L_3(s_i)$.

10. Express $L_B(s_i)$ as the subpopulation means $\mu_B$, standard deviations $\sigma_B$, and densities $p_B$.

11. For graphical representation, plot $s$ versus $L_B(s_i)$ alone, or with the histogram $H$ from Subheading 3.6, **step 6**. An example of mixture modeling is given in **Note 10** and Fig. 4b.

***3.8 qFlow Cytometric Analysis: Non-Normality and Diversity Analysis for Statistically Describing Uni-Population Data***

MATLAB software is used for this data analysis.

1. For each sample, import the number of receptors/cell bincenters as $s$ and frequency as $w$, as described in Subheading 3.6, **step 6**, into MATLAB software.

2. Import the corresponding number of receptors/cell for each sample, as described in Subheading 3.5, **step 8**, into MATLAB software as a vector $v_1$.

3. Determine the Gaussian mean $\mu$ and standard deviation $\sigma$ of $v_1$ using the MATLAB commands "mean($v_1$)"and "std($v_1$)," respectively.

4. Generate a reference Gaussian distribution using the command "$g$ = normpdf($s, \mu, \sigma$),"as given by the equation:

$$g(s|\mu,\sigma) = \frac{1}{\sigma\sqrt{2\pi}}\exp\left\{-\frac{(s-\mu)^2}{2\sigma^2}\right\}$$

5. Generate the probability distribution of $g(s|\mu,\sigma)$, $P_{ref}$, using the command

"Pref = $g$ . /sum($g$)," where sum($g$) is the sum of all elements in vector $g(s|\mu,\sigma)$.

6. Compute scalar statistic K-S values $p_{KS}$ for each sample using the command

"Pks = kstest2($w$, Pref)" (*see* **Note 13**).

7. Compute the quadratic entropy $QE$(*see* **Note 13**) as given by the equation:

$$QE = \sum_{j=1}^{nbins-1} \sum_{i=j+1}^{nbins} (s_i - s_j) \cdot w_i \cdot w_j, 1 \le j < i \le nbins$$

where $s_i$ and $s_j$ are centers of the bins with indices $i$ and $j$ respectively, and $w_i$ and $w_j$ are the frequencies of the bins with indices $i$ and $j$ respectively.

## 4  Notes

1. Trypsin-based cell dissociation using solutions such as TrypLE™ involves cleaving peptide bonds on the C-terminal sides of lysine and arginine [38]. This action may cleave cell surface receptors [35] or stimulate receptor shedding [39–42], and either mechanism would lead to invalid qFlow cytometry results. Therefore, nonenzymatic dissociation solution such as Cellstripper™ is recommended for preserving cell surface receptors when performing qFlow cytometry. For example, we have previously observed TrypLE™-mediated decreases in NRP1 on HUVECs, while cell surface VEGFR1 and VEGFR2 remained relatively unchanged [36]. In this chapter, we observe similar results when we extend the methods to VEGFR3 and Tie2 on HUVECs (Fig. 2a). VEGFR2 and VEGFR3 plasma membrane concentrations on HUVECs are consistent with a previous report, while VEGFR1 is ~30% higher ($p < 0.01$), which may be attributed to donor-specific differences [21]. Interestingly, TrypLE™ treatment results in a ~90% increase in PDGFRβ plasma membrane concentrations on HDF surface ($p < 0.001$), while PDGFRα concentrations remain unchanged (Fig. 1b). Furthermore, the previously reported NRP1 decrease is not specific to HUVECs; we also observe a two orders of magnitude decrease in NRP1 plasma membrane concentrations on HDFs following the TrypLE treatment (Fig. 2b). Altogether, our data and previous reports

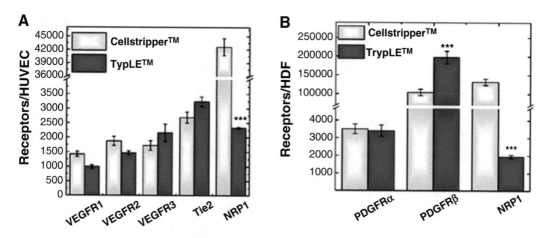

**Fig. 2** Effects of TrypLE™ comparing to Cellstripper™ on receptor quantification of (**a**) HUVECs and (**b**) HDFs. Significance tests were conducted using two-sample t-test where *** indicates $p < 0.001$

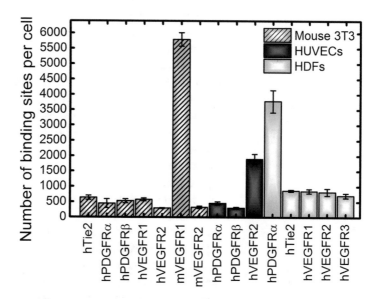

**Fig. 3** Quantification of binding sites for antibodies on mouse 3T3, HUVECs, and HDFs

indicate that if the researcher prefers to use an enzymatic cell dissociation solution, that they check how the solution affects plasma membrane receptor concentrations.

2. Antibody specificity is required for accurate receptor quantification. We labeled mouse 3T3 fibroblasts with either human or mouse-specific antibodies to determine antibody-binding specificity (Fig. 3). Our positive and negative controls give ~5800 mVEGFR1 and ~300 mVEGFR2 per mouse 3T3, respectively (Fig. 3). These trends are in line with prior studies

reporting high mVEGFR1 plasma membrane concentration and little to no mVEGFR2 [29]. As expected, the human antibodies showed low binding to the mouse 3T3s. We observe less than 600 hVEGFR1, hVEGFR2, hPDGFRα, or hPDGFRβ per mouse 3T3 fibroblast, which is similar in value to our negative controls (Fig. 3). Therefore, the cross-reactivity of the human antibody to mouse receptors is very low. When analyzing multiple biomarkers, the researcher should use similar methods, as we have outlined, to ensure that their antibodies have good specificity.

3. SYTOX Blue can be excited with a solid-state laser (407 nm) and its emission collected using a 450/50 band-pass filter. SYTOX Blue live/dead stain is preferable to Propidium Iodide (PI) or 7-AAD because SYTOX Blue emission has little overlap with PE. Live-dead cell staining should always be performed, as we noted above, and if the researcher chooses to use a stain other than SYTOX Blue, they should check the spectral spillover to ensure accurate receptor quantification.

4. This protocol is defined to use the software explicitly listed in Subheading 3; however, alternative software can be used, based on the user's preferences. An alternative flow cytometry software option that we have had good experience with is FCS Express (De Novo Software). The Purdue University Cytometry Laboratory (PUCL) offers a comprehensive listing of free flow cytometry software [43]. Alternatives to Excel for data calculations and storage include any spreadsheet software capable of basic algebraic functions and data storage, such as Accel Spreadsheet. Alternatives to R for performing mixture modeling include any programming language capable of conducting mathematical operations, such as C/C++ or Python. Likewise, other mathematical programming languages such as C/C++ or Python, or statistical analysis software, such as Minitab or SPSS, can be used instead of MATLAB for conducting statistical analyses. Overall, we advise researchers to choose the software based on experience, preference, and availability.

5. A commonly used fibroblast culture medium is Dulbecco's Modified Eagle Medium (DMEM) supplemented with 10% fetal bovine serum (FBS) and 1% of Penicillin Streptomycin (Pen Strep) [21, 44–46]. However, we would like to point out to readers that plasma membrane concentrations can be affected by serum levels in the cell culture medium. We previously found that PDGFRs and NRP1 plasma membrane concentrations decreased when HDFs were cultured in DMEM supplemented with 10% FBS compared to the standard FGM™-2 fibroblast growth medium [21], and the new data shown in Figs. 2–4 are consistent with our previous results. Given the inverse serum concentration-receptor concentration

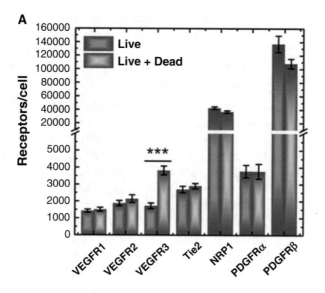

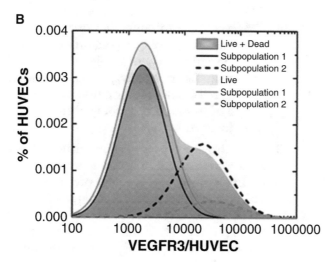

**Fig. 4** Mixture modeling can be applied to unmix VEGFR3 expression on live and dead HUVECs. (**a**) Comparison of plasma membrane receptor concentration between live single cells and live + dead single cells. Significance tests are conducted using two-sample t-test where * indicates $p < 0.05$ and *** indicates $p < 0.001$. (**b**) Gaussian mixture models of VEGFR3 subpopulations on HUVECs. *Dashed lines* represent subpopulations having a higher density and *dotted lines* represent subpopulations having a smaller density

relationship that we consistently observe, we recommend that researchers consider or assay these effects when performing qFlow cytometry.

6. We and others find that plasma membrane receptor concentrations in HUVECs may change if they reach 100% confluence. Indeed, Napione et al. also found that long-confluent HUVECs express two-fold higher VEGFR2 than in sparse cells, and proposed the theory that increased cell

concentrations are linked with the presence of mature cell junctions, which regulate receptor trafficking [30]. Therefore, we recommend researchers harvest HUVECs when they reach 80–85% confluence to avoid receptor changes that made lead to inconsistent data.

7. We find that cell dissociation choice comes with advantages and disadvantages. An important disadvantage of TrypLE as noted above, some receptors are affected by enzymatic treatment. However, a noteworthy advantage is that cells are easily dissociated from flasks with TrypLE; whereas cell dissociation may be incomplete when using Cellstripper. Sometimes, abrupt mechanical force is required to completely dissociate Cellstripper-treated cells. Interestingly, we have not observed significant plasma membrane receptor concentration changes when using abrupt force. However, we do not recommend readers apply the force more than once. Some alternatives are to tap gently on the side of flasks or to place cells in a 37 °C/5% $CO_2$ incubator with Cellstripper for an additional 1–2 min. However, >2 additional min in the incubator is not recommended, as we have observed unpredictable changes in receptor concentrations with significant incubator-Cellstripper treatment. Taken together, we recommend the use of a nonenzymatic cell dissociation solution as a default, if cells do not lift completely—one may employ abrupt force or an additional, short incubation step.

8. Non-labeled receptors will invalidate qFlow results. Therefore, a receptor saturation study is necessary for accurate qFlow cytometry profiling. We have previously determined the optimal concentrations of PE-conjugated monoclonal antibodies for staining each sample ($1 \times 10^5$ cells): 14 μg/mL for VEGFR1 and VEGFR2, 7.1 μg/mL for NRP, and 9.4 μg/mL for PDGFRs [21, 35]. Readers should determine the optimal antibody concentrations for their respective markers by staining cell samples with a series of increasing antibody concentrations and quantify the biomarker levels using qFlow cytometry. The biomarker levels should reach a plateau when the optimal antibody concentration is applied. Therefore, we advise researchers to perform a saturation study to determine the least amount of antibodies needed to achieve consistent and accurate receptor quantification.

9. The photo-physical properties of PE make it an ideal choice for receptor quantification. Its high extinction coefficient lowers error due to photobleach, and its large size imparts the advantageous 1:1 antibody to fluorophore ratio needed to accurately quantify receptors [47]. Several studies have established the use of QuantiBRITE™ PE beads for receptor quantification [21, 27, 29, 35, 36], so the depth of research available further assists

the researcher in troubleshooting and optimizing for their application. Important application notes for QuantiBRITE™ PE beads are described here. They comprise four groups of polystyrene beads conjugated with different PE densities: low (474 PE molecules/bead), medium-low (5359 PE molecules/bead), medium-high (23,843 PE molecules/bead), and high (62,336 PE molecules/bead). The exact PE number may differ from batch to batch and can be found on the flyer in the kit. A representative figure of all four bead populations distinctively displayed on a PE histogram can be seen in Fig. 1b. Other options for qFlow cytometry include Quantum MESF and Quantum Simply Cellular microspheres, which also offer FITC-based quantitative tools. However, the fluorophore sensitivity to photobleach, buffer, pH, etc. should be considered when choosing a fluorescent bead (e.g., FITC vs. PE). Overall, our approach has been optimized for applying QuantiBRITE-PE for quantifying angiogenic receptors; however other tools exist and can be translated to qFlow cytometry via optimization.

10. Live/dead cell staining or a reliable way to exclude dead cells is necessary for accurate receptor quantification. We observe that there is no significant difference in receptor quantitation when all cells are analyzed (live + dead cells) versus when only live cells are analyzed (via SYTOX™ Blue staining) for the following receptors: VEGFR1, VEGFR2, NRP1, Tie2, and PDGFRα (Fig. 4a). However, we do observe significant changes in VEGFR3 plasma membrane concentrations on HUVECs and PDGFRβ on HDFs when dead cells are not excluded (Fig. 4a). In order to further examine VEGFR3 plasma membrane expression on live and dead cells, cell-by-cell analysis was applied as described in Subheading 3.6 and 3.7. We observe that the live-cell population and live + dead mixture population both exhibit twoVEGFR3 subpopulations (Fig. 4b). Two-component lognormal mixture modeling indicates that 97% of live HUVECs have an average of ~1900 VEGFR3/cell, while 3% display an average of ~65,000 VEGFR3/cell (Table 1). Conversely, the high-VEGFR3 subpopulation has

**Table 1**

**Gaussian mixture model parameters for HUVECs labeled with anti-VEGFR3-PE**

| Sample | Mean | | Standard deviation | | Density | |
|---|---|---|---|---|---|---|
| | $\mu_1$ | $\mu_2$ | $\sigma_1$ | $\sigma_2$ | $\pi_1$ | $\pi_2$ |
| Live | 1900 | 65,000 | 3.19 | 2.49 | 0.97 | 0.03 |
| Live + Dead | 1700 | 21,000 | 2.54 | 2.88 | 0.67 | 0.33 |

greater density in the live + dead HUVEC population: 33% display ~21,000 VEGFR3/cell and 67% display ~1700 VEGFR3/cell (Table 1). Increased density of the high-VEGFR3 subpopulation when dead cells are included suggests that the high-VEGFR3 subpopulation is comprised of dead and apoptotic cells. This underlies the necessity to apply live/dead staining for accurate receptor profiling.

11. Cells are distributed based on their sizes on a FSC-A vs. SSC-A dot plot (Fig. 1e). The bigger the cells are, the higher their FSC-A is; the greater the cell granularity, the higher their SSC-A. When cells are prepared carefully, minimizing aggregation and cell lysis (e.g., kept on ice, titrated and/or strained prior to imaging, solutions are buffered, solutions are isotonic, solutions do not include Ca or Mg), cell populations primarily resolve as singlets [34, 35, 48–50]. We observe that higher order cell clusters (e.g., doublets, triplets, etc.) are few and can be distinguished linearly in the FSC-A vs. SSC-A dot plot. Overall, observing best-practices in handling enables easy gating of single-cell populations.

12. Accounting for cell autofluorescence in ensemble qFlow cytometry analysis can be as simple as subtracting the average cell fluorescence of non-labeled cells. However, when performing cell-by-cell analysis accounting for autofluorescence may incorporate error. Indeed, the background subtraction method that we present may result in some negative $PE_{real}$ values, indicating that the noise is larger than the signal. For simplicity, we set the negative values to zero. Figure 5 shows a comparison between cell-by-cell VEGFR1 histogram before and after background subtraction. Overall, the autofluorescence method presented here allows researchers to account for background noise such

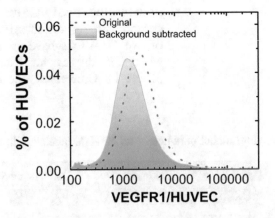

**Fig. 5** Corresponding VEGFR1 cell-by-cell distribution computed from PE fluorescence of labeled HUVECs before (*red dotted line*) and after background subtraction (*gray filled area*)

as cell autofluorescence by shifting the fluorescence signal based on a derived signal-to-noise ratio [21].

13. The high-throughput cell-by-cell data obtained by flow cytometry renders it an ideal tool for studying cell heterogeneity. The quantitative nature of qFlow cytometry adds the additional dimension of quantifying cell heterogeneity, which can be useful for several areas of research, most pressingly cancer medicine [34]. Toward these goals, K-S and QE values are good analytical tools to statistically characterize cellular heterogeneity. The K-S test compares two populations and statistically determines whether they are drawn from the same continuous distribution. In this protocol, the "Pks = kstest2 ($w$, Pref)," command in MATLAB tests the null hypothesis that the number of receptors/cell population ($w$) is drawn from a Gaussian distribution, provided by the reference Gaussian distribution Pref. Pkswill either equal 0 or 1; Pks = 1 indicates that the null hypothesis is rejected at the 5% significance level (the number of receptors/cell population is not Gaussian), whereas Pks = 0 fails to reject the null hypothesis. Therefore, K-S testing offers a statistical method for reporting differences in qFlow cytometry data that is analyzed on a cell-by-cell level. Traditional diversity measurement assumes all differences between groups are equal, whereas QE, introduced by Rao, accounts for the probability differences among those groups [51–54]. The measurement of QE bins a group of cells into a finite number of subsets based on membrane receptors concentrations, and calculates the average dissimilarities between two randomly drawn subsets from the group of cells. QE has been shown to provide a quantitative measure of the diversity of cellular phenotypes in cancer tissue sections for diagnostic applications [55]. It has also been applied to characterize cellular heterogeneity in response to drug treatment [56]. As researchers extract quantitative data from qFlow cytometry studies, we recommend K-S and QE values as good approaches for characterizing non-normality and diversity in heterogeneous cell populations [21, 56].

## Acknowledgments

We would like to thank Dr. Barbara Pilas for her advice and help with flow cytometry. We would also like to thank Spencer B. Mamer and Ali Ansari for their help with editing. Finally, we would like to thank the American Heart Association Grant #16SDG26940002,

American Cancer Society Illinois Division Basic Research Grant #282802, and National Science Foundation CBET Grant #1512598 for funding support.

## Reference

1. Kim BYS, Rutka JT, Chan WCW (2010) Nanomedicine. N Engl J Med 363:2434–2443

2. Agrawal S, Prajapati R (2012) Nanosensors and their pharmaceutical applications: a review. Int J Pharm Sci Technol 4:1528–1535

3. Swierczewska M, Liu G, Lee S et al (2012) High-sensitivity nanosensors for biomarker detection. Chem Soc Rev 41:2641–2655

4. Etzioni R, Urban N, Ramsey S et al (2003) The case for early detection. Nat Rev Cancer 3:243–252

5. Li J, Dobrucki LW, Marjanovic M et al (2015) Enhancement and wavelength-shifted emission of Cerenkov luminescence using multifunctional microspheres. Phys Med Biol 60:727–739

6. Zhang R, Pan D, Cai X et al (2015) alphaVbeta3-targeted copper nanoparticles incorporating an Sn 2 lipase-labile fumagillin prodrug for photoacoustic neovascular imaging and treatment. Theranostics 5:124–133

7. Backer MV, Levashova Z, Patel V et al (2007) Molecular imaging of VEGF receptors in angiogenic vasculature with single-chain VEGF-based probes. Nat Med 13:504–509

8. Ludwig JA, Weinstein JN (2005) Biomarkers in cancer staging, prognosis and treatment selection. Nat Rev Cancer 5:845–856

9. Weis SM, Cheresh DA (2011) Tumor angiogenesis: molecular pathways and therapeutic targets. Nat Med 17:1359–1370

10. Harper SJ, Bates DO (2008) VEGF-A splicing: the key to anti-angiogenic therapeutics? Nat Rev Cancer 8:880–887

11. Arao T, Matsumoto K, Furuta K et al (2011) Acquired drug resistance to vascular endothelial growth factor receptor 2 tyrosine kinase inhibitor in human vascular endothelial cells. Anticancer Res 31:2787–2796

12. Li J, Brown LF, Hibberd MG et al (1996) VEGF, flk-1, and flt-1 expression in a rat myocardial infarction model of angiogenesis. Am J Physiol 270:H1803–H1811

13. Brown LF, Berse B, Jackman RW et al (1995) Expression of vascular permeability factor (vascular endothelial growth factor) and its receptors in breast cancer. Hum Pathol 26:86–91

14. Gerritsen ME, Tomlinson JE, Zlot C et al (2003) Using gene expression profiling to identify the molecular basis of the synergistic actions of hepatocyte growth factor and vascular endothelial growth factor in human endothelial cells. Br J Pharmacol 140:595–610

15. Dougher M, Terman BI (1999) Autophosphorylation of KDR in the kinase domain is required for maximal VEGF-stimulated kinase activity and receptor internalization. Oncogene 18:1619–1627

16. Duval M, Bédard-Goulet S, Delisle C et al (2003) Vascular endothelial growth factor-dependent down-regulation of Flk-1/KDR involves Cbl-mediated ubiquitination. Consequences on nitric oxide production from endothelial cells. J Biol Chem 278:20091–20097

17. Guo Y, Xiao P, Lei S et al (2008) How is mRNA expression predictive for protein expression? A correlation study on human circulating monocytes. Acta Biochim Biophys Sin 40:426–436

18. Bhargava R, Gerald WL, Li AR et al (2005) EGFR gene amplification in breast cancer: correlation with epidermal growth factor receptor mRNA and protein expression and HER-2-status and absence of EGFR-activating mutations. Mod Pathol 18:1027–1033

19. Xu J, Chai H, Ehinger K et al (2014) Imaging P2X4 receptor subcellular distribution, trafficking, and regulation using P2X4-pHluorin. J Gen Physiol 144:81–104

20. Faratian D, Christiansen J, Gustavson M et al (2011) Heterogeneity mapping of protein expression in tumors using quantitative immunofluorescence. J Vis Exp 56:e3334

21. Chen S, Guo X, Imarenezor O et al (2015) Quantification of VEGFRs, NRP1, and PDGFRs on endothelial cells and fibroblasts reveals serum, Intra-Family Ligand, and Cross-Family Ligand Regulation. Cell Mol Bioeng 8:383–403

22. Rocha-Martins M, Njaine B, Silveira MS (2012) Avoiding pitfalls of internal controls: validation of reference genes for analysis by qRT-PCR and Western blot throughout rat retinal development. PloS one 7(e43028)

23. Vigelsø A, Dybboe R, Hansen CN et al (2015) GAPDH and β-actin protein decreases with aging, making Stain-Free technology a superior loading control in Western blotting of human skeletal muscle. J Appl Physiol (1985) 118:386–394

24. Baumgartner R, Umlauf E, Veitinger M et al (2013) Identification and validation of platelet low biological variation proteins, superior to GAPDH, actin and tubulin, as tools in clinical proteomics. J Proteomics 94:540–551

25. Nguyen R, Perfetto S, Mahnke YD et al (2013) Quantifying spillover spreading for comparing instrument performance and aiding in multi-color panel design. Cytometry A 83:306–315

26. Wheeless LL, Coon JS, Cox C et al (1989) Measurement variability in DNA flow cytometry of replicate samples. Cytometry 10:731–738

27. Pannu KK, Joe ET, Iyer SB (2001) Performance evaluation of quantiBRITE phycoerythrin beads. Cytometry 45:250–258

28. Wang L, Abbasi F, Gaigalas AK et al (2006) Comparison of fluorescein and phycoerythrin conjugates for quantifying CD20 expression on normal and leukemic B-cells. Cytometry B Clin Cytom 70:410–415

29. Imoukhuede PI, Dokun AO, Annex BH et al (2013) Endothelial cell-by-cell profiling reveals the temporal dynamics of VEGFR1 and VEGFR2 membrane localization after murine hindlimb ischemia. Am J Physiol Heart Circ Physiol 304:H1085–H1093

30. Napione L, Pavan S, Veglio A et al (2012) Unraveling the influence of endothelial cell density on VEGF-A signaling. Blood 119:5599–5607

31. Weddell JC, Imoukhuede PI (2014) Quantitative characterization of cellular membrane-receptor heterogeneity through statistical and computational modeling. PloS one 9:e97271

32. Weddell JC, Imoukhuede PI. Integrative meta-modeling ranks RTK signaling and identifies connection between nuclear translocation and extracellular ligand concentrations. In: American Institute of Chemical Engineers. San Francisco, CA; 2016

33. Chen S., Ansari A., Sterrett W., et al. (2014). Current state-of-the-art and future directions in systems biology. http://ojs.unsysdigital.com/index.php/pcs/article/view/148

34. Imoukhuede PI, Popel AS (2014) Quantitative fluorescent profiling of VEGFRs reveals tumor cell and endothelial cell heterogeneity in breast cancer xenografts. Cancer Med 3:225–244

35. Imoukhuede PI, Popel AS (2011) Quantification and cell-to-cell variation of vascular endothelial growth factor receptors. Exp Cell Res 317:955–965

36. Imoukhuede PI, Popel AS (2012) Expression of VEGF receptors on endothelial cells in mouse skeletal muscle. PloS one 7:e44791

37. Roxworthy BJ, Johnston MT, Lee-Montiel FT et al (2014) Plasmonic optical trapping in biologically relevant media. PloS one 9:e93929

38. TrypLE™ Express Enzyme (1X), phenol red—Life Technologies, https://www.lifetechnologies.com/order/catalog/product/12605036

39. Miller MA, Meyer AS, Beste MT et al (2013) ADAM-10 and -17 regulate endometriotic cell migration via concerted ligand and receptor shedding feedback on kinase signaling. Proc Natl Acad Sci U S A 110:E2074–E2083

40. Guaiquil VH, Swendeman S, Zhou W et al (2010) ADAM8 is a negative regulator of retinal neovascularization and of the growth of heterotopically injected tumor cells in mice. J Mol Med (Berl) 88:497–505

41. Weskamp G, Mendelson K, Swendeman S et al (2010) Pathological neovascularization is reduced by inactivation of ADAM17 in endothelial cells but not in pericytes. Circ Res 106:932–940

42. Delano FA, Chen AY, Wu K-IS et al (2011) The autodigestion and receptor cleavage in diabetes and hypertension. Drug Discov Today Dis Models 8:37–46

43. Purdue University Cytometry Laboratories Catalog of Free Flow Cytometry Software. http://www.cyto.purdue.edu/flowcyt/software/Catalog.htm

44. Holton SE, Walsh MJ, Bhargava R (2011) Subcellular localization of early biochemical transformations in cancer-activated fibroblasts using infrared spectroscopic imaging. Analyst 136:2953

45. Chan V, Zorlutuna P, Jeong JH et al (2010) Three-dimensional photopatterning of hydrogels using stereolithography for long-term cell encapsulation. Lab Chip 10:2062

46. Qayyum MA, Kwak JT, Insana MF (2015) Stromal-epithelial responses to fractionated radiotherapy in a breast cancer microenvironment. Cancer Cell Int 15:67

47. Lyer S, Bishop J, Abrams B et al (1997) QuantiBRITE: a new standard for PE flourescence quantitation. White Paper, Becton Dickinson Immunocytometry Systems, San Jose, CA, In

48. Houtz B, Trotter J, Sasaki D (2004) Tips on cell preparation for flow cytometric analysis and sorting. BD FACService Technotes 4:3–4

49. Ormerod MG, Imrie PR Flow cytometry. In: Jeffrey W. Pollard and John M. Walker, Animal cell culture. Humana Press, New Jersey, pp 543–558

50. Schmid I, Uittenbogaart CH, Giorgi JV (1994) Sensitive method for measuring apoptosis and cell surface phenotype in human thymocytes by flow cytometry. Cytometry 15:12–20

51. Rao CR Diversity: its measurement, decomposition, apportionment and analysis. Sankhya A44:1–22

52. Rao CR (1982) Diversity and dissimilarity coefficients: A unified approach. Theor Popul Biol 21:24–43

53. Rao CR (2010) Quadratic entropy and analysis of diversity. Sankhya A 72:70–80

54. Botta-Dukát Z (2005) Rao's quadratic entropy as a measure of functional diversity based on multiple traits. J Veg Sci 16:533–540

55. Potts SJ, Krueger JS, Landis ND et al (2012) Evaluating tumor heterogeneity in immunohistochemistry-stained breast cancer tissue. Lab Invest 92:1342–1357

56. Gough AH, Chen N, Shun TY et al (2014) Identifying and quantifying heterogeneity in high content analysis: application of heterogeneity indices to drug discovery. PloS one 9: e102678

# Chapter 9

## Evaluating Nanoparticle Binding to Blood Compartment Immune Cells in High-Throughput with Flow Cytometry

**Shann S. Yu**

## Abstract

Nanoparticles are increasingly being utilized for in vivo applications, where they are implemented as carriers for drugs, contrast agents for noninvasive medical imaging, or delivery vehicles for macromolecular agents such as DNA or proteins. However, they possess many physical and chemical properties that cause them to become rapidly recognized by the immune system as a foreign body, leading to their clearance and elimination, even before they may accumulate to critical concentrations at anatomic and cellular sites of action. The techniques described in this chapter aim to identify potential interactions of test, fluorescently tagged nano-formulations with circulating immune cells, with the goal of predicting potentially problematic formulations that may be rapidly cleared following in vivo administration. The techniques make use of flow cytometry, a method commonly used in immunology to phenotype and identify immune cell subtypes based on their expression of signature surface marker profiles.

**Key words** Flow cytometry, Nanoparticle clearance, Injectable, Blood, Immune system, Reticuloendothelial system

## 1  Introduction

Nanomaterials have seen increasing applications in biotechnology and medicine, and specifically for in vivo use, new vehicles have been developed for the purposes of drug delivery, noninvasive medical imaging, and vaccination. Upon administration, nanoparticles' physical and chemical properties may lead to their identification and rapid clearance by resident or circulating immune cells, or by the reticuloendothelial system [1]. Because of their size, which is often similar to that of viral particles or bacteria, nanoparticles may be directly (and correctly so) recognized as a foreign body by macrophages or other phagocytes. Even the chemical functional groups that decorate the surface of a nano-formulation may promote its coating with blood proteins in a process called opsonization; the resulting activation of the complement pathway, which begins the moment of administration, leads to rapid recognition

Sarah Hurst Petrosko and Emily S. Day (eds.), *Biomedical Nanotechnology: Methods and Protocols*, Methods in Molecular Biology, vol. 1570, DOI 10.1007/978-1-4939-6840-4_9, © Springer Science+Business Media LLC 2017

and clearance from the body [2]. The methods described in this chapter aim to identify some of the potential interactions between nanoparticles and circulating immune cells, as this will allow for optimization of nanoparticle formulations prior to their in vivo application. As highlighted below, in some situations it may be desirable to decrease immune cell interactions, while in other situations a researcher may want to increase immune interactions. The knowledge gained from the methods described in this chapter will enable researchers to achieve these goals.

As just introduced, in some cases nanoparticles are intended for noncirculating targets, so interactions with immune cells and clearance by the immune system would represent a major barrier to the nanoparticles' intended action. Rapid clearance represents a major challenge in terms of cost and, more importantly, in terms of dosing and toxicity. When nanoparticles are cleared too quickly, it makes it difficult to achieve effective concentrations of drugs or contrast agents within a desired organ or cell, and this necessitates the administration of high initial doses to achieve an intended biological effect. To combat this, researchers have developed various approaches to optimize nanoparticle formulations for targeting specific cells and organs. In one approach, a group interested in targeting inflamed endothelial cells screened nanoparticles against individual primary cell lines and scored ideal nanoparticle formulations based on on-target versus off-target cell uptake [3]. In another approach, an in vivo phage display method was used to identify the best ligands for targeting contrast agent-loaded nanoparticles toward inflamed arterial lesions, but this assay did not predict clearance routes [4]. Later, contrast agent-loaded nanoparticles coated with the best ligands were administered in vivo, and noninvasive imaging was used to determine splenic, kidney, or hepatic clearance routes [4, 5]. In several of these examples, the "ideal nanoparticle formulations" still exhibited rapid clearance rates following in vivo administration. Therefore, the identification of potential interactions in circulation and their cellular mediators may promote better prediction of nanoparticles' behavior and clearance routes in vivo.

While in some situations researchers may want their nanoparticles to avoid immune interactions, in other cases direct binding and interactions with circulating cells are desirable. For example, nanomaterials have been injected intravascularly to target circulating monocytes [6, 7], erythrocytes [8], and T cells [9] with the ultimate goals of producing therapeutic effects, or "hitch-hiking"—using these cells' natural trafficking abilities to better reach a desired organ or cell type. The protocol described in this chapter was developed to tackle either of the goals of decreasing or increasing immune interactions. By providing a technique for determining potential cellular targets of systemically administered nanoparticles (potentially related to opsonization of nanoparticles by bloodborne

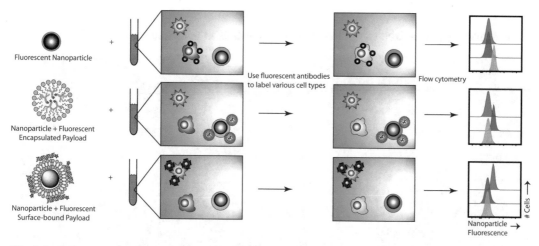

**Fig. 1** Overall schematic of the method. Fluorescent nanoparticles, or nonfluorescent nanoparticles carrying fluorescent payload (drugs, proteins, macromolecules, etc.) may be added to whole blood samples. Eventually, the blood samples will be stained with an antibody cocktail that labels individual cell types with different fluorophores, allowing identification of the individual cell types with flow cytometry. The flow cytometer additionally quantifies the nanoparticle-bound fluorescence signal, allowing identification of main cell types responsible for uptake of the nanoparticles

complement proteins) [2], formulations can be optimized to target specific cell types—including erythrocytes [8], and avoid potential pitfalls that may be encountered following intravascular administration (such as induction of platelet aggregation) [10].

The protocol relies on the use of flow cytometry for the detection and quantification of multiple fluorescent signals on single cells [11] (Fig. 1). In brief, blood samples are incubated with fluorescently tagged nanoparticles (or fluorescently tagged "drugs" loaded onto nanocarriers), and then the blood cells are stained with cocktails of fluorescent antibodies against signature markers present on individual cell types that may be present in the blood (e.g., B cells express B220 or CD19, T cells express CD3) [12]. Together with cell size and granularity, which are quantified by the cytometer using the light scattering characteristics of each cell, general classes of immune cells in the blood are identified, and nanoparticle (or "drug") fluorescence is quantified for each immune cell class, in order to determine the cell types that a test nanoparticle formulation appears to associate with. The main classes of circulating immune cells (B cells, T cells, monocytes, NK cells, granulocytes, and some dendritic cells) may be identified in this fashion using a minimum of seven markers (Table 1).

Here, we describe the collection of fresh mouse or human blood and its incubation with injectable nanoparticle formulations of interest for up to 15 min in vitro prior to flow cytometry analysis. Given that the human heart pumps ~3 L of blood per minute, and the average adult human possesses 4–5 L of blood, the 15-min time

**Table 1**
**Basic signature marker profiles of circulating immune cells in mice and men**

| Immune cell type | Signature markers in humans | Signature markers in mice |
|---|---|---|
| B cells | CD45$^+$ CD3$^-$ CD19$^+$ | CD45$^+$ CD3e$^-$ B220$^+$ |
| T cells | CD45$^+$ CD3$^+$ CD19$^-$ | CD45$^+$ CD3e$^+$ B220$^-$ |
| Monocytes | CD45$^+$ CD3$^-$ CD19$^-$ CD14$^+$ | CD45$^+$ CD3e$^-$ B220$^-$ CD11c$^-$ Ly6C$^{hi}$ Ly6G$^{lo}$ |
| Granulocytes | CD45$^+$ CD14$^-$ CD66$^+$ | CD45$^+$ CD3e$^-$ B220$^-$ CD11c$^-$ Ly6C$^{lo}$ Ly6G$^{hi}$ SSC$^{hi}$ |
| NK cells | CD45$^+$ CD3$^-$ CD56$^+$ | CD45$^+$ NK1.1$^+$ |
| Dendritic cells | CD45$^+$ CD3$^-$ CD19$^-$ CD14$^-$ HLA-DR$^+$ | CD45$^+$ CD3e$^-$ B220$^-$ CD11c$^+$ |

frame selected for this test sufficiently reflects the timeframe necessary for test formulations to make several trips around the circulatory system [13]. With mouse blood (~3 mL total volume, ~15 mL/min), a much shorter test time (<5 min) may be justified to achieve the same ends [14]. Once the incubation period is complete, unbound nanoparticles are removed prior to staining the cells with a selected antibody cocktail and performing subsequent sample analysis on a flow cytometer. The suggested flow cytometry panel in the protocol is designed for the determination of cellular routes of nanoparticle clearance, but the other assays listed are made possible simply by changing the flow cytometry panel (the fluorescently tagged antibodies used) in the assay.

## 2   Materials

1. ACK (ammonium-chlorine-potassium) red blood cell lysis buffer.

2. FACS buffer: HBSS (Hank's buffered salt solution), pH 7.2–7.6, no phenol red. To 1 L of HBSS, add 20 mL of fetal bovine serum and add 2 mL of 1 M EDTA (*see* **Note 1**). Adjust the pH with HCl if necessary.

3. Fluorescently tagged primary antibodies against CD45, CD3 (or CD3e, if mouse cells are used), CD19, CD14, CD16, and HLA-DR (MHC-II, if mouse cells are used; *see* **Note 2**). For panel design notes, *see* **Notes 3** and **4**. A suggested panel has been provided for humans and mice in Table 2.

4. Blood collection tubes: For mouse blood, use K$_2$EDTA-coated blood collection microcentrifuge tubes. For human blood, use

**Table 2**
**Possible 7- or 9-color flow cytometry panels enabling identification of major circulating immune cells in mice and men (Assumes presence of 405, 488, 633 nm laser sources for excitation)**

| Fluorophore | Human panel | Mouse panel |
| --- | --- | --- |
| Pacific Blue/Brilliant Violet 421 | CD45 | CD45 |
| Pacific Orange/Brilliant Violet 510 | CD16 | Ly6G |
| FITC/Alexa Fluor 488 | CD3 | CD3e |
| PE | CD14 | Ly6C |
| PE-Texas Red/ECD/PE-CF594 | Propidium iodide* | Propidium iodide* |
| PerCP-Cy5.5/PE-Cy5 | CD56* | NK1.1* |
| PE-Cy7 | HLA-DR | CD11c |
| APC/Alexa Fluor 633/Alexa Fluor 647 | Nanoparticle | Nanoparticle |
| APC-Cy7/APC-eFluor780 | CD19 | B220 |

*For 7-color panel, remove these markers.

Vacutainer tubes, K$_2$EDTA or heparin-coated. The Vacutainer tubes come evacuated, so that blood collection from human subjects happens automatically once the needle enters the vein.

5. Live/dead stain (optional but recommended; cell death is typically negligible for blood samples analyzed within a few hours of isolation, but dead cells may nonspecifically bind antibodies, leading to false-positive signals): Several options are available; choose DAPI (4′,6-diamidino-2-phenylindole) or propidium iodide (PI) depending on what flow cytometer is to be used. If using DAPI, dissolve in MilliQ water to 5 mg/mL (14 mM) and store in the dark at 4 °C. If using PI, *see* **Note 5**.

6. Lyse/Fix buffer: Prepare 1× working solution of Lyse/Fix buffer by diluting Lyse/Fix buffer 5× with distilled water. (We use the Lyse/Fix buffer from BD Biosciences, but buffer from other vendors may also be suitable for use with this protocol.)

7. Multicolor flow cytometer, preferably equipped with at least 10 channels and a UV laser. For the example data shown in this chapter, a BD LSR-II special-order 18-channel system was used.

8. Fluorescently tagged nanoparticle formulations (*see* **Note 6**).

9. Stretch latex-free tourniquets (for human blood collection).

10. Vacutainer blood-collection needle with tube holder (for human blood collection).

## 3    Methods

### 3.1    Blood Collection and Nanoparticle Incubation

Carry out all steps in this section at 37 °C unless otherwise noted. Perform blood collection from healthy, adult wild-type mice according to an approved protocol from the Institutional Animal Care & Use Committee (IACUC) or equivalent authority, or from consented, healthy adult human volunteers according to an approved protocol from the Institutional Review Board (IRB) or equivalent authority. This method may be also expanded to be applicable to animal disease models and diseased human patients with appropriate IACUC/IRB approval. The reader is assumed to be familiar with some blood collection techniques.

1. For mouse blood, collect blood directly into $K_2$EDTA-coated tubes. To block coagulation, flick the tube several times to mix the blood with the EDTA. 100 µL of blood is sufficient for one nanoparticle formulation, so adjust blood collection volume as necessary.

2. For human blood, ask trained personnel to perform the collection into heparin- or $K_2$EDTA-coated Vacutainer tubes. Each mL of blood is sufficient for five nanoparticle formulations (*see* **Note 7**).

3. Add nanoparticle formulations at desired concentration in a maximum of 10 µL volume into individual wells of a round-bottom 96-well plate. (The appropriate concentration will vary because different materials will interact differently with blood components, and furthermore, large ranges of dose have been reported for various drugs and materials in mice and men (from ~10 mg/kg down to ng/kg depending on formulation; corresponding to ~0.3 µg/µL down to 0.03 pg/µL).)

4. Aliquot whole blood into nanoparticle-containing wells of the plate. Pipet well to mix. Do not discard remainder of the unused blood—these may be used in Subheading 3.3 for the preparation of set-up control samples.

5. Incubate plates at 37 °C on an orbital shaker set at 60 rpm for 15 min (*see* **Note 8**).

6. Pellet cells by centrifugation, 400 rcf × 5 min at 4 °C. Carefully aspirate supernatants, but do not worry if you aspirate away some red blood cells since there are more than enough red blood cells remaining in the pellet.

7. Add 100 µL FACS buffer to the cells (or more, if the wells are deep enough to allow a larger wash volume). Pipet well to mix, then repeat **step 6**.

8. Repeat **steps 6** and **7** for a total of three washes. Aspirate supernatants carefully, and then proceed to the next step.

**3.2 Antibody Cocktail Staining**

Here, blood samples that have already bound fluorescent nanoparticles (or payload) will be stained with an antibody cocktail, with each antibody specific for markers expressed on certain immune cell types. This will enable different immune cell types to be identified based on their marker signature. Carry out all steps in this section at 4 °C and on ice in the dark. Store all cells at 4 °C.

1. Ensure all antibodies are titrated properly prior to use (protocols are widely available online; *see* **Note 9**).

2. Prepare enough antibody cocktail for all samples to be stained. Use optimal antibody concentrations identified from titration (previous step, *see* **Note 9**), and dilute them into FACS buffer. Prepare 50 μL for each sample to be stained.

3. Add the DAPI or PI live/dead stain into the antibody cocktail. Typically, 1:500 dilution of stock solutions into the antibody cocktail provides sufficient stain for 10 million cells.

4. Add 50 μL diluted antibody mix per sample, and pipet well to mix. Incubate plates on ice in the dark for 15–20 min.

5. If erythrocyte binding is important to your assay, skip this step and proceed to the next step. Otherwise, add 150 μL of 1× Lyse/Fix buffer to each sample and pipet to mix. Incubate for 30 min.

6. Pellet cells by centrifugation, 400 rcf × 5 min at 4 °C. Aspirate supernatants; do not worry if you lose some red blood cells in the supernatant.

7. Rinse cells with 200 μL FACS buffer (or more, if the wells are deep enough to allow a larger wash volume). Pipet cells up and down to mix.

8. Repeat **steps 6** and **7** again for a total of two washes.

9. Resuspend cells in 220 μL FACS buffer per sample. They are now ready for analysis on a flow cytometer.

**3.3 Preparation of Setup Controls (For Flow Cytometer Setup)**

Because individual white blood cells may possess multiple fluorescent tags associated with staining for several different antibodies, in addition to possible signal from the nanoparticle and/or its payload, the spectral overlap of these multiple fluorophores may lead to false-positive results. For example, most modern flow cytometers will excite Alexa Fluor 488 and Phycoerythrin (PE) with the same light source, and therefore, cells stained with PE alone may falsely appear Alexa Fluor 488-positive as well. Therefore, appropriate setup of the flow cytometer using single-stained cells is necessary to ensure proper compensation for spectral overlap. For this section, carry out all steps at 4 °C and on ice in the dark (*see* **Notes 10** and **11**).

Even when the goal of your assay is to check erythrocyte binding of nanoparticle formulations, the only purpose of this section is to provide positive control cells for each fluorophore included in the panel. Since all of the antibodies suggested in Tables 1 and 2 will bind immune cells, a red blood cell lysis protocol has been included below to enrich for the immune cells present in the blood.

1. Centrifuge 20 mL of human whole blood or 1 mL of mouse whole blood at 400 rcf × 5 min at 4 °C. Aspirate the supernatant.

2. Add 10 mL (for human blood) or 1 mL (for mouse blood) of ACK red blood cell lysis buffer to the blood samples. Pipet well to mix. Incubate at room temperature for 2 min.

3. Centrifuge lysed blood at 400 rcf × 5 min at 4 °C. Aspirate the supernatant.

4. Repeat **steps 2** and **3** as necessary to eliminate all erythrocytes.

5. Resuspend the remainder of the cells in 2 mL of FACS buffer. Pipet to mix.

6. Aliquot 50 μL of cells into individual polypropylene test tubes. Prepare one tube for each color available in the flow cytometer panel, and an extra tube to be used as unstained controls.

7. To each individual tube, add one of the fluorescently tagged antibodies used as part of the cocktail prepared in Subheading 3.2, **step 2**, using optimal titrations identified in Subheading 3.2, **step 1**. Pipet well to mix. Incubate cells for 15 min (*see* **Note 11**).

8. Dilute all samples with 1 mL FACS buffer.

9. Centrifuge lysed blood at 400 rcf × 5 min at 4 °C. Aspirate the supernatant.

10. Repeat **steps 8** and **9**.

11. Resuspend all cells in 300 μL FACS buffer. Cells are now ready for analysis on a flow cytometer.

*3.4 Flow Cytometry*

Flow cytometers require prior specialized training and orientation to operate properly. Consult with a technician or a flow cytometry core lab manager for help with the software and operation.

1. Using the single-stained setup controls set up in Subheading 3.3, optimize cytometer voltages for the detection of each color represented in the cytometry panel (*see* **Note 12**).

2. Run all samples from Subheading 3.2 on the cytometer, using optimized settings from the previous step.

3. Export all sample data as FCS files.

4. Import data into a suitable flow cytometry analysis software, such as FlowJo (Tree Star, Ashland, OR, USA).

5. Data analysis tips are provided in **Notes 13–17**.

## 4    Notes

1.  1 M EDTA is best prepared by mixing 292.24 g of EDTA in 1 L of MilliQ water (adjusted to pH 8.0 with NaOH). If EDTA crystals remain undissolved, heating the mixture to at least 60 °C will typically enable these to dissolve.

2.  For mouse cells, antibodies against MHC-II must be selected with the haplotype in mind (i.e., depending on the mouse strain used for blood collection, the MHC-II may vary). For most inbred mouse strains the haplotype may be identified in the Mouse Phenome Database maintained by Jackson Laboratories (http://phenome.jax.org). For genetically modified or outbred mouse strains, consult the provider.

3.  The design of flow cytometry panels must be undertaken with the cytometer in mind: the fluorescence detectors available will define what fluorophores are available for your use. We recommend using a cytometer with at least 10 channels available; at the bare minimum, such cytometers today are typically equipped with three lasers: violet (~405 nm), blue (~488 nm), and red (~633 nm). A suitable fluorophore configuration (for 6 markers + live-dead marker + nanoparticle dye) for such a cytometer may utilize the following channels: Pacific Blue, Pacific Orange, FITC, PerCP-Cy5.5, PE, PE-Cy7, APC, and APC-Cy7. Consult with a flow cytometry core lab manager to ensure that staining panels are compatible with the cytometer before proceeding. A suggested panel has been provided in Table 2.

4.  If larger staining panels are desired (to better delineate blood cell types), consider switching to another cytometer with more channels and lasers available; while Alexa Fluor 700 and PE-Texas Red/ECD/PE-CF594 channels may be available on 10-channel cytometers, fluorophores that are detected by these channels typically have significant spillover into other channels, giving false-positive results and making data interpretation messy.

5.  DAPI is recommended as the live/dead stain for most applications, but in case the selected flow cytometer is not equipped to detect DAPI, PI is a suitable alternative. This will typically be detected using the PE-Texas Red/ECD/PE-CF594 channel. Although PI typically has a high emission spillover into other channels (such as into PE), causing PI-positive cells to be falsely identified as positive for fluorophores matched into these channels, PI-positive cells (dead cells) are always excluded from analysis, eliminating this concern.

6.  Either the nanoparticles must be directly tagged or display inherent fluorescence properties; or the payload drugs,

siRNA, plasmid, etc. must be tagged. In the sample data provided in this chapter, we used polymeric carriers with fluorescently tagged siRNA as the nanoparticle "tag." FITC/AlexaFluor 488/Dylight 488, Cy3/PE, or Cy5/AlexaFluor 647/DyLight 633/DyLight 650 are the best options that should be detectable by most flow cytometer systems.

7. Blood volumes for the tests are determined based on two criteria: typical intravenous injection volumes in mice versus in humans, as well as typical immune cell concentrations per mL of blood. Healthy adult mice typically have ~3 mL of blood, to which a maximum injection volume of ~200 μL is typically approvable under IACUC standards (~6–10% blood volume). Therefore, 10 μL of nanoparticles to 100 μL of blood is an acceptable model of the eventual in vivo situation. Lower volumes may result in low yields of some immune cell subtypes. Adult humans typically have approximately 5 L of blood, to which 500 mL of fluid may be safely administered intravenously, although many drugs are administered in a much smaller volume. Therefore, the chosen ratio of 10 μL of nanoparticles to 200 μL of blood is an estimate of the probable clinical situation, although different ratios may also be rationalized.

8. The optimal incubation time for blood with nanoparticles is typically governed by half-life measurements; for some formulations, we have observed circulation half-lives of more than 1 h in mice, in which case the incubation time for this assay is increased to 1 h instead of the suggested 15 min.

9. Antibody titration is recommended to determine optimal amounts of antibody to use because of batch-to-batch variability in antibodies (even from the same provider), and also because in some cases, providers may recommend a certain volume per sample, although similar results may be obtained with a smaller antibody volume. Typical antibody titrations may range from 1:10 to 1:4000 depending on the supplier, antibody clone, and the cells to be stained, so there is no good way to set a ballpark value here. Essentially, antibody titration involves staining similar cell samples (ideally, taken from the same organ and with the same cell concentrations) as the ones that will be used for the actual experiment, using different dilutions of the antibody. Protocols and sample data from a titration are widely available online, and an example may be found here: http://healthsciences.ucsd.edu/research/moores/shared-resources/flow-cytometry/protocols/Pages/antibody-titration.aspx.

10. Set-up controls for flow cytometry are cells or microbeads stained with one of the colors in the multicolor panel. This

enables the cytometer to detect potential sources of fluorescence spillover in a process called compensation. While microbeads are commonly used for compensation due to their practicality and strong antibody binding, we generally prefer using the same cells that are to be analyzed for the experiment, as microbeads poorly model the natural autofluorescence patterns exhibited by the cells.

11. With the goal of the set-up controls in mind (*see* **Note 9**), preparation of a set-up control to match the nanoparticle fluorescence tag requires that some cells are deliberately labeled with the exact same fluorophore as is used on the nanoparticle or on the fluorescent cargo. In the case that the fluorescent tag is available conjugated to a CD45 antibody (CD45-FITC, CD45-Alexa Fluor 488, CD45-Cy3), prepare cells stained with this antibody, as a large proportion of blood cells are positive for CD45. In the case that the fluorescence tag is available as an NHS-ester/succinimide ester, it will suffice to take some cells, wash them with PBS (to remove free proteins in the FACS buffer), and to react them for 5–10 min with the dye on ice in the dark (typically, concentrations of 1 μM or lower for every $10^7$ cells). This reaction can then be quenched by the addition of FACS buffer (such that free dye reacts with the proteins in the buffer instead of with the cells), and then free dye is washed off by multiple cycles of centrifugation and aspiration of supernatants. In summary, the most important thing is to have a pairing of the same cells to be analyzed with the exact fluorophore that will be detected by the cytometer.

12. Voltages/detector gains must be set such that the signal of any fluorophore in its matched detector channel is at least 10× (one decade) stronger than its false-positive signal in off-target detector channels. This can be done by increasing the voltage/gain of the target channel, or by decreasing the voltage/gain of the off-target channels. This process becomes inevitably more complex when fluorophores with high emission bleedover are used, or when more colors are employed.

13. High-quality data is as dependent on the data analysis as it is on user-dependent practices. At the analysis level, proper gating of "junk" signals, which may arise from cell aggregates and dead cells (may nonspecifically bind antibodies and nanoparticles), is the first step toward this (Fig. 2).

14. Erythrocytes can be easily separated from leukocytes and lymphocytes based on forward scatter (FSC), side scatter (SSC), and CD45 staining (Fig. 3).

15. A strategy for delineation of B cells, T cells, monocytes, and other cells has been provided (Fig. 4), which is based off standards published by the Human Immunology Project [12].

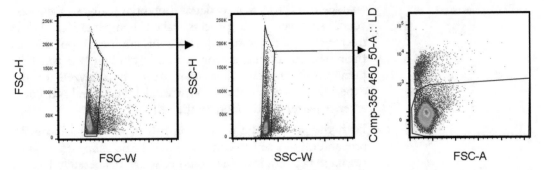

**Fig. 2** Gating strategy for removal of signals arising from cell aggregates and dead cells. Dot plots show all cells collected from a single sample. All cells enclosed within the gates are carried forward for further analysis; cells falling outside the gates are excluded from all downstream analysis. (*Left*) Dot plots show FSC-W (forward scatter signal width) versus FSC-H (forward scatter signal height), (*center*) SSC-W (side scatter signal width) versus SSC-H (side scatter signal height), and (*right*) live-dead stain (LD) versus FSC-A (forward scatter signal area)

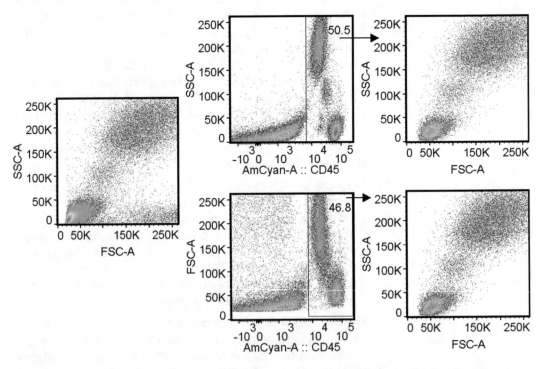

**Fig. 3** Selecting for erythrocytes versus immune cells using FSC, SSC, and CD45 staining. (*Left*) Dot plot of forward scatter versus side scatter signal for individual detected cells, before gating. The same cells were plotted for their CD45 fluorescence signal versus either side scatter (*top center*) or forward scatter (*below center*) signals, in order to demonstrate two essentially identical strategies for gating on non-erythrocytes, which are positive for CD45. (*Right*) Dot plot of forward scatter versus side scatter signal for CD45-positive cells following gating schemes shown in the center. Note differences versus the pre-gating dot plot (*left*). *Inset numbers* indicate percentage of all cells within the dot plot that fall within the gated area

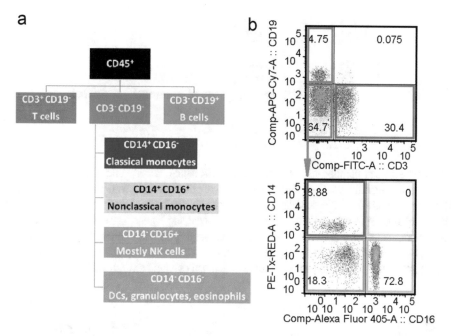

**Fig. 4** (**a**) Human Immunology Project-inspired gating strategy for identification of major immune cell subtypes from CD45$^+$ singlets (next steps following gating scheme in Fig. 3). Adapted from [12]. (**b**) Dot plots of CD45$^+$ immune cells from a healthy human adult donor were gated as in (**a**), with color-coded gates corresponding to the populations named in (**a**). *Inset numbers* indicate percentage of all cells in the plot that occur within the indicated gate. CD3$^-$CD19$^-$ were subjected to further gating based on CD14 and CD16 to identify monocytes and other populations. CD14$^+$CD16$^+$ cells may be present in some donors. The CD14$^-$CD16$^-$ cells can be further analyzed with HLA-DR to identify circulating DCs, while the CD14$^-$CD16$^+$ population can be further gated based on CD56 and CD20 to categorically identify NK cells (CD20$^-$, but may be CD56$^{hi}$ or CD56$^{lo}$), but these subgates are not shown here

16. To identify cell populations that may preferentially internalize nanoparticle formulations [15], gate for each individual cell type as suggested in **Note 13–15**. Then prepare histograms of nanoparticle fluorescence signal for each individual cell type (Fig. 5).

17. Some immune cell populations, such as classical monocytes, may be a minority of all circulating immune cells, and yet take up nanoparticles more avidly than other more abundant cell types (Fig. 6a). Conversely, more abundant cells, such as T cells, may not avidly interact with nanoparticles, but due to their abundance, represent almost as significant a portion of the cells that take up nanoparticles (Fig. 6b). To identify such issues, follow the gating scheme in Fig. 3, and then gate on all nanoparticle-positive cells prior to proceeding with the gating scheme in Fig. 4.

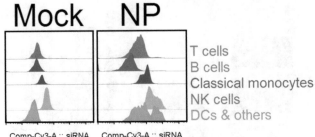

**Fig. 5** Histograms of cell number versus fluorescent siRNA signal, within various cell populations as defined using the gating strategy in Figs. 2–4. Mock refers to whole blood treated with buffer only, while NP refers to whole blood treated with NP-siRNA complexes. Notice right-shift in many of the histograms, suggesting that the NP-siRNA complexes target a broad range of blood cells, including T cells, classical monocytes, NK cells, and DCs. Furthermore, notice how buffer-treated cells exhibit varying baseline levels of fluorescence signal in the siRNA channel—this is typically due to varying levels of autofluorescence by cell type, but may occasionally be caused by improper setup and compensation for fluorescence overlap between antibodies used

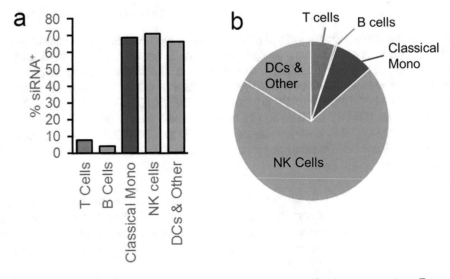

**Fig. 6** Analysis of NP-siRNA delivery to various cell types of the circulating immune system. Two reporting methods are commonly used. (**a**) This analysis, which shows what cell types most avidly interact with the experimental formulation, is performed by selection of immune cell subpopulations based on Figs. 2–4, and then quantification of the percentage of each cell type that presents as siRNA+ via the histograms in Fig. 5. (**b**) Because some avidly interacting cell types may not be abundant in the blood, other cell types may end up making up the majority of nanoparticle-interacting cells. This alternative analysis looks for such issues, and is performed by first gating on all siRNA+ cells within all CD45+ cells, prior to subgating of the immune cells based on Fig. 4. Pie chart therefore shows distribution of all siRNA+ immune cells within the subsets

## References

1. Alexis F, Pridgen E, Molnar LK et al (2008) Factors affecting the clearance and biodistribution of polymeric nanoparticles. Mol Pharm 5:505–515

2. Thomas SN, Van Der Vlies AJ, O'neil CP et al (2011) Engineering complement activation on polypropylene sulfide vaccine nanoparticles. Biomaterials 32:2194–2203

3. Kelly KA, Shaw SY, Nahrendorf M et al (2009) Unbiased discovery of in vivo imaging probes through in vitro profiling of nanoparticle libraries. Integr Biol 1:311–317

4. Nahrendorf M, Jaffer FA, Kelly KA et al (2006) Noninvasive vascular cell adhesion molecule-1 imaging identifies inflammatory activation of cells in atherosclerosis. Circulation 114:1504–1511

5. Gallo J, Kamaly N, Lavdas I et al (2014) CXCR4-targeted and MMP-responsive iron oxide nanoparticles for enhanced magnetic resonance imaging. Angew Chem Int Ed 53:9550–9554

6. Cohen-Sela E, Rosenzweig O, Gao J et al (2006) Alendronate-loaded nanoparticles deplete monocytes and attenuate restenosis. J Controll Release 113:23–30

7. Majmudar MD, Keliher EJ, Heidt T et al (2013) Monocyte-directed RNAi targeting CCR2 improves infarct healing in atherosclerosis-prone mice. Circulation 127:2038–2046

8. Kontos S, Kourtis IC, Dane KY et al (2013) Engineering antigens for in situ erythrocyte binding induces T-cell deletion. Proc Natl Acad Sci U S A 110:E60–E68

9. Ramishetti S, Kedmi R, Goldsmith M et al (2015) Systemic gene silencing in primary T lymphocytes using targeted lipid nanoparticles. ACS Nano 9:6706–6716

10. Shattil SJ, Cunningham M, Hoxie JA (1987) Detection of activated platelets in whole-blood using activation-dependent monoclonal-antibodies and flow-cytometry. Blood 70:307–315

11. Perfetto SP, Chattopadhyay PK, Roederer M (2004) Seventeen-colour flow cytometry: unravelling the immune system. Nat Rev Immunol 4:648–655

12. Maecker HT, Mccoy JP, Nussenblatt R (2012) Standardizing immunophenotyping for the human immunology project. Nat Rev Immunol 12:191–200

13. Guyton AC, Hall JE (2006) Textbook of medical physiology. Elsevier Saunders, Philadelphia

14. Doevendans PA, Daemen JM, De Muinck ED et al (1998) Cardiovascular phenotyping in mice. Cardiovasc Res 39:34–49

15. Kourtis IC, Hirosue S, De Titta A et al (2013) Peripherally administered nanoparticles target monocytic myeloid cells, secondary lymphoid organs and tumors in mice. PLoS One 8: e61646

# Chapter 10

# A Gold@Polydopamine Core–Shell Nanoprobe for Long-Term Intracellular Detection of MicroRNAs in Differentiating Stem Cells

## Chun Kit K. Choi, Chung Hang J. Choi, and Liming Bian

## Abstract

MicroRNAs (miRNAs) represent an emerging class of biomarkers for studying and understanding biological events; the development of viable tools for detecting or monitoring the intracellular expression levels of specific miRNAs is of great interest to life scientists and biomedical engineers. Here, we describe the fabrication of a novel class of core–shell nanoprobes that comprise a gold nanoparticle core and a polydopamine (PDA) shell. Our nanoprobes can be used to specifically track the expression profiles of two miRNA markers of osteogenic differentiation (i.e., osteogenesis), namely, miR-29b and miR-31, in differentiating human mesenchymal stem cells (hMSCs). The newly designed nanoprobes may hold great promise in the noninvasive investigation of the long-term dynamics of cellular events such as stem cell differentiation.

**Key words** MicroRNAs, Intracellular detection, Gold nanoparticles, Polydopamine shell, Osteogenic differentiation, Stem cells

## 1 Introduction

The intracellular detection of biomarkers in living cells has greatly impacted the field of biology and biomedical science by allowing advanced understanding of fundamental cellular events. Particularly, microRNAs (miRNAs), single-stranded noncoding RNAs typically 21–23 nucleotides in length, have recently been identified as essential gene regulators [1, 2] associated with cellular status [3, 4] and diseases in animals [5–7]. Consequently, the development of approaches for monitoring the expression of specific RNA targets may potentially benefit the diagnosis of diseases and pave new ways for medical treatment [8–10].

In the past decade, nanoparticles have been widely used as an efficient platform that allows for the noninvasive intracellular tracking of single or multiple RNA targets, especially for cancer-related RNAs [11–13]. Such nano-sized probes can enter cells without the

Sarah Hurst Petrosko and Emily S. Day (eds.), *Biomedical Nanotechnology: Methods and Protocols*, Methods in Molecular Biology, vol. 1570, DOI 10.1007/978-1-4939-6840-4_10, © Springer Science+Business Media LLC 2017

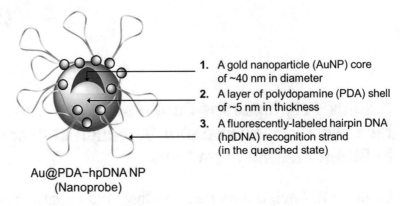

1. A gold nanoparticle (AuNP) core of ~40 nm in diameter
2. A layer of polydopamine (PDA) shell of ~5 nm in thickness
3. A fluorescently-labeled hairpin DNA (hpDNA) recognition strand (in the quenched state)

Au@PDA–hpDNA NP
(Nanoprobe)

**Fig. 1** Schematic illustration of a hairpin-DNA-based (hpDNA) nanoprobe

aid of transfection agents [12], and enable the real-time monitoring of the expression of RNAs at the single-cell level in living cells extracted from blood serum [14]. Here, we introduce a novel class of core–shell nanoprobes that can track osteogenesis-related miRNAs in differentiating human mesenchymal stem cells (hMSCs) over a long observation window [15].

As illustrated in Fig. 1, our novel nanoprobes consist of three major components: (1) a gold nanoparticle (AuNP) of ~40 nm in diameter, a size that favors uptake by mammalian cells [16], as the core; (2) polydopamine (PDA), a bioinspired polymer formed via the self-polymerization of monomeric dopamine [17], as the shell; and (3) fluorescently labeled hairpin DNA (hpDNA) strands, which can be rationally designed for any specific miRNA target and immobilized onto the surface of the gold@polydopamine core–shell nanoparticles (Au@PDA NPs), as the recognition sequences. The PDA shell represents a key element of our nanoprobe design as it provides a surface for the facile immobilization of DNA oligonucleotide strands via π–π interactions and hydrogen bonding [18]. Furthermore, it acts as a secondary quenching entity [19] (apart from the gold core [20]) to help suppress the background, off-target fluorescence signals of the immobilized fluorescently labeled hpDNA recognition strands, thus affording a high signal-to-noise ratio in the context of intracellular detection.

We have shown that our newly designed nanoprobes (termed Au@PDA–hpDNA nanoprobes) can naturally enter living hMSCs, a cell type that is known to have poor transfection efficiency [21]. Following their cellular entry, we have also demonstrated that these nanoprobes can monitor the expression profiles of two miRNAs, namely, miR-29b and miR31, both of which are positive regulators of the osteogenic differentiation of stem cells [22, 23], in differentiating hMSCs and living mouse osteoblasts. Most importantly, our nanoprobes afford long-term tracking of specific miRNAs

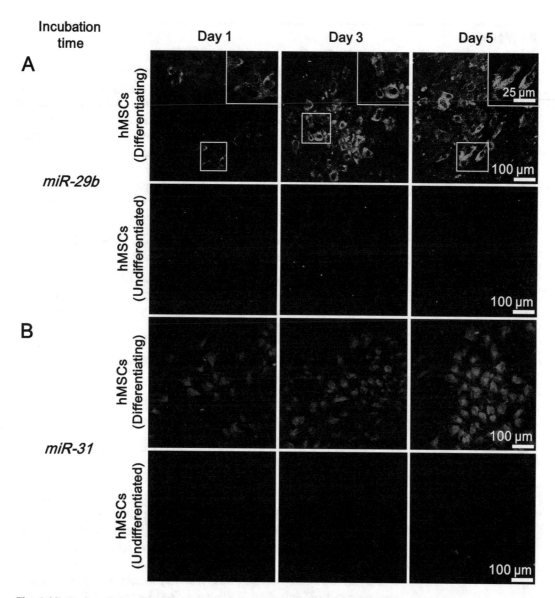

**Fig. 2** Monitoring of the differentiation progress of hMSCs through the intracellular detection of miR-29b and miR-31. (**a**) Upper channel: upon treatment of our Au@PDA–hpDNA nanoprobes targeting miR-29b, a miRNA marker for osteogenic differentiation, hMSCs that are undergoing osteogenic differentiation show gradual increase in their intracellular fluorescence signals (*green*), as revealed by confocal microscopy. Lower channel: in sharp contrast, undifferentiated hMSCs exhibit no appreciable fluorescence response after treatment with the nanoprobes. Scale bar is 100 μm. Inset: high-magnification images of the boxed area. Scale bar is 25 μm. (**b**) Similar results can be obtained when differentiating hMSCs are treated with our nanoprobes targeting miR-31, another marker for osteogenesis, showing increasing intracellular fluorescence signals (*red*) with increasing osteogenic induction time (reprinted from [15] with permission from the American Chemical Society)

without observable degradation in hMSCs for at least 5 days, allowing for the nondestructive monitoring of the differentiation progress of living hMSCs (Fig. 2).

## 2   Materials

Prepare all solutions using deionized water with a resistivity of 18.2 MΩ · cm at 25 °C and analytical grade reagents. All chemicals can be ordered from Sigma-Aldrich (St. Louis, MO, USA) unless otherwise specified. Any DNA strand mentioned in the following text can either be commercially obtained or synthesized using an automated DNA oligonucleotide synthesizer.

### 2.1   Synthesis of the AuNP Core

1. 10 mM hydrogen tetrachloroaurate trihydrate (Au ≥ 48%, $HAuCl_4·;3H_2O$) in deionized water (*see* **Note 1**). Store at 4 °C (*see* **Note 2**).

2. 10 mg/mL trisodium citrate dihydrate (≥99%, $C_6H_5Na_3O_7·2H_2O$) (*see* **Note 2**).

3. Glass apparatus: one 100 mL two-necked round-bottom flask, one coiled condenser, two glass stoppers, and two 25 mL glass vials (all glassware should be pre-rinsed with aqua regia before use) (*see* **Note 3**).

### 2.2   Synthesis of the PDA Shell

1. Reaction buffer: 10 mM Tris–HCl buffer (*see* **Note 4**), pH 8.5. Adjust the pH with concentrated HCl.

2. 0.1 mg/mL dopamine (*see* **Note 5**). Prepare immediately before use.

## 3   Methods

The preparation of Au@PDA–hpDNA nanoprobes is illustrated in Fig. 3a. Basically, a thin-layer PDA shell is coated onto the AuNP core to form Au@PDA NPs. Fluorescently labeled hpDNA recognition strands are then immobilized onto the surface of Au@PDA NPs directly through π–π interactions to form our Au@PDA–hpDNA nanoprobes. The fluorescence signals of the dye-tagged hpDNA strands are quenched by the Au@PDA core-shell structure. In the presence of the target miRNA strands, the immobilized hpDNA recognition strands will open up and hybridize with the target miRNA strands by forming intermolecular hydrogen bonding within the complementary regions. Such specific hybridization weakens the original π–π interactions between the fluorescently labeled hpDNA recognition strands and the PDA shell, thereby triggering the release of the recognition strands from the surface of the nanoparticles and leading to the recovery of their fluorescence [24]. By treating hMSCs with our nanoprobes for 24 h, we can observe the gradually increasing intracellular fluorescence signals in differentiating hMSCs when they are incubated in osteogenic induction medium due to the high expression of the

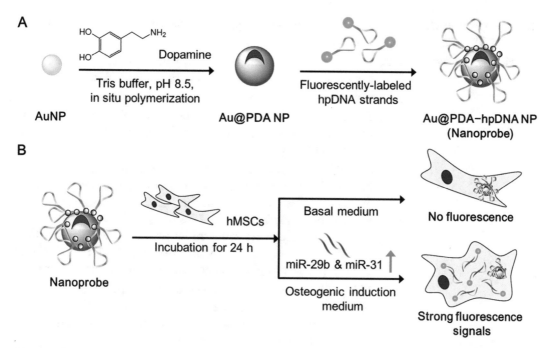

**Fig. 3** (**a**) Preparation of the gold@polydopamine core–shell hairpin-DNA-based nanoprobes (termed Au@PDA–hpDNA NPs or nanoprobes in short). (**b**) Intracellular detection of miRNAs in living human mesenchymal stem cells (hMSCs) (reprinted from [15] with permission from the American Chemical Society)

specific miRNA target inside differentiating hMSCs (Fig. 3b). In contrast, we cannot detect any fluorescence signal when hMSCs are incubated in basal medium.

**3.1 Synthesis of the AuNPs of ~40 nm in Diameter**

A more detailed description of the preparation and characterization of citrate-capped gold nanoparticles (AuNPs) of ~40 nm in diameter can be found in a previous report (*see* ref. 25).

1. Boil 50 mL of water in a two-necked round-bottom flask.

2. Inject 1.214 mL of 10 mM $HAuCl_4$ solution into the reaction flask. Allow the reaction mixture to reboil for 1–2 min.

3. Quickly inject 0.5 mL of 10 mg/mL trisodium citrate solution to the reaction mixture. Observe the color change from pale yellow to dark blue and finally to wine red (*see* **Note 6**).

4. Allow the reaction mixture to boil for an additional 30 min.

5. Cool down the product mixture to room temperature. Collect the resultant AuNP solution in a pre-rinsed glass vial.

6. Characterize the synthesized AuNPs by UV-vis spectroscopy. A typical UV-vis absorption spectrum of the AuNP solution should indicate a maximum peak centered around 530 nm. Based on the UV-vis absorbance, the concentration of the as-prepared AuNP solution can be estimated as previously reported [26]. In a typical synthesis, concentrations of ~0.1 nM AuNP solution are obtained.

**3.2 Synthesis of the Au@PDA NPs with a ~5 nm Shell Thickness**

1. Freshly prepare a 0.1 mg/mL dopamine solution using 10 mM Tris–buffer.

2. Add 10 mL of the 0.1 mg/mL dopamine solution to 10 mL of the ~0.1 nM AuNP solution. Carry out this reaction in a glass vial.

3. Sonicate the reaction mixture for 1 h. Maintain the reaction temperature at room temperature by adding ice into the sonication water bath.

4. Centrifuge the resultant Au@PDA NP solution at $17,350 \times g$ for 10 min in a microcentrifuge to remove any excess dopamine monomer. Wash and disperse the particles in deionized water. Repeat the centrifugation and washing steps twice. Resuspend the nanoparticle pellet in 10 mL of deionized water in the final round of centrifugation. Store the collected nanoparticles inside a pre-rinsed glass vial at room temperature (*see* **Note** 7).

5. Using TEM imaging, the core–shell structure of the Au@PDA NPs can be clearly visualized without any staining by heavy metals; a PDA shell of ~5 nm in thickness can be seen on the surface of the AuNP core (Fig. 4).

**3.3 Synthesis of the Au@PDA–hpDNA Nanoprobes**

Fluorescently labeled hairpin DNA (hpDNA) strands can be rationally designed as the recognition sequences for any specific miRNA target. Carry out the following procedures in a dark environment.

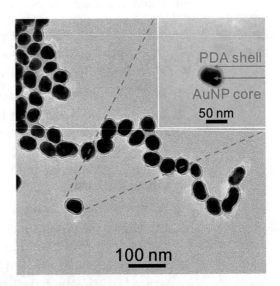

**Fig. 4** A representative TEM image of Au@PDA NPs. Inset: a magnified image of a single Au@PDA NP shows clearly its core–shell structure (reprinted from [15] with permission from the American Chemical Society)

1. Concentrate 2 mL of the as-prepared Au@PDA NPs by 10 times (to ~1 nM) by centrifugation.

2. Add 0.2 mL of a 500 nM hpDNA recognition strand solution to disperse the nanoparticle pellet (*see* **Note 8**). Sonicate the reaction mixture for 5 min. Age the solution for 1 h at room temperature with occasional vortexing if sedimentation of the nanoparticles is noted.

3. Centrifuge the resultant Au@PDA–hpDNA NP solution at 17,350 × *g* for 10 min at 4 °C in a microcentrifuge to remove any excess hpDNA strand. Extract and collect the supernatant for further analysis (*see* **Note 9**). Resuspend the pellet in 2 mL of water to obtain the resultant nanoprobes (in ~0.1 nM). Store the product at 4 °C in the dark (*see* **Note 10**).

4. Conduct fluorescence measurements to confirm the immobilization of the fluorescently labeled hpDNA strands onto the surface of Au@PDA NPs. Effective quenching of the fluorescence signals from the fluorescently labeled hpDNA strands indicates successful immobilization (Fig. 5). Typically, around 250 hpDNA strands can be immobilized onto the surface of each Au@PDA NP to produce the intracellular fluorescence signals as observed in Fig. 2.

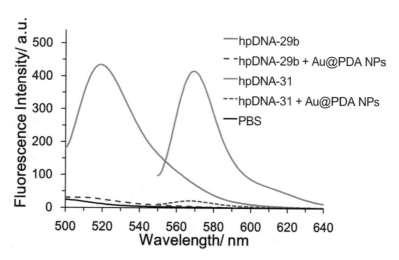

**Fig. 5** Fluorescence emission spectra of 250 nM fluorescently labeled hpDNA recognition strands before and after immobilization onto the surface of Au@PDA NPs. *Green traces* show the efficient quenching of the FITC-labeled hpDNA recognition strands targeting miR-29b (termed hpDNA-29b in the plot), as evidenced by the appreciable reduction of the FITC emission at $\lambda_{max} = 520$ nm. *Red traces* show the efficient quenching of the Cy3-labeled hpDNA recognition strands targeting miR-31 (termed hpDNA-31 in the plot), as evidenced by the appreciable reduction of the Cy3 emission at $\lambda_{max} = 570$ nm (reprinted from [15] with permission from American Chemical Society)

### 3.4 Treatment of hMSCs with the Au@PDA–hpDNA Nanoprobes

Detailed procedures for preparing basal medium to maintain cellular growth as well as osteogenic induction medium to induce osteogenesis of hMSCs can be found elsewhere (*see* ref. 15).

1. Grow hMSCs to confluence with basal medium in a confocal dish.

2. When the cells become confluent, remove the basal medium.

3. Dilute the as-prepared nanoprobes four times (to ~0.025 nM) with either basal medium (as control) or osteogenic induction medium. Inject the working nanoprobe solution into the confocal dish gently.

4. At various time points, the nanoprobes-treated cells are rinsed with PBS thoroughly and then imaged under a confocal laser scanning microscope.

## 4    Notes

1. Prepare the $HAuCl_4$ solution in a glass vial to prevent corrosion of any plastic wall.

2. Store the $HAuCl_4$ solution at 4 °C to prevent evaporation of the reagent.

3. Tris–HCl buffer at a concentration of 5–10 mM can be used for the synthesis of Au@PDA NPs. A lower concentration of the buffer is preferable to prevent the salt-induced aggregation of the citrate-capped gold nanoparticles.

4. A concentrated dopamine solution (e.g., 1 mg/mL) can be prepared first in deionized water. Afterward, use Tris–HCl buffer to dilute the dopamine solution to the working concentration (i.e., 0.1 mg/mL) and induce spontaneous polymerization of dopamine for coating the AuNPs.

5. Glassware or glass vials used for synthesizing or collecting the AuNPs should be pre-rinsed with aqua regia, a mixture of concentrated hydrochloric acid and concentrated nitric acid (v/v, 3:1). Aqua regia is highly corrosive. Handle with extreme caution.

6. The color change is fast and happens within 2 min. The absence of a color change indicates a failed synthesis of the AuNPs.

7. During storage, sedimentation of Au@PDA NPs may be observed. Sonication should be used to help redisperse the particles before further use.

8. For detecting a single specific miRNA in differentiating stem cells, the sequences of the hpDNA recognition strand used for miR-29b and miR-31 are, respectively, 5′-FITC-CCG GGT AAC ACT GAT TTC AAA TGG TGC TA ACC CGG-3′ and

$5'$-CY3-CCG GGT AGC TAT GCC AGC ATC TTG CCT ACC CGG-$3'$, where FITC denotes fluorescein isothiocyanate while Cy3 denotes Cyanine 3. All the recognition sequences can be rationally designed by referring to the mature miRNA sequences of the targets of interest shown in the online miRNA database (www.mirbase.org). Multiplexed nanoprobes that can show fluorescence responses toward multiple specific miRNA targets by a single particle are more preferable than preparing and incubating cells with different nanoprobes against only one single miRNA target to prevent cell-to-cell fluorescence variation due to the heterogeneous rate of nanoparticle uptake among a cell population [12]. To achieve that, multiplexed nanoprobes can be prepared in a similar way as aforementioned. Briefly, a mixture of different hpDNA recognition strand solutions (with the total concentration of each kind of hpDNA strands kept constant at 500 nM) is used to disperse the Au@PDA NP pellet. Note that all the working hpDNA strand solutions are typically prepared in PBS buffer at pH 7.4.

9. Obtain a standard linear calibration curve of the fluorescence signals by using different concentrations of fluorescently labeled hpDNA recognition strands. Perform fluorescence measurements of the collected supernatant and quantify the number of hpDNA strands immobilized onto the Au@PDA NPs based on the calibration curve.

10. Storage at 4 °C can slow down the degradation of immobilized hpDNA recognition strands by DNases in water. Unpublished data show that our nanoprobes that target miR-29b can still show observable fluorescence signals in differentiating stem cells even after 6 months of storage.

## Acknowledgment

This work was supported by an Early Career Scheme grant from the Hong Kong Research Grants Council (Project No. 439913), an Innovation Technology Fund (ITF, Tier 3, ITS/218/13) in Hong Kong, and a Croucher startup allowance. The authors also acknowledge support by the Shun Hing Institute of Advanced Engineering and the Chow Yuk Ho Technology Centre for Innovative Medicine at The Chinese University of Hong Kong.

## References

1. Hobert O (2008) Gene regulation by transcription factors and microRNAs. Science 319:1785–1786

2. Huntzinger E, Izaurralde E (2011) Gene silencing by microRNAs: contributions of translational repression and mRNA decay. Nat Rev Genet 12:99–110

3. Gangaraju VK, Lin H (2009) MicroRNAs: key regulators of stem cells. Nat Rev Mol Cell Biol 10:116–125

4. Ivey KN, Muth A, Arnold J, King FW, Yeh RF, Fish JE et al (2008) MicroRNA regulation of cell lineages in mouse and human embryonic stem cells. Cell Stem Cell 2:219–229

5. Kloosterman WP, Plasterk RH (2006) The diverse functions of microRNAs in animal development and disease. Dev Cell 11:441–450

6. Lu J, Getz G, Miska EA, Alvarez-Saavedra E, Lamb J, Peck D et al (2005) MicroRNA expression profiles classify human cancers. Nature 435:834–838

7. Stefani G, Slack FJ (2008) Small non-coding RNAs in animal development. Nat Rev Mol Cell Biol 9:219–230

8. Tyagi S (2009) Imaging intracellular RNA distribution and dynamics in living cells. Nat Methods 6:331–338

9. Thaxton CS, Georganopoulou DG, Mirkin CA (2006) Gold nanoparticle probes for the detection of nucleic acid targets. Clin Chim Acta 363:120–126

10. Santangelo P, Nitin N, Bao G (2006) Nanostructured probes for RNA detection in living cells. Ann Biomed Eng 34:39–50

11. Seferos DS, Giljohann DA, Hill HD, Prigodich AE, Mirkin CA (2007) Nano-flares: probes for transfection and mRNA detection in living cells. J Amer Chem Soc 129:15477–15479

12. Prigodich AE, Randeria PS, Briley WE, Kim NJ, Daniel WL, Giljohann DA et al (2012) Multiplexed nanoflares: mRNA detection in live cells. Anal Chem 84:2062–2066

13. Ryoo SR, Lee J, Yeo J, Na HK, Kim YK, Jang H et al (2013) Quantitative and multiplexed microRNA sensing in living cells based on peptide nucleic acid and nano graphene oxide (PANGO). ACS Nano 7:5882–5891

14. Halo TL, McMahon KM, Angeloni NL, Xu Y, Wang W, Chinen AB et al (2014) NanoFlares for the detection, isolation, and culture of live tumor cells from human blood. Proc Natl Acad Sci USA 111:17104–17109

15. Choi CK, Li J, Wei K, Xu YJ, Ho LW, Zhu M et al (2015) A gold@polydopamine core–shell nanoprobe for long-term intracellular detection of microRNAs in differentiating stem cells. J Amer Chem Soc 137:7337–7346

16. Chithrani BD, Ghazani AA, Chan WC (2006) Determining the size and shape dependence of gold nanoparticle uptake into mammalian cells. Nano Lett 6:662–668

17. Lee H, Dellatore SM, Miller WM, Messersmith PB (2007) Mussel-inspired surface chemistry for multifunctional coatings. Science 318:426–430

18. Ham HO, Liu Z, Lau KH, Lee H, Messersmith PB (2011) Facile DNA immobilization on surfaces through a catecholamine polymer. Angew Chem Int Ed 50:732–736

19. Qiang W, Li W, Li X, Chen X, Xu D (2014) Bioinspired polydopamine nanospheres: a superquencher for fluorescence sensing of biomolecules. Chem Sci 5:3018–3024

20. Dubertret B, Calame M, Libchaber AJ (2001) Single-mismatch detection using gold-quenched fluorescent oligonucleotides. Nat Biotechnol 19:365–370

21. Eguchi A, Meade BR, Chang YC, Fredrickson CT, Willert K, Puri N et al (2009) Efficient siRNA delivery into primary cells by a peptide transduction domain-dsRNA binding domain fusion protein. Nat Biotechnol 27:567–571

22. Li Z, Hassan MQ, Jafferji M, Aqeilan RI, Garzon R, Croce CM et al (2009) Biological functions of miR-29b contribute to positive regulation of osteoblast differentiation. J Biol Chem 284:15676–15684

23. Baglio SR, Devescovi V, Granchi D, Baldini N (2013) MicroRNA expression profiling of human bone marrow mesenchymal stem cells during osteogenic differentiation reveals Osterix regulation by miR-31. Gene 527:321–331

24. Yang R, Jin J, Chen Y, Shao N, Kang H, Xiao Z et al (2008) Carbon nanotube-quenched fluorescent oligonucleotides: probes that fluoresce upon hybridization. J Amer Chem Soc 130:8351–8358

25. Choi CK, Xu YJ, Wang B, Zhu M, Zhang L, Bian L (2015) Substrate coupling strength of integrin-binding ligands modulates adhesion, spreading, and differentiation of human mesenchymal stem cells. Nano Lett 15:6592–6600

26. Haiss W, Thanh NT, Aveyard J, Fernig DG (2007) Determination of size and concentration of gold nanoparticles from UV-vis spectra. Anal Chem 79:4215–4221

# Chapter 11

# Antibody-Conjugated Single Quantum Dot Tracking of Membrane Neurotransmitter Transporters in Primary Neuronal Cultures

Danielle M. Bailey, Oleg Kovtun, and Sandra J. Rosenthal

## Abstract

Single particle tracking (SPT) experiments have provided the scientific community with invaluable single-molecule information about the dynamic regulation of individual receptors, transporters, kinases, lipids, and molecular motors. SPT is an alternative to ensemble averaging approaches, where heterogeneous modes of motion might be lost. Quantum dots (QDs) are excellent probes for SPT experiments due to their photostability, high brightness, and size-dependent, narrow emission spectra. In a typical QD-based SPT experiment, QDs are bound to the target of interest and imaged for seconds to minutes via fluorescence video microscopy. Single QD spots in individual frames are then linked to form trajectories that are analyzed to determine their mean square displacement, diffusion coefficient, confinement index, and instantaneous velocity. This chapter describes a generalizable protocol for the single particle tracking of membrane neurotransmitter transporters on cell membranes with either unmodified extracellular antibody probes and secondary antibody-conjugated quantum dots or biotinylated extracellular antibody probes and streptavidin-conjugated quantum dots in primary neuronal cultures. The neuronal cell culture, the biotinylation protocol and the quantum dot labeling procedures, as well as basic data analysis are discussed.

**Key words** Quantum dot, Single particle tracking, Antibody, Neurotransmitter transporter

## 1 Introduction

Single particle tracking (SPT) applications have continued to expand in the field of molecular biology, shedding light on how individual proteins are organized and regulated. SPT has allowed for individual molecules to be visualized at subdiffraction limited spatial resolutions, providing critical information about targets in the entire cell (Fig. 1) [1]. In contrast to ensemble averaging approaches, the use of SPT has led to the discovery of heterogeneous modes of motion of membrane proteins. Since nanoscience has been implemented into biomedical research, SPT in endogenous systems is now a useful tool to elucidate underlying mechanisms related to disease states.

Sarah Hurst Petrosko and Emily S. Day (eds.), *Biomedical Nanotechnology: Methods and Protocols*, Methods in Molecular Biology, vol. 1570, DOI 10.1007/978-1-4939-6840-4_11, © Springer Science+Business Media LLC 2017

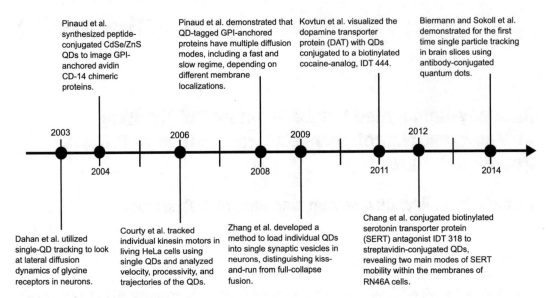

**Fig. 1** Examples of QD SPT applications starting with the first single-QD tracking experiment in neurons and ending with the first single-QD tracking experiment in brain slices [13, 17–23]

Quantum dots (QDs), semiconductor nanocrystals that have a core nanocrystal passivated with a wider bandgap shell, make excellent fluorescent probes for SPT applications [2]. Commercially available QDs with CdSe/CdS/ZnS core/shell configurations are most commonly used in biological applications [3]; they are capped with native nonpolar surface ligands that are solubilized. For example, an amphiphilic polymer can be intercalated into nonpolar ligands to preserve the original architecture and photoluminescence properties of the particles. Once the QDs are solubilized, streptavidin, antibodies, peptides, single-stranded DNA, and small molecule ligands can be incorporated onto the surface for biological targeting and SPT experiments [4]. The core/shell design gives QDs their excellent fluorescent properties, including high brightness that increases their signal-to-noise ratios, excellent photostabilities, narrow emission spectra, and large Stokes shifts [2]. One unique feature of QDs is their characteristic "blinking," or fluorescence fluctuations, which can be used to confirm single QDs are being analyzed, although trajectory reconstruction is necessary during analysis [5]. Transporters have been visualized in heterologous cell lines using QDs, but they have not yet been utilized in endogenously expressing primary neuronal systems, making this protocol a valuable resource (*see* **Note 1**). Our method is generalizable as many extracellular antibodies are already commercially available. In this protocol, biotinylated and unmodified antibody probes are used with streptavidin-functionalized and antibody-conjugated QDs, respectively, to detect individual neurotransmitter transporters. Monoclonal extracellular antibodies are more specific to the target of interest than polyclonal antibodies; thus, they are preferred for use in SPT experiments [6].

In a typical SPT experiment, QD probes are first bound to their targets, and a time-lapse image series is then acquired with a fluorescence microscope at video rates as the probe is shuttled around the cell membrane by the neurotransmitter transporter. Next, the images are processed in an image analysis program, such as Image-J, IDL, or MATLAB, using tracking algorithms that are readily available [7–9]. The density of QD labeling should be sufficiently low so that each QD spot is spatially separated further than the diffraction limit. The localization of each centroid is estimated by fitting individual bright spots to a 2D Gaussian distribution [7]. Localization data (x, y) are then used to generate trajectories, and multiple parameters can be determined, such as mean square displacement (MSD), diffusion coefficient, confinement index, and instantaneous velocity. This protocol details how to carry out SPT experiments using antibodies and streptavidin-conjugated or antibody-conjugated QDs targeted to membrane neurotransmitter transporters in primary neuronal cultures. The basic analysis of the single particle trajectory data will also be discussed.

## 2  Materials

### 2.1  Neuronal Culture

1. Sprague-Dawley rats, postnatal day 0–2.
2. 24-well plates.
3. No. 0 bottom glass coverslips.
4. Forceps.
5. 70% ethanol for sterilization.
6. Fire-polished Pasteur pipettes.
7. Matrigel.
8. Trypsin.
9. HEPES buffer.
10. DNase.
11. Hanks buffer salt solution (HBSS).
12. Plating media: Minimum essential media (MEM, 500 mL), glucose (2.5 g), $NaHCO_3$ (1 M), transferrin (50 mg), L-glutamine (0.2 M), insulin (12.5 mg/mL), fetal bovine serum (FBS) (50 mL).
13. 4-AraC: MEM (500 mL), glucose (2.5 g), $NaHCO_3$ (1 M), transferrin (50 mg), L-glutamine (0.2 M), B-27 (10 mL), AraC stock (1.12 mg/mL), FBS (25 mL).
14. H+20: HBSS + 20% FBS.
15. Dissociation solution: HBSS + 12 mM $MgSO_4$ + DNase (1 kilo-unit/mL).

| | |
|---|---|
| **2.2  QD Labeling** | 1. Streptavidin-conjugated 605 or 655 quantum dots. |
| | 2. Secondary antibody-conjugated 605 or 655 quantum dots. |
| | 3. Extracellular monoclonal antibody. |
| | 4. EZ-Link micro sulfo-NHS-LC biotinylation kit. |
| | 5. Bovine serum albumin (BSA) or dialyzed FBS. |
| | 6. Fluorobrite live-cell imaging media (to reduce background). |
| | 7. Tyrode's solution: 150 mM NaCl, 4 mM KCl, 2 mM $MgCl_2$, 2 mM $CaCl_2$, 10 mM HEPES buffer, 10 mM glucose in a final volume of 1 L deionized water, pH 7.35. |
| | 8. 1.5 mL microcentrifuge tubes. |
| | 9. 15 mL conicals. |
| **2.3  Instrument and Analysis** | 1. Fluorescence microscope with frame rate $\geq 10$ frames per second (custom-built epifluorescence, Andor EMCCD camera, 600/30 bandpass filter). |
| | 2. Microscope heating chamber. |
| | 3. Microscope software (MetaMorph, Micromanager, or similar). |
| | 4. Image analysis software (Image-J or FIJI). |
| | 5. Data analysis software (Excel or MATLAB). |

---

# 3  Methods

| | |
|---|---|
| **3.1  Neuronal Culture** | This protocol gives a general outline of neuronal culture dissociation and plating. |

1. Dissect the midbrain or hippocampus from post-natal day 0–2 Sprague-Dawley rats. Cut tissue into 4–6 small pieces for further dissociation and place in 5 mL H + 20.

2. Wash the tissue three times with HBSS. Add trypsin with DNase to the tissue and incubate for 10 min to digest the tissue and degrade the released DNA.

3. Fire-polish three glass Pasteur pipettes so that the diameters get progressively smaller. The largest diameter should have rounded edges, but should not decrease in size. The next two should be smaller in diameter, with the smallest being around 0.5 mm.

4. Wash the tissue three times with 5 mL of H + 20 to block the trypsin. Wash the tissue three more times with 5 mL of HBSS. Add 3 mL of the dissociation solution to prepare for the titration of the tissue.

5. Manually titrate the dissociation solution with tissue 5 times with the largest pipette to begin breaking up tissue. Repeat for the two other pipette sizes to dissociate the tissue into cells,

keeping the tissue in the same dissociation solution. Do not attempt to wash.

6. Centrifuge the dissociation solution at $700 \times g$ at 4 °C for 5 min.

7. Aspirate the supernatant and resuspend the cell pellet with warm plating media using the smallest-sized (about 0.5 mm diameter) pipette. Add enough volume to plate 100 μL of the cell solution onto the desired number of coverslips.

8. Plate 100 μL of cell solution onto matrigel-coated glass coverslips in a culture plate. Add 1 mL of plating media after 1–2 h.

9. Add 1 mL 4-AraC the following day after cell plating to inhibit glial growth.

**3.2 Antibody Biotinylation**

Streptavidin/biotin binding has long been utilized in biological studies due to its formation affinity constant of $10^{15}$ L mol$^{-1}$, making it one of the strongest noncovalent interactions reported [10]. In cases where the biotinylation of antibodies is possible, this protocol can be used in conjunction with streptavidin-conjugated QDs for labeling. Antibodies have numerous primary amino ($NH_2$) groups available that are ideal for coupling reactions involving sulfo-N-hydroxysulfosuccinimide (sulfo-NHS). The EZ-Link Sulfo-NHS-LC-Biotin kit from Thermo Scientific makes the linking chemistry easy, quick, and straightforward. For all biotinylation reactions, the detailed protocol from Thermo Scientific was followed with no modification [11]:

1. Calculate the millimoles of Sulfo-NHS-Biotin to add using the following formula:

$$\text{mL protein} \times \frac{\text{mg protein}}{\text{mL protein}} \times \frac{\text{mmol protein}}{\text{mg protein}} \times \frac{50 \text{ mmol biotin}}{\text{mmol protein}}$$

$$= \text{mmol biotin}$$

A 50-fold molar excess ensures 1–4 biotin groups per antibody.

2. Calculate the microliters of 9 mM Sulfo-NHS-LC-Biotin to add using the mmol biotin calculated in 1:

$$\text{mmol biotin} \times \frac{557 \text{ mg}}{\text{mmol biotin}} \times \frac{200 \text{ μL}}{1.0 \text{ mg}} = \text{μL biotin solution}$$

3. Dissolve 50–200 μg of antibody in 200–700 μL of phosphate-buffered saline (PBS).

4. Right before use, puncture one microtube of Sulfo-NHS-LC-Biotin and add 200 μL of water, giving 9 mM biotin, and add the calculated amount (**step 2**) to the antibody.

5. Incubate on ice for 2 h or room temperature for 30–60 min.

6. Centrifuge this solution in the Zeba Spin Desalting Column from the kit by placing it in a 15 mL conical at $1000 \times g$ for 2 min. Discard the storage buffer and mark the side that has the resin slanted upward. Place this side facing out for all further centrifugation steps.

7. Equilibrate the column with 1 mL of PBS and centrifuge again at $1000 \times g$ for 2 min, discarding the buffer. Repeat 2–3 times.

8. Place the column in a new 15 mL conical and add antibody solution to the center of resin until absorbed. Centrifuge again at $1000 \times g$ for 2 min. The collected solution is the purified antibody.

9. Store the final product in microcentrifuge tubes in 5 µL aliquots at $-20\,^{\circ}\mathrm{C}$.

## 3.3 Single Quantum Dot Labeling of Neurotransmitter Transporters in Live Neuronal Cultures

These protocols are optimized for a two-step labeling procedure utilizing unmodified primary antibody and secondary antibody-conjugated QDs (Method A) or a biotinylated antibody and streptavidin-functionalized QDs (Method B). In both methods, the antibodies are incubated with the neurons, washed and finally incubated with a 0.05–0.15 nM QD solution. After washing multiple times, the QDs are imaged using a custom-built epifluorescence microscope with an oil-immersion 60× or 100× objective lens (Fig. 2). Video-microscopy rates (≥10 frames per second) and nanometer QD localization accuracy (<20 nm) are achieved with an electron multiplying charge-coupled device (EMCCD) camera (Andor). Single particles are then localized using imaging software, such as Image-J, to map trajectories (Fig. 3). These protocols are generalizable for any protein associated with a specific, extracellular antibody, examples including the GABAa receptor [12], the glycine receptor [13], the dopamine D1 receptor [14], and the glutamate transporter [15]. While this protocol is optimized for primary neuronal cultures, it is also applicable to heterologous expression systems.

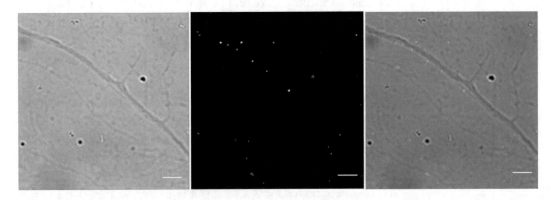

**Fig. 2** QD labeling of neurotransmitter transporters in cultured primary neurons. Bright field, fluorescence, and overlay are shown. Scale bars = 10 µm

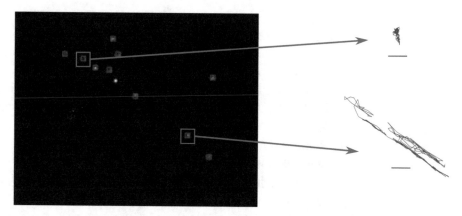

**Fig. 3** Single particles are localized using Image-J and trajectories are subsequently mapped. Scale bar = 2 μm

*Method A: Unmodified Primary Extracellular Antibody + Secondary Antibody-Conjugated QDs*

1. Culture dissociated neurons for 2 weeks on No. 0 glass coverslips.

2. Prepare a solution of 5–10 μg/mL primary antibody in warm fluorobrite media with 4% bovine serum albumin (BSA) to be added to cells (*see* **Note 2**). Invert multiple times to ensure mixing.

3. Prepare a solution of 0.05–0.15 nM secondary antibody-conjugated QDs (depending on desired QD density) in warm 4% BSA fluorobrite media. Invert multiple times to ensure mixing.

4. Remove the culture media and add the antibody solution to the neurons. Incubate at 37 °C for 5–10 min (*see* **Note 3**).

5. Gently wash two times with warm fluorobrite media. Add the QD solution and incubate at 37 °C for 3–5 min.

6. Gently wash five times with Tyrode's solution. Leave in Tyrode's (*see* **Note 4**).

7. Mount the coverslip onto a heated (37 °C) microscope stage.

8. Acquire time-lapse images at a frame rate of 10 frames per second or faster using an appropriate filter for the QD used. A 600/30 bandpass filter was used for 605 QDs (*see* **Note 5**).

*Method B: Biotinylated Antibody + Streptavidin-Conjugated QDs*

This method can be used when biotinylation of the antibody is possible without losing functionality.

1. Culture dissociated neurons for 2 weeks on No. 0 glass coverslips.

2. Prepare a solution of 5–10 µg/mL biotinylated antibody in warm fluorobrite media with 2–4% dialyzed FBS (dFBS) to be added to cells (*see* **Note 2**). Invert multiple times to ensure mixing.

3. Prepare a solution of 0.05–0.15 nM streptavidin QDs in warm 2–4% dFBS fluorobrite media. Vortex for 5 s to break up aggregates.

4. Remove the culture media and add the antibody solution to the neurons. Incubate at 37 °C for 10–20 min (*see* **Note 3**).

5. Gently wash two times with warm dFBS fluorobrite. Add the QD solution and incubate at 37 °C for 5 min.

6. Gently wash five times with Tyrode's solution. Leave in Tyrode's (*see* **Note 4**).

7. Mount the coverslip onto a heated (37 °C) microscope stage.

8. Acquire time-lapse images at a frame rate of 10 frames per second or faster using an appropriate filter for the QD used. A 600/30 bandpass filter was used for 605 QDs (*see* **Note 5**).

**3.4 Data Analysis and Diffusion Models**

The goal of SPT trajectory analysis is to extract quantitative parameters of motion and consequently elucidate the type of motion and diffusive behavior each particle undergoes. Table 1 defines a set of particularly useful motion parameters associated with each trajectory. As the diffusion is a stochastic process, a pool of multiple displacements, $\Delta r(p_i, p_{i+1})$, is necessary to attain a more complete understanding of the particle dynamic behavior. The most popular means of analyzing a pool of multiple displacements within a single

**Table 1**

Biologically useful quantitative parameters derived from a single particle trajectory. The example 2D trajectory drawn on the left consists of $N$ points $p_1(x_1, y_1)$ through $p_N (x_N, y_N)$. The table on the right provides mathematical definitions for several quantitative measures commonly encountered in the literature. $\Delta r$ is defined as the Euclidean norm between two *consecutive trajectory points*

| Parameter | Definition |
|---|---|
| Total displacement | $\Delta r_{\text{total}} = \sum_{i=1}^{N-1} \Delta r(p_i, p_{i+1})$ |
| Net displacement | $\Delta r_{\text{net}} = \Delta r(p_i, p_{i+1})$ |
| Confinement ratio | $z_{\text{conf}} = \Delta r_{\text{net}} / \Delta r_{\text{total}}$ |
| Instantaneous angle | $\alpha_i = \arctan{(y_{i+1} - y_i)} / (x_{i+1} - x_i)$ |
| Instantaneous velocity | $v_i = \Delta r(p_i, p_{i+1}) / \Delta t$ |
| Mean square displacement | $\text{MSD}(n) = \frac{1}{N-1} \sum_{i=1}^{N-n} \Delta r^2(p_i, p_{i+n})$ |

**Table 2**
**Different modes of diffusion defined by the analytical forms of the MSD versus time curves [16]**

| Diffusion Model | MSD Definition | Parameters |
|---|---|---|
| Linear | $4Dt$ | $D$ (diffusion coefficient) |
| Anomalous | $4D_\alpha t^\alpha$ | $D_\alpha$, $\alpha$ (confinement coefficient) |
| Directed | $4Dt + (Vt)^2$ | $D$, $V$ (velocity) |
| Restricted | $L^2\left(1 - A_1 e^{-4A_2 Dt/L^2}\right)$ | $D$, $L$ (length of the confinement domain) |

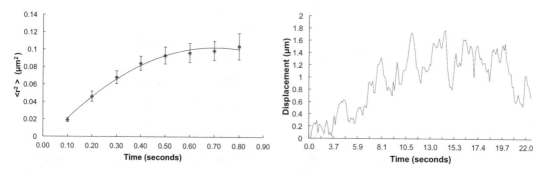

**Fig. 4** Example MSD and displacement plots showing confined motion for a single particle trajectory (*see* Fig. 3)

trajectory is by computing the MSD, which gives a measure of the area explored by a particle at any given time interval. By fitting the MSD curve over time with the appropriate equations, it is thus possible to identify the mode of motion the particle undergoes. Table 2 provides analytical forms of the MSD curve that describe linear (free) diffusion, anomalous subdiffusion, directed motion, and restricted (corralled) diffusion [16]. Once x and y positions have been identified from a single trajectory, MSD can be calculated using the following formula [18]:

$$\text{MSD}\,(n\Delta t) = (N - n)^{-1} \sum_{i=1}^{N-n} \left[(x_{i+n} - x_i)^2 + (y_{i+n} - y_i)^2\right]$$

where $N$ is the total number of frames, n$\Delta$t is the time interval in which the MSD is calculated, and $x_i$ and $y_i$ are the positions of the particles in the trajectories (Fig. 4). Here, we provide a stepwise protocol for calculating the MSD:

1. Display the trajectory data as a numeric array of particle xy-position over time t:

| $x_1$ | $y_1$ | $t_1$ |
|---|---|---|
| $x_2$ | $y_2$ | $t_2$ |
| $x_3$ | $y_3$ | $t_3$ |
| $x_4$ | $y_4$ | $t_4$ |
| $x_5$ | $y_5$ | $t_5$ |

2. Determine the minimum lag time as $\Delta t = t_2 - t_1$ (i.e., temporal resolution of acquisition).

3. Extract from the trajectory the following displacements at a lag time of $\Delta t$:

$$\Delta r_{21} = (\Delta x_{21} = x_2 - x_1, \Delta y_{21} = y_2 - y_1)$$
$$\Delta r_{32} = (\Delta x_{32} = x_3 - x_2, \Delta y_{32} = y_3 - y_2)$$
$$\Delta r_{43} = (\Delta x_{43} = x_4 - x_3, \Delta y_{43} = y_4 - y_3)$$
$$\Delta r_{54} = (\Delta x_{54} = x_5 - x_4, \Delta y_{54} = y_5 - y_4)$$

4. Extract the following displacements at a lag time of $2\Delta t$:

$$\Delta r_{31} = (\Delta x_{31} = x_3 - x_1, \Delta y_{31} = y_3 - y_1)$$
$$\Delta r_{42} = (\Delta x_{42} = x_4 - x_2, \Delta y_{42} = y_4 - y_2)$$
$$\Delta r_{53} = (\Delta x_{53} = x_5 - x_3, \Delta y_{53} = y_5 - y_3)$$

5. Repeat this calculation for a given trajectory at increasing lag times that are multiples of $\Delta t$. Square and average individual displacements to yield MSD values for each lag time. For convenience, the example MATLAB code for implementing this calculation is provided in Fig. 5.

```
%   x: vector of x positions converted from pixels to microns;
%   y: vector of y positions converted from pixels to microns;
%   code below is applicable to trajectories without gaps due to blinking;

tau = 0.1; % time in s between trajectory points, i.e. temporal resolution of acquisition
data = sqrt(x.^2 + y.^2);
N = length(data); % number of data points in the trajectory
ndt = floor(N/4); % ndt should be up to 25% of number of data points

msd = zeros(ndt, 1); % vector storing a pool of MSD values calculated for each dt
msdpts = zeros(ndt, 1); % vector with calculated MSD values
dt = zeros(ndt, 1); % dt lag time vector
sem = zeros(ndt, 1); % vector with SEM values for each MSD point

% calculate msd for all dt's

for i = 1:ndt;
  msd = (data(1+i:end) - data(1:end-i)).^2;
  dt(i) = i * tau;
  sem(i) = std(msd) / sqrt(length(msd));
  msdpts(i) = mean(msd);
end
```

**Fig. 5** Example MATLAB code demonstrating how to calculate MSD from individual x and y positions

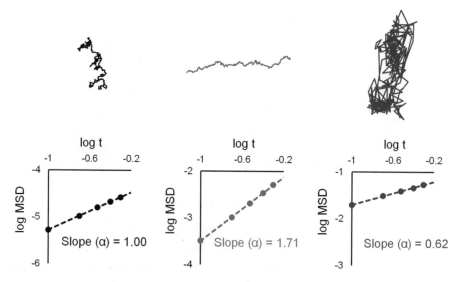

**Fig. 6** Graphical guide for determining the trajectory motion type based on $\alpha$ parameter. Individual MSD versus time curves are transformed to the log-log plot; the $\alpha$ parameter, i.e., the slope of the linear regression line (*dashed*) of the resulting plot, is then used to classify the motion type of each trajectory as restricted ($\alpha < 1$), Brownian ($\alpha = 1$), Brownian + directed ($1 < \alpha < 2$), or "pure" active transport ($\alpha = 2$)

6. To determine the motion type a tagged molecule undergoes, transform the MSD versus $\Delta t$ curve to its log-log form:

$$\log MSD\,(t) = \log 4D + \alpha \log t$$

7. Calculate the $\alpha$ value from the slope of the log-log curve and determine the mode of motion:

   $\alpha < 1$—anomalous subdiffusion
   $\alpha = 1$—Brownian motion
   $\alpha > 1$—anomalous superdiffusion
   $\alpha = 2$—active transport

8. The $\alpha$ exponent gives a measure of the degree the motion of a QD-tagged molecule is influenced by the local environment (Fig. 6). Smaller values of the $\alpha$ exponent correspond to either increased binding or a higher density of obstacles (organelles, lipid rafts, or cytoskeleton) in the diffusion path. Larger $\alpha$ values are usually indicative of active cellular transport events.

# 4  Notes

1. QDs are bound to an extracellular epitope of transporter proteins in an antibody-dependent manner. The motion of the transporter-bound QDs is monitored on the surface of cultured neurons. As cultured neurons are considered a low expression system with less than 1–2 transporters per $\mu m^2$ at

the cell membrane, we estimate QD labeling at 1:1 stoichiometry. However, in a higher expression system antibody-mediated QD labeling might lead to protein crosslinking and promote undesirable target protein internalization. In such cases, we recommend preforming QD-primary antibody complexes at 1:1 stoichiometry and only then labeling target proteins. An alternate route is generating monovalent antibody fragments with a single-antigen binding site.

2. Premix the antibody or QD solutions to reduce aggregation of each. Dilute the antibody and QDs in the final volume that will be added to the cells.

3. If the QDs are internalizing:

    (a) Image immediately after labeling.

    (b) Decrease the incubation time with QDs to 3 min.

    (c) Label cells at 4 °C to further minimize internalization.

4. If the QDs are binding nonspecifically:

    (a) Increase the percentage of BSA or dFBS when incubating with the QDs.

    (b) Include an additional blocking reagent (e.g., casein, newborn calf serum, dehydrated fat-free milk).

    (c) Increase the number of wash steps.

    (d) Perform a control experiment with QDs only.

    (e) Perform a control experiment with a blocking agent (e.g., a peptide sequence for the protein).

5. If signal-to-noise ratio is low:

    (a) Use a low-background buffer, like DMEM Fluorobrite.

    (b) Adjust the pinhole size, gain, and amplitude.

    (c) Increase the excitation intensity.

## Acknowledgments

The authors would like to gratefully acknowledge Dr. Qi Zhang for his guidance in neuronal cultures and imaging parameters, as well as for the use of his fluorescence microscope. The authors would also like to thank Dr. Jerry Chang for useful discussions regarding SPT experiments and imaging. DMB was supported by the NSF Graduate Research Fellowship Program.

## References

1. Chang JC, Rosenthal SJ (2013) A bright light to reveal mobility: single quantum dot tracking reveals membrane dynamics and cellular mechanisms. J Phys Chem Lett 4:2858–2866

2. Rosenthal SJ et al (2002) Targeting cell surface receptors with ligand-conjugated nanocrystals. J Am Chem Soc 124:4586–4594

3. Valizadeh A et al (2012) Quantum dots: synthesis, bioapplications, and toxicity. Nanoscale Res Lett 7:480

4. Bilan R, Fleury F, Nabiev I, Sukhanova A (2015) Quantum dot surface chemistry and functionalization for cell targeting and imaging. Bioconjug Chem. doi:10.1021/acs.bioconjchem.5b00069

5. Chang JC, Kovtun O, Blakely RD, Rosenthal SJ (2012) Labeling of neuronal receptors and transporters with quantum dots. Wiley Interdiscip Rev Nanomed Nanobiotechnol 4:605–619

6. Abcam Protocol (2015) A comparison between polyclonal and monoclonal. http://www.abcam.com/protocols/a-comparison-between-polyclonal-and-monoclonal

7. Chang JC, Rosenthal SJ (2013) Quantum dot-based single-molecule microscopy for the study of protein dynamics. Methods Mol Biol 1026:71–84

8. Crocker JC, Grier DG (1996) Methods of digital video microscopy for colloidal studies. J Colloid Interface Sci 179:298–310

9. Blair D, Dufresne E (2013) The matlab particle tracking code repository. http://site.physics.georgetown.edu/matlab/index.html

10. Christopoulos K (1991) The biotin-(strept) avidin system: principles and applications in biotechnology. Clin Chem 37:625–636

11. Thermo Scientific Protocol EZ-Link Micro Sulfo-NHS-LC Biotinylation Kit. https://tools.thermofisher.com/content/sfs/manuals/MAN0011568_EZ_Micro_Sulfo_NHS_LC_Biotinylation_UG.pdf

12. Bannai H et al (2009) Activity-dependent tuning of inhibitory neurotransmission based on GABAAR diffusion dynamics. Neuron 62:670–682

13. Dahan M et al (2003) Diffusion dynamics of glycine receptors revealed by single-quantum dot tracking. Science 302:442–445

14. Ladepeche L et al (2013) Single-molecule imaging of the functional crosstalk between surface NMDA and dopamine D1 receptors. Proc Natl Acad Sci U S A 110:18005–18010

15. Murphy-Royal C et al (2015) Surface diffusion of astrocytic glutamate transporters shapes synaptic transmission. Nat Neurosci 18:219–226

16. Saxton MJ, Jacobson K (1997) Single-particle tracking: applications to membrane dynamics. Annu Rev Biophys Biomol Struct 26:373–399

17. Pinaud F, King D, Moore H-P, Weiss S (2004) Bioactivation and cell targeting of semiconductor CdSe/ZnS nanocrystals with phytochelatin-related peptides. J Am Chem Soc 126:6115–6123

18. Courty S, Luccardini C, Bellaiche Y, Cappello G, Dahan M (2006) Tracking individual kinesin motors in living cells using single quantum-dot imaging. Nano Lett 6:1491–1495

19. Pinaud F et al (2009) Dynamic partitioning of a glycosyl-phosphatidylinositol-anchored protein in glycosphingolipid-rich microdomains imaged by single-quantum dot tracking. Traffic 10:691–712

20. Zhang Q, Li Y, Tsien RW (2009) The dynamic control of kiss-and-run. Science 329:1448–1453

21. Kovtun O et al (2011) Visualization of the cocaine-sensitive dopamine transporter with ligand-conjugated quantum dots. ACS Chem Nerosci 2:370–378

22. Chang JC et al (2012) Single molecule analysis of serotonin transporter regulation using antagonist-conjugated quantum dots reveals restricted, p38 MAPK-dependent mobilization underlying uptake activation. J Neurosci 32:8919–8929

23. Biermann B et al (2014) Imaging of molecular surface dynamics in brain slices using single-particle tracking. Nat Commun 5:3024

# Chapter 12

# Spectroscopic Photoacoustic Imaging of Gold Nanorods

## Austin Van Namen and Geoffrey P. Luke

### Abstract

Photoacoustic imaging is a rapidly developing tool capable of achieving high-resolution images with optical contrast at imaging depths up to a few centimeters. When combined with targeted nanoparticle contrast agents, sensitive detection of molecular signatures is possible. In this chapter, we discuss the achievements and future directions of nanoparticle-augmented photoacoustic imaging. We present a method to synthesize silica-coated gold nanorods, which are highly stable, signal amplifying photoacoustic contrast agents, and also describe spectroscopic image acquisition and processing steps to provide a specific map of nanoparticle distribution in vivo.

**Key words** Photoacoustic imaging, Spectroscopy, Gold nanorods, Molecular imaging

## 1 Introduction

Photoacoustic (PA) imaging is an emerging modality that offers unique advantages over existing soft tissue imaging technology. It uses laser-induced ultrasound to address the need for real-time, high-resolution imaging at clinically relevant depths. As a hybrid modality, PA imaging integrates the benefits from the high contrast and specificity of optical spectroscopy with fine spatial resolution from ultrasound detection. These attributes have made PA imaging a promising alternative or complementary modality in a diverse array of fields such as oncology, neurology, cardiology, and tissue engineering [1–5]. Innovative approaches to build on the promise of the technology are currently being explored in imaging algorithms, instrumentation, and molecular targeted contrast agents [6–10]. In this chapter, we present a brief introduction to the photoacoustic effect, image acquisition techniques, spectroscopic imaging, and contrast agents. Then, we provide a detailed protocol for synthesis of silica-coated gold nanorods and methods to acquire and process PA images to localize the particles in vivo.

The PA effect is the generation of sound waves resulting from absorption of time-varying electromagnetic energy. The process is

Sarah Hurst Petrosko and Emily S. Day (eds.), *Biomedical Nanotechnology: Methods and Protocols*, Methods in Molecular Biology, vol. 1570, DOI 10.1007/978-1-4939-6840-4_12, © Springer Science+Business Media LLC 2017

## 1.1 The Photoacoustic Effect and Image Formation

most often initiated with energy deposited by a nanosecond pulsed laser in the near-infrared spectrum. The use of near-infrared light allows for deep penetration in biological tissue. Absorbed by endogenous or exogenous photoabsorbers, energy is converted to heat through vibrational and collisional relaxation. The photoabsorber's energy releases through a rapid thermoelastic expansion in the incident matter. This heat deposited in the surrounding tissue is linearly related to a rise in the local maximum initial pressure, $p_0$. Pressure $p_0$ depends on a number of parameters and is treated with a simplified equation:

$$P_0 = \mu_a \Gamma F$$

where $\mu_a$ is the optical absorption coefficients of the photoabsorbers, $F$ is the fluence at the photoabsorber, and $\Gamma$ is the tissue's Gruneisen parameter, which dictates the energy conversion from heat to pressure [10]. The linear dependence of $p_0$ on photoabsorber concentration quantifies and localizes physiological parameters.

Optical absorption, proportional to initial pressure as in the above equation, is the dominant form of contrast in the formation of the PA image. The initial pressure increase and subsequent relaxation travel unscattered through tissue in the form of a broadband acoustic wave. Ultrasound receivers measure the propagation of the acoustic wave when it reaches the surface and use time of flight to calculate the initial acoustic source distribution and map absorption properties. Image formation is dependent on the collection technique, initial energy deposition, and geometry of the US transducers. For example, photoacoustic tomography (PAT) is a complex imaging method that reconstructs the photoacoustic signal from multiple ultrasound transducer elements to form a 2D or 3D image from a single laser pulse. Another common technique, photoacoustic microscopy (PAM), raster-scans a single ultrasound transducer over a 2D field. Each laser pulse results in a single depth-resolved 1D line in the image. Both of these techniques result in a map of optical absorption in tissue. In either case, spectroscopic imaging approaches and targeted contrast agents can augment the images to extract additional functional and molecular information.

## 1.2 Spectroscopic Photoacoustic Imaging

Spectral imaging provides physiological and molecular information by retrieving signals from multiple tissue chromophores and contrast agents. Spectroscopic PA imaging has been used to image blood vessels, quantify oxygen saturation, identify melanoma, and detect lipids in vessels [11–14]. Photoacoustic image contrast can be selectively enhanced for tissue chromophores by tuning the excitation wavelength to the absorption spectra of specific components. For example, the absorption spectrum of blood in the optical window is highly dependent upon its oxygen saturation, a consequence of spectral differences between the two blood chromophores, deoxyhemoglobin and oxyhemoglobin. Using an assortment of wavelengths, it is possible to quantify concentrations of the two photoabsorbers and estimate oxygen saturation. In this manner, PA imaging can provide functional and anatomical information about processes such as angiogenesis or tissue inflammation [15]. Deriving absolute chromophore concentrations from signals at multiple wavelengths is essential for accurate imaging and an area of current innovation [16, 17]. In addition to imaging endogenous chromophore concentrations, spectroscopic techniques provide a means of detecting and quantifying the accumulation of targeted contrast agents used in PA molecular imaging [7].

*1.3  Contrast Agents*

Exogenous contrast agents expand the versatility of PA imaging. Constructs that absorb strongly in the near-infrared spectrum can be coupled with targeting mechanisms for molecular imaging. Signaling agents including dyes, noble metal nanoparticles, carbon nanostructures, or liposome encapsulations have been used with PA imaging to target and enhance contrast [9, 18–23]. Targeting can be achieved using small biological molecules such as peptides, aptamers, or antibodies. Assembling a signaling compound with a targeting ligand provides a built-to-order photoabsorber that can bind to process biomarkers or uniquely identifiable cells. Exogenous contrast agents continue to expand the applications of PA imaging by enabling patient-specific molecular diagnosis of disease, image-guided release of targeted nanosized drug carriers, and real-time monitoring of therapy outcomes.

This chapter presents the synthesis of silica-coated gold nanorods (Fig. 1), and describes PA image acquisition and spectral processing using these contrast agents. The methods outlined here have been adapted from well-established protocols of particle synthesis and surface modification [24–26]. Plasmonic nanoparticles make excellent PA contrast agents because of their large optical absorption and easily customizable surface [25–29]. The addition of the silica shell to the gold nanorods provides enhanced stability by preventing nanoparticle reshaping and amplified PA signal generation through more efficient heat transfer to the surrounding tissue [26, 30, 31]. When combined with spectroscopic imaging techniques, this enables highly sensitive tracking of nanoparticles in vivo.

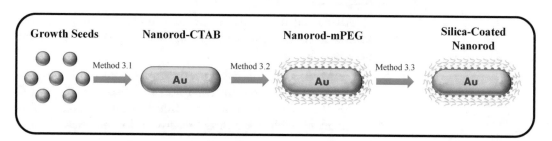

**Fig. 1** Schematic showing the three-step growth and functionalization of silica-coated gold nanorods

## 2   Materials

### 2.1   Gold Nanorod Synthesis

1. Aqua regia: Carefully mix three parts hydrochloric acid (37 %) and one part nitric acid (70 %) (v/v) in a pyrex glass bottle. Prepare this solution in a chemical fume hood while wearing personal protective equipment including safety goggles, lab coat, and acid-resistant gloves. Place the lid on the bottle, but do not close the lid tightly, as this could lead to a pressure build-up that would rupture the glass (*see* **Note 1**).

2. Deionized ultrafiltered water (DIUF, 18.2 M$\Omega$ cm) (*see* **Note 2**).

3. 200 mM cetyl trimethyl ammonium bromide (CTAB): Dissolve 5.47 g CTAB in 75 mL DIUF at 35 °C to yield a final concentration of 200 mM (*see* **Notes 3** and **4**).

4. 1 mM Gold(III) chloride trihydrate ($HAuCl_4$): Dissolve 29.5 mg HAuCl4 in 75 mL water to generate a 1 mM stock solution.

5. 10 mM sodium borohydride ($NaBH_4$): Dissolve 1.89 mg $NaBH_4$ in 5 mL of ice-cold DIUF to yield a final concentration of 10 mM.

6. 4 mM silver nitrate ($AgNO_3$): Dissolve 3.40 mg $AgNO_3$ in DIUF to yield a final concentration of 4 mM.

7. 78.8 mM ascorbic acid (AA): Dissolve 13.9 mg AA in 1 mL DIUF to yield a final concentration of 78.8 mM.

8. Hot plate with magnetic stirrer.

9. 250 mL glass beakers.

10. 20 mL glass scintillation vials.

11. Magnetic stir bars.

12. Centrifuge.

13. Ultraviolet-visible spectrophotometer (UV-Vis).

14. Oven.

### 2.2   Pegylation of Gold Nanorods

1. Hot plate with magnetic stirrer.

2. 20 mL glass scintillation vials.

3. Magnetic stir bars.

4. CTAB-coated nanorods: Suspended in DIUF water at a concentration yielding a peak optical density (OD) of 15 (1-cm path length).

5. 0.2 mM 5 kDa methyl-polyethylene glycol-thiol (mPEG-thiol): Weigh 20 mg mPEG-thiol and dissolve in 20 mL DIUF. This solution should be prepared immediately before use to avoid oxidization of the reactive thiol moieties

6. DIUF water.

7. Benchtop sonicator.

8. Filtered centrifuge tubes (100 kDa).

**2.3  Silica-Coating of Gold Nanorods**

1. Pegylated gold nanorods: Suspended in DIUF water to peak OD=15.

2. DIUF water.

3. Isopropanol (IPA).

4. 1.4 % Tetraethyl orthosilicate (TEOS) in IPA: mix 21 μL TEOS with 1.479 mL IPA to yield a final concentration of 1.4 % v/v.

5. 3 % Ammonium Hydroxide (NH$_4$OH, 29 %) in IPA. Mix 45 μL NH$_4$OH with 1.455 mL IPA to yield a final concentration of 3 % v/v.

6. Dulbecco's phosphate buffer solution (DPBS).

7. 20 mL glas scintillation vials.

8. Magnetic stir bars.

9. Hot plate with magnetic stirrer.

10. Digital pH meter.

11. UV-Vis spectrophotometer.

12. Transmission electron microscope (TEM).

13. Centrifuge.

14. Filtered centrifuge tubes (100 kDa).

15. Sterile 5 mL syringe.

16. Sterilizing syringe filter (0.45 μm).

**2.4  Image Acquisition**

1. Photoacoustic and ultrasound imaging system.

2. Linear array ultrasound transducer.

3. Heated electrocardiogram (ECG) pad.

4. ECG coupling gel.

5. Athletic tape.

6. Clear ultrasound gel.

7. 50 mL centrifuge tubes.

8. Scoopula.

9. Centrifuge.

10. Cotton swabs.

11. Tumor-bearing nude mouse.

12. 27-gauge needle.

13. 1 mL syringe.

14. Isoflurane Vaporizor.

15. Pressurized O$_2$ canister.

16. Isoflurane.

17. Sterile silica-coated gold nanorods.

**2.5 Spectroscopic Processing**

1. Matlab.

## 3 Methods

**3.1 CTAB-Coated Gold Nanorod Synthesis**

1. Perform this step in a chemical fume hood (*see* **Note 1**). Place magnetic stir bars in glass beakers. Fill all glass beakers half-full with aqua regia and carefully tilt and rotate the beakers to ensure aqua regia has contacted all surfaces. Allow beakers and stir bars to sit with the aqua regia for 30–60 min to ensure they are thoroughly cleaned. Remove aqua regia from the beakers and discard in accordance with institutional policies. Wash three times with DIUF water, discarding it in the same manner as the aqua regia, and dry in an oven (*see* **Note 2**).

2. Prepare nanorod growth solution by mixing 50 mL of 200 mM CTAB with 50 mL of 1 mM $HAuCl_4$, 2 mL of 4 mM $AgNO_3$, and 0.7 mL of 78.8 mM ascorbic acid in a 250-mL glass beaker containing a magnetic stir bar. Stir at 300 rpm at 30 °C (*see* **Note 5**).

3. In a separate 20-mL glass scintillation vial containing a magnetic stir bar, prepare a seed solution containing 5 mL of 200 mM CTAB and 5 mL of 0.5 mM $HAuCl_4$ (0.5 mM $HAuCl_4$ can be prepared by diluting the 1 mM stock solution 1:1 in DIUF). Stir at 700 rpm at 30 °C.

4. Quickly add 0.6 mL of ice-cold 10 mM $NaBH_4$ to the seed solution and stir for 2 minutes (*see* **Notes 6** and **7**).

5. Add 120 μL of the seed solution from **step 4** of Subheading 3.1 to the growth solution and continue stirring at 300 rpm for 1 minute before stopping stirring and keeping the temperature at 30 °C (*see* **Note 8**).

6. Cover the solution and allow it to age overnight at 30 °C.

7. Confirm the formation of gold nanorods by diluting a sample 10× in DIUF and measuring its extinction spectrum with a UV-Vis spectrophotometer (*see* **Note 9**).

8. Transfer the nanorods to a centrifuge tube. Centrifuge $(18,000 \times g, 45 \text{ min})$. The nanorods will form a pellet at the bottom of the tube. Decant the supernatant and add DIUF to disperse the nanorods to match the previous volume. Repeat the centrifugation and decanting process. After the second wash step, resuspend the nanorods in DIUF at a concentration that yields a peak optical density (OD) of 15 (1-cm path length) as measured with the UV-Vis spectrophotometer.

While the nanorods are typically stable for several days at room temperature in this solution, we recommend proceeding to the steps outlined in Subheading 3.2 immediately to avoid any possible nanorod degradation.

### 3.2 Pegylation of Gold Nanorods

1. Mix the OD-15 nanorods with an equal volume of freshly prepared 0.2 mM mPEG-thiol in a glass beaker containing a magnetic stir bar while stirring at 700 rpm.

2. Sonicate the solution for 5 min then cover with parafilm and let sit overnight.

3. Transfer the nanorods into the filtered centrifuge tubes. Centrifuge at $3000 \times g$ for 10 min to form a nanorod pellet. Remove excess mPEG-thiol with the supernatant (*see* **Note 10**).

4. Resuspend the pegylated nanorods in DIUF water to an OD of 6 (*see* **Note 11**).

### 3.3 Silica-Coating of Gold Nanorods

1. Stir 5 mL of the OD-6 pegylated gold nanorods at 700 rpm in a 20 mL glass beaker scintillation vial.

2. Add 1.2 mL of the 1.4 % $NH_4OH$ in IPA solution to the stirring nanorods.

3. Confirm that the pH is 10.9–11 (*see* **Note 12**).

4. Add 1.2 mL of the 3 % TEOS in IPA solution to the sample.

5. Allow the solution to react uncovered for 2–3 h.

6. Track the peak absorption wavelength using UV-vis spectrophotometer measurements of a $10\times$ diluted sample to confirm adsorption of silica onto the gold nanorods (*see* **Note 13**).

7. Transfer the nanorods to a filtered centrifuge tube (100 kDa filter, 15 mL) and centrifuge to form a pellet (*see* **Note 14**). Remove the waste from the centrifuge tube and resuspend the pellet with 15 mL DIUF water. Repeat the centrifugation to form a pellet of concentrated silica-coated gold nanorods.

8. Apply a single drop of concentrated silica-coated gold nanorods to a copper TEM grid and allow it to dry completely. Imaging of the nanoparticles with TEM can confirm the size and morphology of the silica shell (Fig. 2a–d).

9. Acquire final UV-vis spectra of the nanoparticles (diluted to OD of approximately 1) to be used with the spectroscopic image processing algorithms (Fig. 2e).

10. Resuspend the silica-coated gold nanorods in DPBS to reach an OD of 20.

11. Sterilize the nanorods with a 0.45 micron syringe filter in a cell culture hood (*see* **Note 15**).

12. Store the silica-coated nanorods at 4 °C. They should be stored for not more than 3–4 days prior to in vivo use.

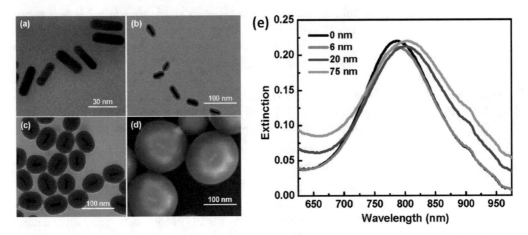

**Fig. 2** TEM images of gold nanorods (**a**) without a silica shell and with a (**b**) 6 nm, (**c**) 20 nm, and (**d**) 75 nm thick shell. (**e**) The corresponding UV-Vis spectra show a gradual redshift in the absorption peak with the adsorption of silica. Adapted with permission from [30]

*3.4 Image Acquisition*

In this section, we refer to acquiring the images with a Verasonics Vevo LAZR imaging system, but other imaging systems may also be used with this protocol (*see* **Note 16**).

1. Centrifuge the ultrasound coupling gel to remove all bubbles ($3000 \times g$, 3 min).

2. Anesthetize the tumor-bearing mouse (*see* **Note 17**) (1.5– 2 % isoflurane, 1 L/min $O_2$).

3. Apply a small (less than 5-mm) drop of ECG coupling gel to each electrode on the ECG pad.

4. Place the mouse on the ECG pad so that the tumor is easily accessible from above for imaging. Set the ECG pad to maintain 37 °C body temperature.

5. Tape each of the mouse's four paws to the ECG electrodes with the athletic tape (*see* **Note 18**).

6. Apply a small drop of ultrasound gel to the imaging area and carefully spread it with a cotton swab to remove air bubbles on the skin surface (*see* **Note 19**).

7. Extract a large volume of ultrasound gel with the scoopula, taking great care to avoid generating any air bubbles. With the scoopula upside down over the tumor, slide a cotton swab along the inside ridge of the scoopula to dislodge the gel in one piece (*see* **Note 19**).

8. Lower the ultrasound transducer onto the gel using the real-time B-Mode ultrasound image for guidance. An offset of approximately 8 mm–1 cm between the surface of the transducer and the mouse skin is optimal for light delivery with this

imaging probe. This can be measured by noting the depth of the skin surface on the ultrasound image.

9. Shift the imaging stage to find the desired field of view.

10. Acquire ultrasound and spectroscopic photoacoustic images in the same imaging plane using several optical wavelengths (*see* **Note 20**). These images constitute the "before" images that can be used as a baseline to compare nanoparticle delivery.

11. Inject 200 μL of the concentrated, sterile gold nanoparticles in the tail vein of the mouse with the 27-gauge needle (*see* **Note 21**).

12. Repeat the imaging steps either immediately while the mouse is still positioned or at a later time point to allow for greater nanoparticle accumulation in the tumor.

**3.5 Spectroscopic Processing**

The overall goal of spectroscopic processing is to match the acquired photoacoustic spectrum in each pixel to the expected optical absorbers to estimate their relative concentrations (*see* **Note 22**). We know from Subheading 1.1 that the generated photoacoustic signal is proportional to the laser fluence and the optical absorption. Hidden within this simple linear relationship are two nonlinear forward problems [7]. First, the light must travel through highly scattering and heterogeneous tissue to arrive at the optical absorber. Second, the photoacoustic wave must travel to the surface of the tissue, during which time frequency-dependent attenuation occurs. Therefore, two inverse problems must be solved to estimate the concentrations of absorbers. An estimate to the solution of the acoustic inverse problem is done by the Vevo system during image formation. This in effect maps out the location and amplitude of the total absorbed energy in the tissue. What it does not do, however, is distinguish between the fluence and the optical absorption coefficient. In order to do this, an estimate of the fluence must be obtained at each point in the image. There are a handful of methods to estimate this, including a Monte Carlo simulation, a diffusion approximation of light propagation, and Beer's law. The Monte Carlo simulation works well if the geometry and components of the tissue are known. The diffusion approximation is particularly well suited for full-angle tomography systems where the boundary conditions are well defined. Both of these methods require a considerable amount of processing power, and are thus not yet realizable in real time. Beer's law, on the other hand, is a relatively simple calculation. It states that the fluence, $F$, decays exponentially as a function of tissue depth:

$$F = F_0 e^{-\mu_{\text{eff}} z}$$

where $F_0$ is the fluence incident on the surface of the skin, $z$ is the depth in tissue, and $\mu_{\text{eff}}$ is a term that combines the optical absorption, $\mu_a$ with the reduced optical scattering, $\mu_s'$:

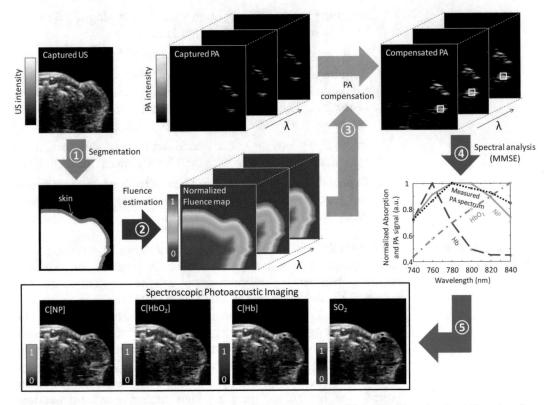

**Fig. 3** Overall scheme for acquiring and processing spectroscopic PA images to extract relative absorber concentrations. This involves (*1*) segmenting the ultrasound image to find tissue boundaries, (*2*) estimating the fluence with Beer's law, (*3*) correcting the original PA images for depth- and wavelength-dependent optical fluence, (*4*) spectrally unmixing the absorber concentrations with the linear least squares method, and (*5*) displaying functional spectroscopic PA images overlaid on anatomical ultrasound images. Adapted with permission from [33]

$$\mu_{\text{eff}} = \sqrt{3\mu_a\left(\mu_a + \mu_s'\right)}$$

This allows for an estimate of the fluence to be made at each spatial location and enables the decoupling of the optical absorption coefficient. The details of the spectroscopic image processing procedure are shown in Fig. 3 and described stepwise below.

1. First, segment the ultrasound image to identify the tissue borders. This can be automated by first applying a spatial median filter (to suppress the characteristic ultrasound speckle) and thresholding to highlight the mouse skin and form a binary image.

2. Directly calculate the depth of each pixel in the tissue based on the segmented ultrasound image.

3. Estimate tissue properties from the literature to apply Beer's law to calculate the fluence distribution map [32]. Repeat this for

each optical wavelength. Normalize the captured photoacoustic images (which are proportional to the total absorbed energy) by the estimated fluence in each pixel to generate a compensated photoacoustic image, which is proportional to $\mu_a$.

4. Spectrally unmix the compensated photoacoustic image to provide estimates of relative absorber concentrations in the tissue. In this example, each pixel can conceivably contain a combination of three distinct absorbers: gold nanorods, deoxyhemoglobin, and oxyhemoglobin. A fourth absorber, melanin, should also be considered if the region of interest contains the skin. A matrix of the molar absorption spectra, $\varepsilon$, should be constructed, with the rows corresponding to optical wavelengths and the columns corresponding to distinct absorbers. Then, a $3 \times 1$ vector of concentrations, $C$, in each pixel can be solved in a minimum mean squared error sense using the equation:

$$C = \varepsilon^+ \, PA_{comp}$$

where $PA_{comp}$ is the compensated photoacoustic signal in the pixel and $\varepsilon^+$ is the pseudoinverse of $\varepsilon$:

$$\varepsilon^+ = \left(\varepsilon^T \varepsilon\right)^{-1} \varepsilon^T$$

5. Display the concentration of gold nanorods, deoxyhemoglobin, and oxyhemoglobin as an image overlaying the anatomical ultrasound image (Fig. 4) (see **Note 23**).

# 4  Notes

1. Aqua regia is a powerful oxidizing solution that generates chlorine and nitrogen oxide gas. Any steps in this protocol that describe the preparation or use of aqua regia for cleaning glassware and stir bars should be performed in a chemical fume hood.

2. Unopened capped scintillation vials can be used without cleaning with aqua regia.

3. Unless noted otherwise, all solutions are made with DIUF water.

4. CTAB will only dissolve at 25 °C or above and will crystalize if it returns to room temperature. However, higher temperatures will lead to faster growth, larger nanorods, and a more polydisperse solution.

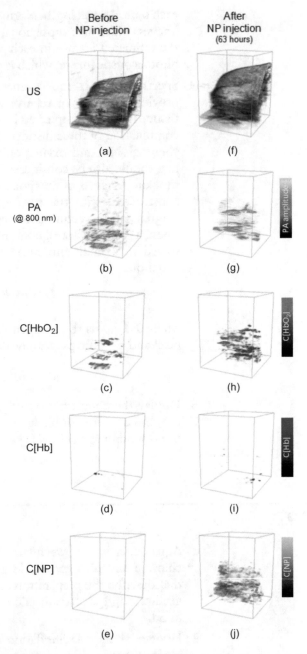

**Fig. 4** Three-dimensional ultrasound and spectroscopic photoacoustic images (**a–e**) before and (**f–j**) 63 h after injection of silica-coated gold nanorods. The nanorods (yellow) clearly accumulate in and around the tumor and lead to enhanced PA signal after the injection. Adapted with permission from [34]

5. This solution of CTAB-HAuCL$_4$ should be bright yellow-orange from the Au$^{3+}$ ions. The ascorbic acid acts as a mild reducing agent, turning the solution colorless.

6. NaBH$_4$ reacts with water at room temperature. Therefore, it is also best to make this solution immediately before seed formation.

7. The seed solution should instantly change to a dark yellow-brown color. A violet solution is indicative of growth of large nanospheres or aggregation of the seeds. If this occurs, the seeds should be discarded and a new seed solution should be made.

8. The growth solution will gradually turn from colorless to burgundy red over the course of 10–20 min. This corresponds to the growth of the nanorods and the development of surface plasmon resonance.

9. Absorption peaks near 530 and 800 nm will gradually develop. The 530-nm peak corresponds to the plasmon resonance in the transverse direction of the nanorods and also to contaminating nanospheres. The 800-nm peak corresponds to the plasmon resonance in the longitudinal direction. In general, narrow peaks and a high ratio between the heights of the peaks indicate more mono-disperse nanorods.

10. CTAB cannot be filtered through the centrifugation filters. Therefore, it is important to adequately wash the nanorods to remove excess CTAB in **step 8** of Subheading 3.1.

11. The addition of a PEG layer to the surface of the nanorods can yield a slight (2–3 nm) blue-shift in their absorption peak.

12. In general, a higher pH leads to a faster reaction, which can result in a more porous silica coating. In addition, too high of pH can lead to self-seeding of silica nanoparticles and aggregation.

13. Each 1-nm layer of silica shifts the peak absorption wavelength toward the near infrared by approximately 1 nm.

14. Add the silica-coated nanorods to a centrifugation filter that has been filled half-way with DIUF water. The IPA and NH$_4$OH can damage the filter and inhibit filtration.

15. If any aggregation has occurred in the particles, they will accumulate in the filter. Thus, it is imperative to optimize the synthesis procedures prior to filtering.

16. A wide variety of imaging systems could be used to collect spectroscopic photoacoustic images of the nanoparticles, including several commercial solutions. The critical feature that is required is a laser that is tunable across the spectral range of interest (typically in the red to near infrared region). This is commonly accomplished with a Nd:YAG 2nd harmonic pumped optical parametric oscillator, which has a wavelength range of 680–950 nm. The laser should also provide sufficient

fluence (up to 20 mJ/cm$^2$ to stay within safety limits) for a high signal-to-noise ratio in the images.

17. The images in this chapter were acquired in a xenograft mouse model in which imaging was performed 3 weeks after subcutaneous inoculation of A431 cells (ATCC) in the flank of an immunodeficient Nu/Nu mouse (Charles River). However, the overall procedure is readily adaptable to other anatomical locations and disease models. The key requirement is that the imaging is performed in soft tissue to avoid acoustic reflections in bone or air.

18. Apply tape to the paw first, then tape the paw to the ECG pad. Otherwise, the ECG gel is likely to make adhesion difficult.

19. Air bubbles severely degrade the quality of the ultrasound and photoacoustic image and introduce unwanted artifacts. Therefore, if air bubbles are clearly visible (seen as bright points in the gel or on the surface of the mouse skin with shadowing underneath) it is recommended to repeat the gel coupling procedure.

20. A minimum of three wavelengths are required to successfully unmix the gold nanoparticle spectrum from deoxy- and oxy-hemoglobin spectra. Errors in spectral unmixing can be suppressed if a greater number of wavelengths are selected. Imaging wavelengths selected by photoacoustic researchers are typically evenly spaced across the spectrum of interest. However, performance can be improved if the spectral characteristics of the absorbers are taken into account (*see* Ref. 16).

21. The tail vein injection is a difficult procedure that often requires a substantial amount of practice. Stimulating vasodilation by dipping the tail in warm water or applying oil of wintergreen to the skin can help improve the likelihood of success.

22. While quantification of tissue components and nanoparticle concentrations would be highly desirable, it remains an open problem in the field and is highly dependent on transducer geometry and tissue components [7].

23. If non-silica-coated pegylated gold nanorods are used for PA imaging, their deposition in tissue can be confirmed histologically though silver staining, as silver stain deposits on gold to enable visualization by light microscopy. Another option is to use hyperspectral microscopy on unstained tissue sections to optically detect nanoparticles in excised tissue [19]; this approach may be utilized for either silica-coated or non-silica-coated nanorods.

## Acknowledgments

The methods described in this chapter build on the excellent work of and discussions with Dr. Stanislav Emelianov of Georgia Institute of Technology, Dr. Kimberly Homan of Nanohybrids Inc., Dr. Yun-Sheng Chen of Stanford University, and Dr. Seungsoo Kim of Siemens Healthcare.

## References

1. Mallidi S, Luke GP, Emelianov S (2011) Photoacoustic imaging in cancer detection, diagnosis, and treatment guidance. Trends Biotechnol 29:213–221

2. Taruttis A, van Dam GM, Ntziachristos V (2015) Mesoscopic and macroscopic optoacoustic imaging of cancer. Cancer Res 75:1548–1559

3. Hu S, Wang LV (2010) Neurovascular photoacoustic tomography. Front Neuroenergetics 2:10

4. Su JL, Wang B, Emelianov SY (2009) Photoacoustic imaging of coronary artery stents. Opt Express 17:19894–19901

5. Nam SY, Ricles LM, Suggs LJ, Emelianov SY (2014) Imaging strategies for tissue engineering applications. Tissue Eng Part B Rev 21:88–102

6. Li C, Wang LV (2009) Photoacoustic tomography and sensing in biomedicine. Phys Med Biol 54:R59

7. Cox B, Laufer JG, Arridge SR, Beard PC (2012) Quantitative spectroscopic photoacoustic imaging: a review. J Biomed Opt 17:061202

8. Beard P (2011) Biomedical photoacoustic imaging. Interface Focus 1:602–631

9. Luke GP, Yeager D, Emelianov SY (2012) Biomedical applications of photoacoustic imaging with exogenous contrast agents. Ann Biomed Eng 40:422–437

10. Wang LV (ed) (2009) Photoacoustic Imaging and Spectroscopy. CRC Press, Boca Raton

11. Jose J, Manohar S, Kolkman RGM, Steenbergen W, van Leeuwen TG (2009) Imaging of tumor vasculature using Twente photoacoustic systems. J Biophotonics 2:701–717

12. Li M-L, Oh J-T, Xie X et al (2008) Simultaneous molecular and hypoxia imaging of brain tumors in vivo using spectroscopic photoacoustic tomography. Proc IEEE 96 (3):481–489

13. Staley J, Grogan P, Samadi AK et al (2010) Growth of melanoma brain tumors monitored by photoacoustic microscopy. J Biomed Opt 15:040510–040513

14. Sethuraman S, Amirian JH, Litovsky SH, Smalling RW, Emelianov SY (2008) Spectroscopic intravascular photoacoustic imaging to differentiate atherosclerotic plaques. Opt Express 16:3362–3367

15. Ku G, Wang X, Xie X, Stoica G, Wang LV (2005) Imaging of tumor angiogenesis in rat brains in vivo by photoacoustic tomography. Appl Optics 44:770–775

16. Luke GP, Nam SY, Emelianov SY (2013) Optical wavelength selection for improved spectroscopic photoacoustic imaging. Photoacoustics 1:36–42

17. Luke GP, Emelianov SY (2014) Optimization of in vivo spectroscopic photoacoustic imaging by smart optical wavelength selection. Opt Lett 39:2214–2217

18. Levi J, Kothapalli SR, Ma T-J et al (2010) Design, synthesis, and imaging of an activatable photoacoustic probe. J Am Chem Soc 132 (32):11264–11269

19. Luke GP, Myers JN, Emelianov SY, Sokolov KV (2014) Sentinel lymph node biopsy revisited: ultrasound-guided photoacoustic detection of micrometastases using molecularly targeted plasmonic nanosensors. Cancer Res 74:5397–5408

20. De la Zerda A, Zavaleta C, Keren S et al (2008) Carbon nanotubes as photoacoustic molecular imaging agents in living mice. Nat Nanotechnol 3:557–562

21. Wilson K, Homan K, Emelianov S (2012) Biomedical photoacoustics beyond thermal expansion using triggered nanodroplet vaporization for contrast-enhanced imaging. Nat Commun 3:618

22. Hannah A, Luke G, Wilson K, Homan K, Emelianov S (2013) Indocyanine green-loaded photoacoustic nanodroplets: dual contrast nanoconstructs for enhanced photoacoustic and ultrasound imaging. ACS Nano 8:250–259

23. Pan D, Pramanik M, Senpan A et al (2010) Near infrared photoacoustic detection of sentinel lymph nodes with gold nanobeacons. Biomaterials 31:4088–4093

24. Nikoobakht B, El-Sayed MA (2003) Preparation and growth mechanism of gold nanorods (NRs) using seed-mediated growth method. Chem Mater 15:1957–1962

25. Niidome T, Yamagata M, Okamoto Y et al (2006) PEG-modified gold nanorods with a stealth character for in vivo applications. J Control Release 114:343–347

26. Chen YS, Frey W, Kim S et al (2010) Enhanced thermal stability of silica-coated gold nanorods for photoacoustic imaging and image-guided therapy. Opt Express 18:8867–8878

27. Jain PK, Lee KS, El-Sayed IH, El-Sayed MA (2006) Calculated absorption and scattering properties of gold nanoparticles of different size, shape, and composition: applications in biological imaging and biomedicine. J Phys Chem B 110:7238–7248

28. Kumar S, Aaron J, Sokolov K (2008) Directional conjugation of antibodies to nanoparticles for synthesis of multiplexed optical contrast agents with both delivery and targeting moieties. Nat Protoc 3:314–320

29. Joshi PP, Yoon SJ, Hardin WG, Emelianov S, Sokolov KV (2013) Conjugation of antibodies to gold nanorods through Fc portion: synthesis and molecular specific imaging. Bioconjug Chem 24:878–888

30. Chen Y-S, Frey W, Kim S et al (2011) Silica-coated gold nanorods as photoacoustic signal nanoamplifiers. Nano Lett 11:348–354

31. Chen Y-S, Frey W, Aglyamov S, Emelianov S (2012) Environment-dependent generation of photoacoustic waves from plasmonic nanoparticles. Small 8:47–52

32. Jacques SL (2013) Optical properties of biological tissues: a review. Phys Med Biol 58: R37

33. Kim S, Chen Y-S, Luke GP, Emelianov SY (2011) In vivo three-dimensional spectroscopic photoacoustic imaging for monitoring nanoparticle delivery. Biomed Opt Express 2:2540–2550

34. Kim S, Chen Y-S, Luke G, Emelianov S (2014) In-vivo ultrasound and photoacoustic image-guided photothermal cancer therapy using silica-coated gold nanorods. IEEE Trans Ultrason Ferroelectr Freq Control 61:891–897

# Chapter 13

# Dual Wavelength-Triggered Gold Nanorods for Anticancer Treatment

## Dennis B. Pacardo, Frances S. Ligler, and Zhen Gu

## Abstract

Gold nanomaterials with light-responsive properties can be exploited as light-triggered delivery vehicles to enhance the therapeutic efficacy of anticancer drugs. Additionally, different wavelengths of light can be utilized to achieve the combined effects of light-triggered release of therapeutics and light-induced localized heating, which results in improved anticancer efficacy. Herein, we describe methods to develop gold nanorod (AuNR) complexes that provide drug delivery or photothermal therapy when activated by ultraviolet (UV) or near-infrared (NIR) wavelengths of light, respectively. The surface functionalization of AuNRs with three key components is presented. The first component, cyclodextrin, serves to encapsulate drugs of interest. The second component, dextran-phenyl-azo-benzoic acid (DexAzo), serves as a capping agent that undergoes a conformational change upon UV light activation to expose the drugs for release. The third component is a folic acid-based targeting ligand that provides efficient delivery of the AuNR complexes to cancer cells. The dual wavelength activation of these drug-loaded AuNR complexes, which enables one to achieve highly efficient anticancer therapy through the combined effects of UV-triggered drug release and NIR-induced hyperthermia, is also described.

**Key words** Cancer therapy, Drug delivery, Triggered release, Gold nanorods, Photothermal therapy, Combination therapy

## 1 Introduction

Specially designed nanocarriers can provide "on-demand" or triggered drug release, resulting in increased efficacy for cancer treatment [1, 2]. These programmable nanomedicines can be delivered precisely to tumor sites by modifying their surfaces with targeting ligands, which can also enable enhanced intracellular transport [3]. In addition, these drug nanocarriers can be designed to contain stimuli-reactive moieties embedded in their structure that activate the release of cytotoxic agents in response to internal or external cues. For example, nanocarriers can be designed to respond to internal triggers in tumor microenvironments such as acidic pH, enzymatic activity, and low redox potential [4, 5]. Alternatively, external stimuli such as the application of ultrasound,

Sarah Hurst Petrosko and Emily S. Day (eds.), *Biomedical Nanotechnology: Methods and Protocols*, Methods in Molecular Biology, vol. 1570, DOI 10.1007/978-1-4939-6840-4_13, © Springer Science+Business Media LLC 2017

magnetic field, heat, X-ray, or light can generate programmed release of therapeutic cargo for cancer treatment [4, 5]. Designing nanocarriers to response to external stimuli, rather than internal stimuli, offers the advantage that users have both spatial and temporal control over the release of payloads. In addition, nanocarriers can be designed to respond to two or more triggers for synergistic and/or sequential drug delivery, which has been demonstrated to improve therapeutic efficacy [4, 6]. In this chapter, we briefly describe recent advances in the field of cancer nanomedicine with dual stimuli-responsive nanomaterials and then describe methods to prepare one specific formulation that enables both drug delivery and hyperthermia upon activation with two different wavelengths of light.

Dual stimuli-responsive nanomaterials use a combination of two different triggers for site-specific drug delivery [6]; gold nanorods (AuNRs) are a subset of these nanomaterials that offer intrinsic optical and electronic properties suitable for dual-modality function using light of different wavelengths as triggers. AuNRs possess characteristic transverse and longitudinal surface plasmon resonance with absorbance bands in the UV and NIR wavelengths, respectively, enabling laser irradiation at tissue-transparent wavelength regions (700–1100 nm) [7]. The ability of AuNRs to strongly absorb and scatter light was initially utilized for dual-mode imaging via fluorescence and surface-enhanced Raman scattering using different excitation wavelengths of 543 nm and 633 nm, respectively [8]. In another example, AuNRs coated with mesoporous silica shells were functionalized with the photosensitizer hematoporphyrin to combine photodynamic and photothermal therapies in a single treatment [9]. Photodynamic therapy is a treatment that allows a photosensitizer to produce singlet oxygen when exposed to a certain wavelength of light causing damage to targeted cells, whereas photothermal therapy employs the conversion of light energy into heat to destroy targeted cells. Singlet oxygen was generated upon irradiation at 633 nm while NIR-laser treatment at 808 nm produced localized heating for photothermal therapy [9]. This dual-modality photodynamic and photothermal therapy was applied in vivo which resulted in improved cancer therapy as indicated by dramatic decrease in tumor volume [9].

We recently demonstrated that metal-based nanocarriers such as AuNRs could be functionalized to produce dual light-based activation using exposure to UV and NIR wavelengths [10]. The dual-mode AuNR complex, illustrated in Fig. 1, exploits the intrinsic ability of AuNRs to generate heat upon exposure to NIR light to yield photothermal therapy, as well as their ease of surface functionalization to incorporate UV-activated moieties for drug delivery [10]. In this three-component system, the AuNR surface is functionalized with cyclodextrin (CD) for efficient encapsulation

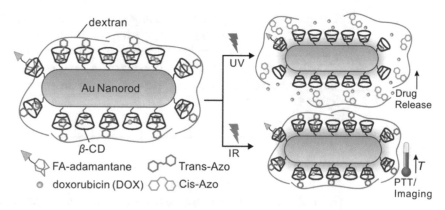

**Fig. 1** Schematic illustration of the dual modality AuNR complex functionalized with cyclodextrin for DOX encapsulation, DexAzo (please define DexAzo) as a capping agent, and Folic Acid (FA)–adamantane as a targeting ligand. Exposure to UV light triggers "on-demand" drug release from the AuNR complex and exposure to NIR light initiates the production of heat by the AuNR complex for targeted photothermal therapy. Reproduced from [10] with permission from the Royal Society of Chemistry

of small molecules and hydrophobic anticancer drugs such as doxorubicin (DOX) through guest–host interaction. The second component of the AuNRs complex is a dextran-phenyl-azo-benzoic acid (DexAzo) moiety, which undergoes *trans*-to-*cis* isomerization upon exposure to UV light. While in the *trans* conformation, the DexAzo serves as a capping agent through its interaction with the CD immobilized on the surface of the AuNR. With UV activation, DexAzo assumes the *cis* conformation, which "uncaps" the CD moiety, thereby triggering the release of DOX. The third component of the dual-mode AuNR system is the folic acid (FA)-based targeting ligand, which enables the nanocarriers to selectively bind and enter targeted cancer cells rather than noncancerous cells. This dual wavelength-activated AuNR complex provides increased therapeutic efficacy in cell studies due to the combined effects of UV-triggered cytotoxic drug delivery and NIR light-induced hyperthermia [7]. For cancer therapy, the multifunctional AuNR complex provides distinct advantages such as biocompatibility, target specificity, and light-activated drug release and heat treatment to induce cancer cell apoptosis [9].

In this chapter, we describe the step-by-step process to generate AuNRs using seed-mediated methods [10–12] and the subsequent surface functionalization of these AuNRs for dual wavelength-triggered drug delivery for cancer therapy. The synthesis of the UV-active DexAzo compound [10, 13] is also described, as well as the procedures to synthesize the folic acid-based targeting ligand. Furthermore, the methods to encapsulate DOX in the AuNR complexes, monitor its UV-triggered release, and enable NIR-induced photothermal therapy are presented in full detail. This AuNRs system represents a novel strategy for cancer therapeutics with high drug-loading capacity due to CD functionalization, specific cancer cell targeting via folic acid ligand, and spatiotemporal control using dual wavelength activation. The ease of AuNRs synthesis and functionalization, as presented in this chapter, showed a promising new approach in anticancer drug delivery.

## 2  Materials

All the solutions are prepared using ultrapure, deionized water (18 mΩ cm,) and analytical grade reagents. All reagents are used as received. The amount of reagents used may be scaled as needed. These procedures require access to fume hood and biosafety cabinet as well as knowledge and training on cell culture protocols.

### 2.1  Preparation of Solutions for CTAB–AuNRs Synthesis

1. 100 mM cetyltrimethylammonium bromide (CTAB) solution: Weigh 3.65 g CTAB (*see* **Note 1**) and transfer to a 250 mL beaker. Add 100 mL water and mix the solution using a mechanical stirrer with gentle heating to approximately 30 °C. The solution should be thoroughly mixed after about 5 min of stirring. Transfer the 100 mM CTAB solution to 50 mL centrifuge tubes. Store in a water bath set to 30 °C (*see* **Note 2**).

2. 10 mM gold metal precursor solution ($Au^{3+}$): Weigh 19.7 mg of $HAuCl_4$ $3H_2O$ in a 5 mL vial. Add 1 mL water and vortex the solution.

3. 10 mM sodium borohydride ($NaBH_4$) solution: Weigh 7.6 mg of $NaBH_4$ in a 20 mL vial and then dissolve by adding 20 mL water. Place the solution in an ice bath for approximately 10 min (*see* **Note 3**).

4. 10 mM silver nitrate ($AgNO_3$) solution: In a 5 mL vial, weigh 8.5 mg of $AgNO_3$ and thoroughly mix with 5 mL water using a vortexer (*see* **Note 4**).

5. 100 mM ascorbic acid solution: Using a 20 mL glass vial, dissolve 17.6 mg of ascorbic acid in 1 mL water. Vortex the solution thoroughly until all the solids are dissolved (*see* **Note 5**).

### 2.2  Functionalization of CTAB–AuNRs

1. 10 mM 11-mercaptoundecanoic acid (MUA) solution: weigh 109 mg MUA in a 50 mL centrifuge tube, then add 50 mL 50:50 ethanol:water ($EtOH:H_2O$) solution. Ensure complete dissolution of the solid MUA using magnetic stirrer.

2. Prepare the following reagents and materials and use as described in the methods section:
   (a) 720 mg (2-hydroxypropyl)-β-cyclodextrin (CD).
   (b) 50 mL N,N′-dimethylformamide (DMF).

(c) 210 mg N,N′-dicyclohexylurea (DCC).

(d) 125 mg 4-dimethylaminopyridine (DMAP).

(e) Polyvinylidene difluoride (PVDF) 45-μm syringe filter.

(f) Regenerated cellulose dialysis tubing with molecular weight cutoff (MWCO) of 12,000 Da.

## 2.3 Synthesis of Dextran-4-Phenyl-Azo-Benzoate (DexAzo) and Targeting Ligand

There is no solution preparation step for this section but the following reagents and materials should be prepared prior to the synthesis of DexAzo and targeting ligand:

1. 500 mg Dextran from *Leuconostoc mesenteroides* (MW 35,000–40,000).

2. 679 mg 4-phenyl-azo-benzoic acid.

3. 15 mL Dimethyl sulfoxide (DMSO).

4. 487.5 mg N,N′-carbonyldiimidazole (CDI).

5. 5 mg Folic acid–polyethylene glycol (PEG)–maleimide (MW 2000).

6. 2.5 mg 1-adamantanethiol.

7. Dialysis tubing benzoylated, MWCO 2000.

## 2.4 Drug Encapsulation and Cell Studies and Instruments for Dual Wavelength Activation of AuNRs

For this process, knowledge of cell culture protocols is required as well as access to biosafety cabinets and incubators. The following reagents and instruments are needed:

1. 10 mg Doxorubicin hydrochloride (DOX).

2. 500 mL Dulbecco's Modified Eagle Medium (DMEM) culture medium containing fetal bovine serum (10% v:v), penicillin (100 U/mL), and streptomycin (100 mg/mL).

3. LIVE/DEAD assay solution containing calcein AM and ethidium homodimer-1.

4. 500 mL phosphate-buffered saline (PBS).

5. Multiplate reader for fluorescence intensity measurements.

6. UV lamp (such as Dymax BlueWave 75).

7. 800 nm NIR Ti-sapphire pulsed laser (such as Coherent, Chameleon, with pulses of 100 fs and 200 MHz).

# 3 Methods

## 3.1 Seed-Mediated Synthesis of CTAB–AuNRs

### 3.1.1 Preparation of Gold Nanoparticle Seeds (AuNPs)

1. In a 15 mL centrifuge tube, transfer 7.5 mL of 100 mM CTAB solution using a pipette.

2. Add 250 μL of the 10 mM of $Au^{3+}$ solution to the CTAB solution using a pipette.

3. Thoroughly mix the solution by gentle inversion of the centrifuge tube approximately ten times.

4. After mixing the CTAB–$Au^{3+}$ solution, immediately add 600 μL of ice-cold 10 mM $NaBH_4$ solution.

5. Gently mix the solution by inverting the centrifuge tube at least ten times.

6. Allow the seed formation to occur by incubating the solution at room temperature for 1 h. The solution will turn from yellow ($Au^{3+}$) to brown ($Au^0$).

*3.1.2  Synthesis of CTAB–AuNRs*

1. Prepare the nanorod growth solution in a 50 mL centrifuge tube by adding 47.5 mL of 100 mM CTAB solution.

2. Add 2.0 mL of 10 mM $Au^{3+}$ to the CTAB solution using a pipette.

3. Gently mix the CTAB-$Au^{3+}$ solution by repeated inversion of the centrifuge tube approximately ten times.

4. Immediately add 300 μL of newly made 10 mM $AgNO_3$ using a pipette and mix the solution by gentle inversion.

5. Add 320 μL of freshly prepared 100 mM ascorbic acid.

6. Invert the centrifuge tube no less than ten times to completely mix all the components of the nanorod growth solution. The solution will change from light yellow to colorless after the addition of ascorbic acid.

7. Incubate the solution at room temperature for 3 h to allow the formation of CTAB–AuNRs. Approximately 5 min after the addition of ascorbic acid, the colorless solution will turn into a purple color indicating the formation of nanorods.

8. After 3 h of incubation, perform UV–visible spectrophotometry to examine the extinction spectrum of CTAB–AuNRs (*see* **Note 6**). The typical UV–vis spectrum for AuNRs is shown in Fig. 2a [14].

9. Purify the generated CTAB–AuNRs by centrifugation at 14,000 × *g* for 30 min. This process will precipitate the nanorods as a pellet.

10. After centrifugation, pipette out or decant the supernatant to separate the newly formed CTAB–AuNRs from the solution containing excess CTAB (*see* **Note 7**).

11. Redisperse the CTAB–AuNR precipitate by adding deionized water to reflect the same total volume as the initial nanorod solution.

12. Perform the centrifugation and washing process two more times to ensure complete removal of excess CTAB in the nanorod solution.

13. The purified CTAB–AuNRs may be immediately used for subsequent surface functionalization. Alternatively, the purified nanorods may be stored in a 4 °C refrigerator for approximately 6 months prior to functionalization.

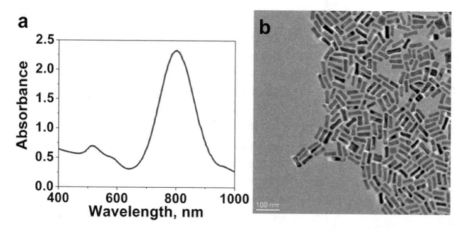

**Fig. 2** Characterization of CTAB–AuNRs using (**a**) UV–vis spectrophotometry showing extinction peaks at 510 nm and 800 nm, corresponding to the transverse and longitudinal surface plasmon bands, respectively; (**b**) representative TEM image of the as-synthesized CTAB–AuNRs. Reproduced from [14] with permission from Springer

14. Verify the formation of the AuNRs using transmission electron microscopy (TEM). Figure 2b [14] shows a representative TEM image of CTAB–AuNRs.

***3.2 Functionalization of CTAB–AuNRs***

The following reactions should be performed in a fume hood.

1. Mix equal volumes of purified CTAB–AuNRs (25 mL) and 10 mM MUA solution in 50:50 EtOH–$H_2O$ (25 mL) in a 50 mL centrifuge tube.

2. In a rotating mixer, mix the solution for 24 h at room temperature to facilitate the ligand exchange reaction.

3. After 24 h, centrifuge the nanorods at 14,000 × g to precipitate out the MUA–AuNRs and to remove the excess ligands in solution. Redisperse the MUA–AuNRs with 50 mL water and repeat the washing process at least two times to remove excess CTAB and MUA.

4. After the last washing and centrifugation process, decant or pipette out the supernatant.

5. Redisperse the precipitated MUA–AuNRs in 50 mL DMF and transfer the solution to a 100 mL round bottom flask.

6. Add 720 mg of CD and dissolve thoroughly by stirring the solution in a magnetic stir plate set at 400 rpm.

7. Immediately add 210 mg of DCC and 125 mg of DMAP to enable the coupling reaction between MUA and CD.

8. Allow the reaction to occur for 24 h at room temperature with constant stirring.

9. After 24 h, isolate the CD–AuRs by centrifugation at 14,000 × *g* for 30 min.

10. Pipette out or decant the supernatant and redisperse the precipitated CD–AuNRs with 50 mL deionized water.

11. Collect the CD–AuNR solution in a syringe and attach a PVDF 45 μm filter to the syringe; filter the sample into a glass vial to remove excess DCC in solution.

12. Use a pipette to transfer the filtered CD–AuNRs to a dialysis membrane (MWCO 12 K) and dialyze the solution against large excess of deionized water (approximately 4 L) for 24 h.

13. Store the purified CD–AuNRs in a refrigerator at 4 °C for future use. The CD–AuNRs can be stored for approximately 6 months.

### 3.3 Synthesis of Capping Agent, Dextran-4-Phenyl-Azo-Benzoate (DexAzo)

The following reactions should be done in a fume hood.

1. In a 50 mL round bottom flask, weigh 500 mg of dextran.

2. Add 15 mL DMSO and stir the mixture for 5–10 min to completely dissolve the dextran powder.

3. Add 679 mg of phenyl-azo-benzoic acid to the dextran solution while stirring constantly.

4. Add 487.5 mg of CDI to the dextran solution.

5. Transfer the reaction mixture to a constant temperature oil bath set at 80 °C.

6. Allow the reaction to proceed for 20 h with constant stirring.

7. After 20 h of reaction, add 200 mL absolute EtOH to precipitate the DexAzo product.

8. Pipette out the supernatant, then wash the precipitated DexAzo with 150 mL EtOH to get rid of unreacted starting materials. Repeat the washing process two more times to further purify the DexAzo product.

9. Transfer the purified DexAzo product to a glass vial and dry for 24 h in a desiccator connected to a vacuum.

10. Store the DexAzo product, a reddish powder, wrapped in aluminum foil, in a drying chamber containing silica gel.

### 3.4 Synthesis of the Targeting Ligand, Folic Acid–PEG–Maleimide–Adamantane (FA)

1. Weigh 5 mg of folic acid–PEG–maleimide in a glass vial and dissolve by adding 1.0 mL DMSO.

2. Add to the same solution 2.5 mg of 1-adamantanethiol.

3. Allow the coupling reaction to occur at room temperature with constant stirring for 24 h (*see* **Note 8**).

4. Transfer the FA solution to a dialysis membrane with MWCO of 2000 Da using a pipette.

5. Dialyze the FA solution against excess amount of deionized water (approximately 4 L) for 24 h with constant stirring at room temperature to remove excess starting materials.

6. Refrigerate the purified FA solution for storage and future use.

### 3.5 Encapsulation of DOX in CD–AuNRs

This section involves the use of pipettes for transferring small volumes of liquids.

1. In a 1.5 mL microcentrifuge tube, transfer 500 μL of purified CD–AuNRs.

2. Add 20 μL of 5 mg/mL DOX solution followed by 10 μL of the purified FA solution and mix the solution thoroughly using a vortexer.

3. Add 100 μL of 1 mg/mL DexAzo solution to act as capping agent for the encapsulated DOX.

4. Incubate the DOX-loaded CD–AuNR solution in the dark by wrapping the microcentrifuge tube with aluminum foil completely.

5. Place the foil-covered sample in a rotating mixer for 24 h at room temperature.

6. After 24 h, remove the foil from the microcentrifuge tube. Centrifuge the sample at 14,000 × g for 30 min to pellet the DOX-loaded CD–AuNRs containing FA and capped by DexAzo.

7. Pipette out the supernatant and redisperse the DOX-loaded AuNRs in 500 μL deionized water.

8. Repeat the centrifugation and washing process (**steps 6** and **7** in Subheading 3.5) twice to ensure removal of excess DOX in the final formulation. After the final centrifugation and supernatant removal, do not add any additional buffer.

9. Store the precipitated DOX-loaded CD–AuNRs at 4 °C for future use.

### 3.6 Demonstration of Dual Wavelength Activation of AuNRs for Drug Release and Photothermal Therapy

#### 3.6.1 UV-Triggered DOX Release

1. Remove the vial containing the DOX-loaded CD–AuNRs obtained in **step 9** of Subheading 3.5 from the refrigerator. Add 1 mL deionized water to the vial to resuspend the DOX-loaded CD–AuNRs.

2. Transfer a 20 μL aliquot of the solution to a 384-well fluorescence plate and measure the initial fluorescence intensity ($I_0$) of DOX before UV irradiation, using 470 nm excitation and 595 nm emission on a plate reader. This measurement records the amount of doxorubicin initially present in the solution due to diffusion (*see* **Note 9**).

3. Focus the UV lamp on the remaining 980 μL solution of DOX-loaded CD–AuNRs and irradiate for 5 s (*see* **Note 10**).

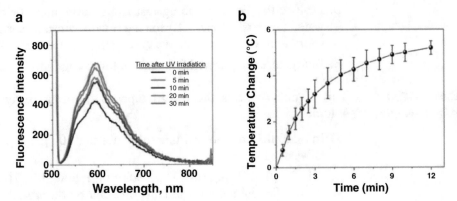

**Fig. 3** (a) The UV-activated release of DOX from CD–AuNRs can be monitored by measuring fluorescence intensity, as shown by the representative graph provided here. As DOX is released from the nanoparticles, its fluorescence becomes unquenched and the signal intensity increases. This figure displays the release of DOX at different time points ranging from 5 min to 30 min after the initial 5 s exposure to UV light. (**b**) The heat generated by AuNRs can be monitored using a digital thermometer. Shown here is a representative increase in temperature upon NIR irradiation. Reproduced from [10] with permission from the Royal Society of Chemistry

4. Remove a 20 μL aliquot from the UV-irradiated solution with a pipette and measure the fluorescence intensity ($I_t$) using 470 nm excitation and 595 nm emission.

5. Continue to monitor the DOX release from the AuNRs by taking 20 μL aliquots of the solution and measuring the fluorescence intensity at different time points: 5, 10, 20, and 30 min, as shown in Fig. 3a [10].

6. For control experiments (no UV treatment), repeat the previous steps except **step 3**.

*3.6.2 NIR-Induced Temperature Increase Using CD–AuNRs (See Note 11)*

1. Laser safety goggles is required to prevent eye damage as a result of exposure to NIR light.

2. Remove a second vial containing DOX-loaded CD–AuNRs (prepared following steps up to **step 9** in Subheading 3.5) from the refrigerator. Add 2 mL deionized water to the purified DOX-loaded CD–AuNRs, redisperse the formulation using a vortexer, and transfer the suspension to a glass vial.

3. Place the vial in a heating chamber set to 37 °C to mimic physiological temperature.

4. Set up the digital thermometer in the solution.

5. With the 800 nm laser off, use the visible wavelength focusing beam to focus the laser beam to the center of the AuNRs solution but away from the thermometer tip.

6. Set the NIR laser to 800 nm, corresponding to the longitudinal surface plasmon band of the AuNRs, and irradiate the solution at 1.5 W/cm$^2$ for 12 min.

7. Monitor the increase in temperature of the solution during the NIR irradiation process using the digital thermometer, as shown in Fig. 3b [10].

**3.7 Evaluating Synergistic Effects of DOX-Loaded CD–AuNRs in Cell Studies**

For this process, knowledge of cell culture protocols is required as well as access to biosafety cabinets and incubators.

1. In a glass-bottom culture dish suitable for confocal microscopy, culture HeLa cells ($1 \times 10^5$ cells/mL) using DMEM culture medium.

2. Incubate for 24 h at 37 °C under an atmosphere of 5% $CO_2$ and 90% relative humidity.

3. After 24 h, remove the culture medium and replace with DOX-loaded AuNR formulation with 2.0 μM DOX concentration dissolved in DMEM culture medium (*see* **Note 12**).

4. Incubate for 4 h at 37 °C under an atmosphere of 5% $CO_2$ and 90% relative humidity.

5. After 4 h, remove the culture medium and replace with fresh DMEM solution.

6. To illustrate the effect of UV irradiation on DOX release, cover the right half of the confocal dish with aluminum foil (dark), as illustrated in Fig. 4a, while leaving the left half uncovered. Place the sample under the UV lamp and irradiate for 5 s.

7. With the right half of the sample dish still covered with aluminum foil, expose the left half of the sample to NIR irradiation for 10 min using 800 nm light.

8. After the dual wavelength treatment, unwrap the foil from the sample, remove the culture medium, and add 1 mL PBS. Rinse gently by tilting the sample. Remove the PBS and repeat the washing procedure twice more.

9. Incubate the HeLa cells with LIVE/DEAD assay solution containing 20 μL EthD-1 stock solution and 5 μL calcein AM stock solution in 10 mL PBS (*see* **Note 13**).

10. Visualize the effects of dual wavelength treatment on HeLa cells using a confocal laser-scanning microscope. The cells on the right half of the sample should display mostly green fluorescence (indicating the cells are alive) whereas the cells on the left half of the sample should display mostly red fluorescence (indicating the cells are dead), as shown in Fig. 4b (*see* **Note 14**).

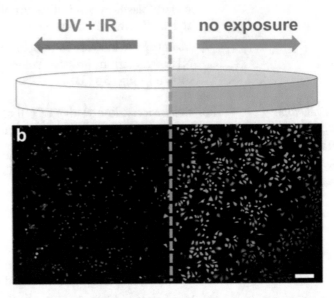

**Fig. 4** (**a**) Schematic illustration of the dual wavelength treatment of HeLa cells incubated with DOX-loaded CD–AuNRs wherein the right half of the sample is covered by aluminum foil (and thus receives no exposure to UV or NIR light), while the left half of the sample is exposed to UV and NIR light treatments. (**b**) Fluorescence microscopy images of live (*green*, not exposed) and dead (*red*, exposed) HeLa cells after dual wavelength treatment. Reproduced from [10] with permission from the Royal Society of Chemistry

## 4  Notes

1. Presence of iodide impurity in commercially available CTAB reagents can prevent nanorod formation [15]. In this regard, we used the CTAB reagent from Sigma with product number H9151.

2. After preparation of CTAB solution, gentle heating using water bath at 30 °C will prevent precipitation of solid CTAB.

3. The reducing agent, $NaBH_4$, should be freshly prepared and stored in ice bath prior to use in the synthesis of AuNRs. $NaBH_4$ reacts rapidly when dissolved in water producing heat in the process. In this regard, the low temperature in an ice bath slows this reaction and allowed the $NaBH_4$ solution to function as a reducing agent when added to gold solution.

4. Prepare the $AgNO_3$ solution right before starting the synthesis of CTAB–AuNRs, since it is prone to degradation upon exposure to light. If prepared ahead of time, the $AgNO_3$ solution should be stored in dark/amber bottle.

5. The ascorbic acid solution should be freshly prepared right before the start of CTAB–AuNRs synthesis to enable the mild reduction of $Au^{3+}$ during the AuNR growth. The solution changes from orange to colorless after the addition of ascorbic acid solution. If the ascorbic acid solution was prepared way before the synthesis, its reducing ability will significantly decrease leading to unsuccessful nanorods synthesis.

6. AuNR formation can be monitored by measuring the extinction spectrum from 300 to 1000 nm with a UV–visible spectrophotometer. AuNRs have two distinct absorbance peaks at 510 nm and 800 nm, indicating transverse and longitudinal plasmon bands, respectively.

7. Purification and removal of excess CTAB is important because free CTAB in solution have cytotoxic effects on cells [16].

8. The folic acid–PEG–maleimide reacts with adamantanethiol via a "click" reaction to form a thiol–maleimide covalent bond.

9. When DOX-loaded CD–AuNRs were resuspended in deionized water, some unencapsulated DOX were released and/or diffused in solution. This was measured initially before UV treatment to accurately determine the DOX released as a result of UV irradiation. The release of DOX upon UV irradiation was due to the "uncapping" of CD as a result of the *trans*-to-*cis* isomerization of DexAzo (Fig. 1). In the *trans* conformation, DexAzo interacts with the CD through the formation of a guest–host complex thereby "capping" the CD with the DOX inside. Upon UV treatment, DexAzo isomerizes to *cis* conformation, which does not interact with CD, thereby "uncapping" the CD and releasing the encapsulated DOX.

10. For your safety, use UV-blocking eye protection goggles when performing UV-triggered release experiments. Furthermore, the use of UV-blocking clothing and gloves are also recommended to prevent skin exposure.

11. This experiment is performed to ensure that the CD–AuNRs produce sufficient heat upon exposure to NIR light to induce photothermal cell death. The NIR-induced localized heating of CD–AuNRs should reach temperatures between 41 and 43 °C to initiate cell death [7].

12. The DOX concentration in CD–AuNRs was determined by first generating a standard calibration curve and linear equation based on Beer's Law using different DOX solutions of known concentration. Then, the absorbance of the DOX-loaded CD–AuNRs was measured and the DOX concentration was calculated from the equation.

13. For the LIVE/DEAD cell assay, the green fluorescence was due to calcein AM, indicating live cells while red fluorescence was due to Etd-1, indicating dead cells.

14. Perform control experiments using only UV light treatment on one cell sample and only NIR light irradiation on another cell sample.

## References

1. Mura S, Nicolas J, Couvreur P (2013) Stimuli-responsive nanocarriers for drug delivery. Nat Mater 12:991–1003

2. Davis ME, Chen Z, Shin DM (2008) Nanoparticle therapeutics: an emerging treatment modality for cancer. Nat Rev Drug Discov 7:771–782

3. Brannon-Peppas L, Blanchette JO (2012) Nanoparticle and targeted systems for cancer therapy. Adv Drug Deliv Rev 64:206–212

4. Pacardo DB, Ligler FS, Gu Z (2015) Programmable nanomedicine: synergistic and sequential drug delivery systems. Nanoscale 7:3381–3391

5. Wang Y, Shim MS, Levinson NS, Sung H-W, Xia Y (2014) Stimuli-responsive materials for controlled release of theranostic agents. Adv Funct Mater 24:4206–4220

6. Cheng R, Meng F, Deng C, Klok H-A, Zhong Z (2013) Dual and multi-stimuli responsive polymeric nanoparticles for programmed site-specific drug delivery. Biomaterials 34:3647–3657

7. Hauck TS, Jennings TL, Yatsenko T, Kumadras JC, Chan WCW (2008) Enhancing the toxicity of cancer chemotherapeutics with gold nanorod hyperthermia. Adv Mater 20:3832–3838

8. Wang Z, Zong S, Yang J, Li J, Cui Y (2011) Dual-mode probe based on mesoporous silica coated gold nanorods for targeting cancer cells. Biosens Bioelectron 26:2883–2889

9. Terentyuk G, Panfilova E, Khanadeev V, Chumakov D, Genina E, Bashkatov A, Tuchin V, Bucharskaya A, Maslyakova G, Khlebtsov N, Khlebtsov B (2014) Gold nanorods with a hematoporphyrin-loaded silica shell for dual-modality photodynamic and photothermal treatment of tumors in vivo. Nano Res 7:325–337

10. Pacardo DB, Neupane B, Rikard SM, Lu Y, Mo R, Mishra SR, Tracy JB, Wang G, Ligler FS, Gu Z (2015) A dual wavelength-activatable gold nanorod complex for synergistic cancer treatment. Nanoscale 7:12096–12103

11. Merrill NA, Sethi M, Knecht MR (2013) Structural and equilibrium effects of the surface passivant on the stability of au nanorods. ACS Appl Mater Interfaces 5:7906–7914

12. Sau TK, Murphy CJ (2004) Seeded high yield synthesis of short au nanorods in aqueous solution. Langmuir 20:6414–6420

13. Wondraczek H, Heinze T (2008) Efficient synthesis and characterization of new photoactive dextran esters showing nanosphere formation. Macromol Biosci 8:606–614

14. Pacardo DB, Neupane B, Wang G, Gu Z, Walker GM, Ligler FS (2015) A temperature microsensor for measuring laser-induced heating in gold nanorods. Anal Bioanal Chem 407:719–725

15. Smith DK, Miller NR, Korgel BA (2009) Iodide in CTAB prevents gold nanorod formation. Langmuir 25:9518–9524

16. Alkilany AM, Nagaria PK, Hexel CR, Shaw TJ, Murphy CJ, Wyatt MD (2009) Cellular uptake and cytotoxicity of gold nanorods: molecular origin of cytotoxicity and surface effects. Small 5:701–708

# Chapter 14

## Photolabile Self-Immolative DNA-Drug Nanostructures

### Xuyu Tan and Ke Zhang

## Abstract

It is often desirable to simultaneously target different cellular pathways to improve the overall efficacy of a drug or to circumvent drug resistance in therapeutic treatments. Nucleic acid therapy has been considered attractive for such combination therapies due to its possible synergistic effects with traditional chemotherapy, especially for targets that do not yet have small molecule inhibitors. However, the co-delivery of nucleic acids and chemotherapeutics typically involves the use of inherently cytotoxic/immunogenic, polycationic carrier systems, for which the benefit is often overshadowed by adverse side effects. Herein, we detail the construction and characterization of a DNA-drug nanostructure that consists almost entirely of payload molecules. Upon triggering with light, the nanostructure collapses via an irreversible, self-immolative process and releases free oligonucleotides, drug molecules, and small molecule fragments. We demonstrate that the nanostructures can be used as a dual-delivery agent in vitro without a carrier system and that the released model drug (camptothecin, CPT) exhibits similar levels of cytotoxicity as unmodified drugs toward cancer cells.

**Key words** Oligonucleotide, Antisense, Combination therapy, Camptothecin, Light activation, Self-immolative, Spherical nucleic acid

## 1 Introduction

Spherical nucleic acids (SNAs), which consist of a densely packed DNA shell surrounding a nanoscopic core (typically gold, but other inorganic and organic nanoparticles can also be used), exhibit many unusual properties compared with their linear and circular counterparts, including significantly enhanced endocytosis [1] and increased nuclease stability [2]. These properties have been demonstrated to stem from the dense packing and orientation of the oligonucleotides on the surface of the nanoparticles, and are core-independent [3]. Therefore, it is possible to take advantage of the unique structure and properties of SNAs as DNA-drug co-delivery vehicles (Scheme 1).

Many anticancer drugs are highly hydrophobic, thus their solubility in aqueous solution and hence their efficacy is limited.

Sarah Hurst Petrosko and Emily S. Day (eds.), *Biomedical Nanotechnology: Methods and Protocols*, Methods in Molecular Biology, vol. 1570, DOI 10.1007/978-1-4939-6840-4_14, © Springer Science+Business Media LLC 2017

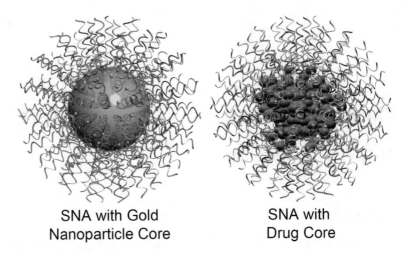

SNA with Gold
Nanoparticle Core

SNA with
Drug Core

**Scheme 1** Schematic representations of a gold-cored SNA and a drug-cored SNA

DNA can be used to improve the solubility of the drug, while the drug helps arrange the DNA into a high-density form that imparts the conjugate structure improved stability and enhanced cellular uptake. Such carrier-free, single-entity agents offer unique benefits including a precise drug loading ratio [4] and spatiotemporally controlled release [5]. They allow for independent access to gene and drug targets, thereby providing a rational approach to take on the challenge of multidrug resistance, which single-target therapeutic strategies cannot adequately address [6]. In a proof-of-concept study, we link therapeutic oligonucleotides with camptothecin (CPT), an anti-cancer drug, via a trivalent photolabile self-immolative linker. The conjugate is amphiphilic, assembling into SNA-like nanostructures with cores that consist almost entirely of CPT in solution under the appropriate conditions. These nanoparticles enter cancer cells and burst-release drug molecules upon light irradiation.

## 2    Materials

Prepare all the solutions and buffers using Nanopure™ water (prepared by purifying deionized water to attain a resistance of at least 18 MΩ cm at 25 °C). Prepare and store all the solutions and buffers at room temperature unless indicated otherwise. All the chemicals listed below are purchased from Sigma-Aldrich Co., Fisher Scientific Inc., or VWR LLC and used directly without purification unless indicated otherwise.

**2.1  Synthesis of CPT-Conjugate 8**

The following chemicals/reagents are required (listed in the order that they appear in Subheading 3):

1. Ethyl 4-hydroxybenzoate.
2. Formaldehyde (37% wt in water).
3. *t*-butyldimethylsilyl chloride.
4. Imidazole.
5. Diisobutylaluminium hydride (1 M in hexane).
6. Filter aid, Celite Hyflo Super-cel®.
7. Lithium hydroxide.
8. 5-hydroxy-2-nitrobenzyl alcohol.
9. 18-crown-6.
10. Propargyl bromide.
11. Triethylamine.
12. Triphosgene.
13. Amberlyst-15 resin.
14. Camptothecin.
15. 4-dimethylaminopyridine.

**2.2  DNA Purification, Modification, and Characterization**

1. DNA purification buffer: 0.1 M triethylammonium acetate (TEAA), pH 7.0. To prepare a 1 M stock solution of TEAA, add about 800 mL of water to a 1-L glass beaker. Weigh 101.2 g of triethylamine and transfer it to the beaker. Weigh 60.1 g of acetic acid and transfer it slowly to the beaker with stirring. Mix and adjust the pH with triethylamine or acetic acid to 7.0. Then, add water to the 1-L mark (*see* **Note 1**). Before HPLC purification, use 100 mL of stock solution to make 1 L of 0.1 M TEAA buffer.
2. DNA modification reagents: Azido-dPEG®$_4$-NHS ester (Quanta Biodesign Co.); 0.1 M sodium bicarbonate aqueous solution.
3. DNase I buffer: 10 mM Tris(hydroxymethyl)aminomethane (Tris base), 2.5 mM $MgCl_2$, 0.5 mM $CaCl_2$, pH 7.5.
4. DNA nuclease: DNase I.
5. MALDI matrix for DNA: 10 mg/mL sinapinic acid in 70/30 (vol%) acetonitrile/water with 0.1% trifluoroacetic acid.

**2.3  Stock Solutions for Click Chemistry (See Note 2)**

1. Solution 1: 10 mg CuBr in 700 μL DMSO.
2. Solution 2: 54 mg tris(benzyltriazolylmethyl)amine (TBTA, *see* **Note 3**) in 1 mL DMSO.
3. Solution 3: 20 mg sodium ascorbate in 1 mL $H_2O$.
4. Premixed solution 4: Mix solutions 1 and 2 in a 1:2 (vol:vol) ratio.

**2.4  PBS Buffers**

1. Phosphate buffered saline ($1\times$ PBS): 138 mM NaCl, 2.7 mM KCl, 10 mM $Na_2HPO_4$, 1.8 mM $KH_2PO_4$, pH 7.4.

2. $0.05\times$ PBS: 6.9 mM NaCl, 0.135 mM KCl, 0.5 mM $Na_2HPO_4$, 0.09 mM $KH_2PO_4$.

3. $0.05\times$ PBS with 2 mM $MgCl_2$: Measure 0.2 mL of 1 M $MgCl_2$ solution and dilute to 100 mL with PBS $0.05\times$ solution.

**2.5  Agarose Gel Electrophoresis**

1. DNA loading buffer: 30% glycerol solution in water (*see* **Note 4**).

2. Tris-borate-EDTA buffer ($0.5\times$ TBE): 44.5 mM Tris base, 44.5 mM boric acid, 1 mM EDTA.

3. 1% agarose gel with TBE buffer ($0.5\times$): Weigh 0.4 g of agarose into a 200-mL Erlenmeyer flask, and add TBE buffer ($0.5\times$) to give a total weight of 40 g. Microwave the mixture at max power (800–1000 W) for 1 min and pour the hot solution into the cast to solidify.

4. 5% agarose gel with $0.05\times$ PBS buffer: Weigh 2.0 g of agarose in a 200-mL Erlenmeyer flask, and add $0.5\times$ TBE buffer to give a total weight of 40 g. Microwave the mixture at max power (800–1000 W) for 1 minute and pour the hot solution into the cast to solidify (*see* **Note 5**).

# 3  Methods

Carry out all the procedures at room temperature unless indicated otherwise. Nanopure™ water is used for all the solutions and buffers.

**3.1  Synthesis of Self-Immolative Precursor 4 (See Note 6, See Scheme 2)**

1. To 70 mL of cooled 12% NaOH solution (0.21 mol, 2.33 eq.), add 15 g of ethyl 4-hydroxybenzoate (0.09 mol, 1 eq.) and stir under cooling by ice water. Add 60 mL of 37% formaldehyde in water (0.74 mol, 8.22 eq.). Continue to stir the mixture at 50–55 °C for 2–3 days (*see* **Note 7**) and monitor the reaction by TLC (Hex:EtOAc=2:3). Quench the reaction by adding 210 mL of 1 M HCl solution, and extract the product three times using 100 mL of ethyl acetate. Combined solvent is removed after drying over anhydrous $Na_2SO_4$. Purify the crude product using silica gel column chromatography (Hex: EtOAc=4:1 to 2:1) to give **1** as a white solid (5.85 g, 28.6%). $^1$H–NMR (400 MHz, d-acetone): δ 7.85 (s, 2 H), 4.82 (s, 4 H), 4.26–4.31 (q, 2 H), 1.31–1.34 (t, 3 H). $^{13}$C–NMR (400 MHz, d-acetone): δ 166.06, 157.95, 129.31, 128.42, 125.04, 64.82, 60.36, 14.08.

2. Weigh 3.00 g of compound **1** (13.27 mmol, 1 eq.), 7.20 g of tert-butyldimethylsilyl chloride (TBSCl, 47.77 mmol, 3.60 eq.), and 3.25 g of imidazole (47.74 mmol, 3.60 eq.)

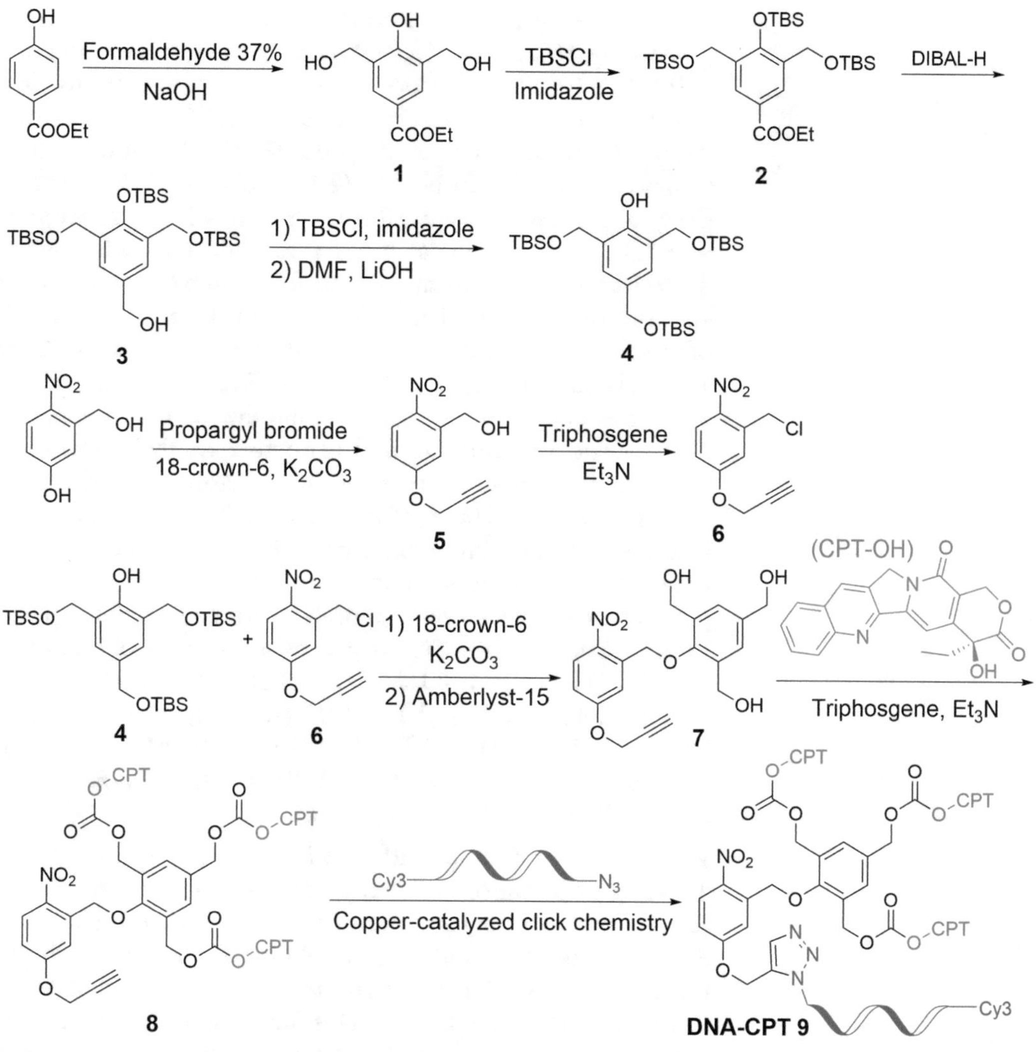

**Scheme 2** Synthetic sequence of DNA-CPT conjugate

and transfer them into a 50-mL round-bottom flask. Add 20 mL of N,N-dimethylformamide (DMF) to the flask and stir the mixture for 1–2 days (*see* **Note 8**). Monitor the reaction by TLC (Hex:EtOAc=19:1), and then quench it by adding 20 mL of water to the mixture. Extract the product three times using 30 mL of hexane. Combined solvent is removed after drying over anhydrous $Na_2SO_4$. Purify the crude product by silica gel column chromatography (Hex:EtOAc=80:1) to give **2** as a colorless oil (6.30 g, 83.5%). $^1$H–NMR (400 MHz, CDCl$_3$): δ 8.10 (s, 2 H), 4.71 (s, 4 H), 4.32–4.34 (q, 2 H), 1.35–1.38 (t, 3 H), 1.02 (s, 9 H), 0.95 (s, 18 H), 0.19 (s, 6 H), 0.09 (s, 12 H). $^{13}$C–NMR (400 MHz, CDCl$_3$): δ 166.97, 152.45, 132.14, 128.08, 124.19, 60.70, 26.22, 26.15, 19.08, 18.62, 14.48, −3.04, −5.05.

3. Dissolve 3.98 g of compound **2** (7 mmol, 1 eq.) in 10 mL of dry tetrahydrofuran (*see* **Note 9**), and transfer the solution to a 100-mL Schlenk flask. Cool the solution to −78 °C using a slurry of dry ice and acetone. While the flask is connected to a stream of flowing nitrogen gas, add 28 mL of 1 M diisobutylaluminium hydride in hexane (28 mmol, 4 eq.) dropwise to the Schlenk flask using a 10-mL syringe. Stir the mixture at −78 °C for 2 h. To quench the reaction, add 1.12 mL of water dropwise to the mixture at −78 °C, followed by the addition of 1.12 mL of 15% NaOH aqueous solution and another 2.8 mL of water. Slowly warm the mixture to room temperature. After an emulsion is formed, add 15 g of Filter aid, Celite Hyflo Super-cel®, and continue to stir for another 30 min. Then, add 30 mL of hexane, and transfer the mixture evenly into two 50-mL centrifuge tubes. Vortex the tubes vigorously and centrifuge at a speed of $1100 \times g$ for 10 min. Collect the hexane layer and repeat the extraction step three times. Combined solvent is removed after drying over anhydrous $Na_2SO_4$. Purify the crude product by silica gel column chromatography (Hex: EtOAc=80:1 to 20:1) to give **3** as a colorless oil (3.30 g, 89.5%). $^1$H–NMR (400 MHz, CDCl$_3$): δ 7.37 (s, 2 H), 4.71 (s, 4 H), 4.64 (s, 2 H), 1.02 (s, 9 H), 0.94 (s, 18 H), 0.16 (s, 6 H), 0.09 (s, 12 H). $^{13}$C–NMR (400 MHz, CDCl$_3$): δ 147.89, 134.33, 132.10, 125.29, 65.95, 60.85, 26.25, 26.21, 19.00, 18.68, −3.11, −5.03.

4. Weigh 1.36 g of compound **3** (2.58 mmol, 1 eq.), 0.43 g of TBSCl (2.85 mmol, 1.1 eq.), and 0.195 g of imidazole (2.86 mmol, 1.1 eq.) and transfer them to a 20-mL glass vial. Add 4 mL of DMF and stir the solution for 4 h. Quench the reaction by adding 8 mL of water and extract the product three times using 10 mL of hexane. Dry the combined organic layer over anhydrous $Na_2SO_4$ and remove the solvent under reduced pressure. Dissolve the crude mixture in 4 mL of DMF. Add 0.19 g of LiOH (7.93 mmol, 3 eq.) to the solution and stir the mixture overnight [7]. Quench the reaction by adding 8 mL of saturated $NH_4Cl$ aqueous solution and extract the product three times using 10 mL of hexane. Combined solvent is removed after drying over anhydrous $Na_2SO_4$. Purify the crude product by silica gel column chromatography (Hex:EtOAc=80:1 to 60:1) to give **4** as a colorless oil (0.95 g, 69.8%). $^1$H–NMR (400 MHz, CDCl$_3$): δ 8.16 (s, 1 H), 7.06 (s, 2 H), 4.84 (s, 4 H), 4.63 (s, 2 H), 0.93 (s, 27 H), 0.11 (s, 12 H), 0.08 (s, 6 H). $^{13}$C–NMR (400 MHz, CDCl$_3$): δ 152.37, 132.29, 125.99, 124.08, 65.27, 63.28, 26.26, 26.12, 18.70, 18.54, −4.92, −5.19.

### 3.2 Synthesis of the Photo-Cleavable Precursor 6 (See Scheme 2)

1. Weigh 1.015 g of 5-hydroxy-2-nitrobenzyl alcohol (6 mmol, 1 eq.), 0.911 g of $K_2CO_3$ (6.6 mmol, 1.1 eq.), and 0.087 g of 18-crown-6 (0.33 mmol, 0.055 eq.) and dissolve them in 10 mL of anhydrous DMSO. Stir the mixture at 70 °C for 30 min, then add 500 μL of propargyl bromide (6.6 mmol, 1.1 eq.). Monitor the reaction by TLC (Hex:EtOAc=4:1). After 4 h, quench the reaction with 13.2 mL of 1 M HCl, and add 7 mL of water. Extract the product three times using 20 mL of ethyl acetate. Combined solvent is removed after drying over anhydrous $Na_2SO_4$. Purify the crude product by silica gel column chromatography (Hex:EtOAc=4:1 to 2:1) to give **5** as a white solid (1.039 g, 81.6%). $^1H$–NMR (400 MHz, d-acetone): δ 8.16–8.18 (d, 1 H), 7.55–7.56 (d, 1 H), 7.08–7.11 (dd, 1 H), 5.02–5.03 (d, 2 H), 4.97–4.98 (d, 2 H), 4.70–4.71 (m, 1 H), 3.17–3.19 (t, 1 H). $^{13}C$–NMR (400 MHz, d-acetone): δ 162.23, 142.42, 140.65, 127.47, 114.03, 113.24, 78.09, 77.27, 61.18, 56.27.

2. Weigh 0.215 g of compound **5** (1.04 mmol, 1 eq.) and measure 160 μL of triethylamine (1.14 mmol, 1.10 eq.), and transfer them to a 20-mL glass vial containing 4 mL of dry dichloromethane (DCM). After the solids are dissolved, rapidly add 0.112 g of triphosgene (1.13 mmol, 1.09 eq.) to the solution, and seal the vial immediately by capping. Allow the mixture to stir for 30 min. Confirm the completion of the reaction by TLC (Hex:EtOAc=4:1). Remove the solvent under reduced pressure. Purify the crude product by silica gel column chromatography (Hex:EtOAc=7:1) to give **6** as a pale yellow oil (0.200 g, 85.5%). $^1H$–NMR (400 MHz, $CDCl_3$): δ 8.12–8.14 (d, 1 H), 7.25–7.26 (d, 1 H), 6.99–7.02 (dd, 1 H), 4.98 (s, 2 H), 4.79–4.80 (d, 2 H), 2.59–2.60 (t, 1 H). $^{13}C$–NMR (400 MHz, $CDCl_3$): δ 161.60, 141.46, 135.60, 128.30, 117.79, 114.70, 77.66, 77.02, 56.61, 43.80.

### 3.3 Synthesis of Light-Triggered, Self-Immolative Linker 7 (See Scheme 2)

1. Weigh 0.150 g of compound **4** (0.28 mmol, 1 eq.) and 0.120 g of compound **6** (0.53 mmol, 1.89 eq.), and dissolve them in 2 mL of anhydrous DMSO, followed by the addition of 80 mg of $K_2CO_3$ (0.58 mmol, 2.07 eq.) and 10 mg of 18-crown-6 (0.038 mmol, 0.14 eq.). Allow the reaction mixture to stir at 100 °C for 2 h. Monitor the reaction by TLC (Hex:EtOAc =19:1 and 7:1). Thereafter, add 2 mL of water to the reaction mixture, and extract the product three times with 5 mL of diethyl ether. Combined solvent is removed after drying over anhydrous $Na_2SO_4$. Dissolve the crude product in 5 mL of methanol and add an excess amount of Amberlyst-15 resin. Stir the suspension overnight and filter the resin. Remove the solvent and purify the crude product by silica gel column chromatography (Hex:EtOAc=1:2) to give **7** as a white solid

(26.0 mg, 24.4%). $^1$H–NMR (400 MHz, d-acetone): δ 8.25–8.27 (d, 1 H), 7.80–7.81 (d, 1 H), 7.43 (s, 2 H), 7.18–7.22 (dd, 1 H), 5.42 (s, 2 H), 5.04–5.05 (d, 2 H), 4.70–4.71 (m, 4 H), 4.62–4.64 (m, 2 H), 3.22–3.23 (t, 1 H); hydroxyl protons (3 H): 4.12–4.18 (m) and 2.80–2.84 (m). $^{13}$C–NMR (400 MHz, d-acetone): δ 162.33, 153.08, 140.31, 138.78, 138.21, 135.10, 127.62, 126.71, 113.99, 113.87, 78.04, 77.45, 72.82, 63.88, 63.75, 59.38, 59.25, 56.42.

### 3.4 Synthesis of the DNA-CPT Conjugate 9 (See Scheme 2)

1. Weigh 7.6 mg of CPT (21.8 μmol), 11.4 mg of 4-dimethylaminopyridine (DMAP, 93.3 μmol), and 3.7 mg of triphosgene (12.5 μmol), and dissolve them in 1.5 mL of anhydrous DCM. Allow the solution to stir for 15 min. Thereafter, add 2.6 mg of linker 7 (7.0 μmol) to the reaction mixture and gently shake the solution overnight. Dry the solution in vacuo and suspend the solids in acetonitrile/H$_2$O (1:1 vol:vol) under sonication. Remove insoluble components by centrifugation at a speed of 275 × $g$ for 5 min. Collect the supernatant, remove insoluble particles using a 0.2 micron nylon filter, and use reverse-phase HPLC (50%/50% acetonitrile/H$_2$O to 100% acetonitrile in 30 min) to isolate the CPT-conjugate 8 (ca. 30% yield, *see* **Note 10**).

2. Measure 0.28 μmol of amino-modified DNA (*see* **Note 11**) using a UV-vis spectrophotometer (*see* **Note 12**), and dilute the solution in 500 μL of 0.1 M sodium bicarbonate solution. Dissolve 1.4 mg of Azido-dPEG®$_4$-NHS ester (Quanta Biodesign Co., 3.6 μmol) in 36 μL of acetonitrile. Mix the Azido-dPEG®$_4$-NHS ester solution with the DNA solution. Shake the mixture for 2 h, and remove excess NHS ester and unreacted DNA by reverse-phase HPLC (100% H$_2$O to 100% acetonitrile in 48 min).

3. Weigh 0.2 mg of compound 8 (0.13 μmol) and dissolve it in 90 μL of DMSO, to which add 50 μL of azide-modified DNA aqueous solution (5–25 OD). After brief stirring, add 10 μL of premix solution 4 to the reaction mixture. Then, add 50 μL of solution 3. Shake the mixture on a thermomixer at 40 °C for 1–1.5 h (*see* **Note 13**). Thereafter, remove the solvent in vacuo. Dissolve the crude product in water, filter it with a 0.2 μm nylon syringe filter, and isolate the conjugated product 9 by reverse-phase HPLC (from 100% H$_2$O to 100% acetonitrile in 48 min) (*see* **Note 14**).

### 3.5 Characterization of DNA-CPT Nanoparticles

1. To form DNA-CPT nanoparticles, add PBS with 5 mM MgCl$_2$ to a vial containing lyophilized conjugate powder to give a final concentration of 5 OD/mL (*see* **Note 15**). To visualize the morphology of the particles, transmission electron microscopy

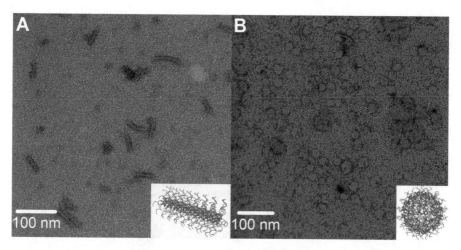

**Fig. 1** TEM images of (**a**) DNA-CPT SNAs made with a short DNA strand (NH$_2$-TTTTT-Cy3) in PBS, and (**b**) DNA-CPT SNAs having an anti-HER2 strand (20-mer) in PBS buffer with 5 mM MgCl$_2$

(TEM) is used. Deposit 4 µL of sample solution (5 OD/mL) on formvar-coated copper grids. After 1.5 min, remove excess sample by wicking using filter paper. Stain the sample by depositing 4 µL of 2% uranyl acetate onto the grid. Allow the stain to stay for 1 min, then wick excess stain with filter paper. Image the sample using TEM with an acceleration voltage of 80 kV (*see* **Note 16**, *see* Fig. 1).

2. To obtain a quantitative relationship of the optical density (OD) at 260 nm and the concentration of the conjugate, measure 200 µL of DNA-CPT nanoparticle solution (1 OD/mL) and treat with prolonged UV light irradiation (>1 h, 10 mJ·s$^{-1}$ cm$^{-2}$, 365 nm) using a handheld UV lamp, which leads to complete decomposition of the conjugate. Subject the UV-treated solution to HPLC analysis, and quantify the peak correlating to the CPT using a standard curve generated with free CPT of known concentrations (*see* **Note 17**).

3. To obtain a quantitative drug release profile upon light triggering, prepare five vials of 10 µL of DNA-CPT solution in PBS buffer (31 nmol/mL) and expose each vial to UV light (365 nm, 10 mJ/s·cm$^2$) for different durations of time (0, 1, 2, 4, and 8 min, respectively). Thereafter, dilute the samples by adding an additional 90 µL of PBS buffer, and inject the samples into reverse-phase HPLC for analysis (*see* **Note 18**). The area under the curve for the peak correlating to the untreated conjugate (measured at 350 nm) is regarded as 0% degraded. The decomposition is determined by measuring the reduction of the conjugate peak for samples treated with UV light. Repeat the measurements three times and average the results (Fig. 2).

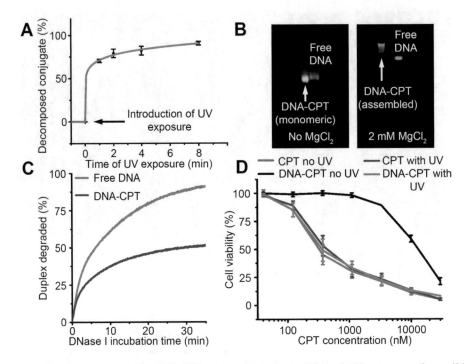

**Fig. 2** (**a**) Release kinetics of the DNA-CPT conjugate as a function of UV exposure times (365 nm, 10 mJ·s$^{-1}$ cm$^{-2}$). (**b**) Agarose gel electrophoresis images showing the effect of Mg$^{2+}$ on the self-assembly of DNA-CPT. (**c**) Kinetics of the enzymatic degradation of DNA-CPT SNAs and free DNA duplexes (100 nM). (**d**) Cytotoxicity measurement of CPT and DNA-CPT SNAs in the absence or presence of UV activation

4. To study the effect of Mg$^{2+}$ ions on the self-assembly of the DNA-drug conjugate to form micellar nanoparticles, first make a 5% agarose gel with 0.05× PBS buffer (*see* **Note 19**). Electrophorese the DNA-CPT conjugate and the corresponding free dye-labeled DNA (~1 OD/mL) in 0.05× PBS buffer with a voltage of 100 V for 30 min (*see* **Note 20**). Repeat the experiment in the presence of 2 mM MgCl$_2$ with a voltage of 100 V for 60 min (*see* **Note 21**). Acquire multiplex gel images using two separate channels (*see* **Note 22**). If the conjugate assembles into micellar nanoparticles, its rate of gel migration should be much slower compared with free DNA (Fig. 2).

5. To study the nuclease stability of the DNA-CPT conjugates, a reported method has been adopted [2]. Synthesize a complementary strand to the conjugate bearing a dabcyl quencher. Add an excess of the complementary dabcyl-DNA (1.5 eq.) to preformed DNA-CPT nanostructures (containing an equivalent of 1000 nM free DNA) or free Cy3-DNA (1000 nM) in DNase I buffer (*see* **Note 23**). Gently shake the solutions overnight. Thereafter, withdraw aliquots of the samples and dilute into 1 mL of 100 nM solutions using DNase I buffer.

Next, place the diluted sample into a fluorimeter, rapidly add 0.1 unit of DNase I and mix thoroughly, and immediately start to monitor Cy3 fluorescence every 3 s for 1 h (excitation: 512 nm, emission: 570 nm). Repeat the measurements three times and average the results. The micellar nanoparticles should exhibit enhanced nuclease resistance (Fig. 2).

6. To test the cell-killing efficacy of the DNA-CPT nanoparticles, first culture SK-BR-3 cells in RPMI 1640 medium supplemented with 10% FBS and 1% Antibiotic-Antimycotic at 37 °C in 5% $CO_2$ atmosphere. To assess the cytotoxicity of DNA-CPT nanostructures, the 96-well plate was first seeded with a density of 5000 cells/well, and incubate cells with different concentrations of DNA-CPT nanoparticles (40–30,000 nM conjugated CPT. control: free CPT) for 6 h after the following day. Subject cells to UV irradiation ($10 \ mJ \cdot s^{-1} \ cm^{-2}$, 365 nm) for 4 min (control: no UV irradiation). Check cell viability 72 h postirradiation by MTT cytotoxicity assay. Repeat the measurements three times and average the results (Fig. 2).

# 4  Notes

1. 1 M TEAA stock solution can be stored at room temperature for months.

2. To obtain best yields, it is important to prepare fresh CuBr and sodium ascorbate stock solutions immediately prior to use. TBTA and premixed solution 4 can be stored at −20 °C for months.

3. We have investigated both TBTA and tris(3-hydroxypropyltriazolylmethyl)amine (THPTA) as the ligand, and we find that TBTA gives better yields for this click reaction.

4. Commercial DNA loading buffer is often colored, which affects the gel imaging. A glycerol aqueous solution (30 wt%) is used directly as the loading buffer.

5. Agarose solution (5%) solidifies very quickly (in 3–5 min), and therefore it is important that it be poured into the cast immediately after microwaving.

6. The synthesis of the self-immolative precursor 4 is adapted from the literature report [8]. In every TBS-protection step, an excess of TBSCl is used to achieve maximized yield, which leads to the overprotection of the phenolic hydroxyl group. The phenolic TBS group can be selectively removed using a reported method [7].

7. This reaction is temperature sensitive. At low temperatures, the reaction is too slow, but increasing the temperature leads to the increased rates of side reactions. This results in the poor yield of this reaction. Monitor the reaction regularly using TLC to determine when to stop it.

8. It takes more time to protect the phenolic hydroxyl group.

9. It is critical to use dry THF since water can react with DIBAL-H and deactivate the reagent.

10. The yield is calculated from the HPLC chromatogram. The area-under-the-curve correlating to the product is about 30% of the overall area (measured at 350 nm where CPT absorbs).

11. The conjugation strategy is general with regard to the oligonucleotide sequence. We have used an anti-HER2 antisense sequence: $5'$-$NH_2$-TTTTTCTCCATGGTGCTCAC-Cy3-$3'$ (the 5 T sequence serves as a spacer). A $5'$ amino modifier is used to attach a terminal azide group through amidation chemistry. The choice for the drug component has two requirements: (1) the drug must be highly hydrophobic, and (2) the drug must have a nucleophilic functional group ($-OH$, $-NH_2$, etc) for coupling with a self-immolative linker. Many common drugs, such as paclitaxel and CPT, satisfy these requirements.

12. For low-volume (1–2 μL) measurements, a Nanodrop™ instrument is suggested.

13. The reaction mixture is cloudy at the beginning. As the reaction proceeds, the solution becomes clear, which is an indication that a reaction has occurred.

14. Surprisingly, the reaction yield is improved while the solvent is removed in vacuo using a CentriVap™ concentrator.

15. We find that DNA-CPT conjugate is able to self-assemble in the presence of 2–5 mM $Mg^{2+}$, which may act as a bridging to stabilize the nanostructure because it may reduce the electrostatic repulsion between DNA strands.

16. The size and shape of the micellar nanoparticles are dependent upon the length of the DNA and the concentration of divalent ions (e.g., $Mg^{2+}$, $Ca^{2+}$).

17. The amount of CPT cannot be accurately determined in the conjugate form since the extinction coefficient of conjugated CPT is different from molecular CPT. Therefore, to precisely quantify the conjugate, free CPT is released from the conjugate by prolonged UV treatment. Given the DNA strand used (TTTTTCTCCATGGTGCTCAC-Cy3), 9.3 nmol of CPT and 3.1 nmol of DNA are present in 1 OD (260 nm) of DNA-CPT (1 mL).

18. Additional PBS buffer is added to minimize the error caused by the syringe injection.

19. TBE buffer is not used since EDTA chelates with $Mg^{2+}$. Thus, $0.05 \times$ PBS is used as the running buffer since the electrical conductivity of $1 \times$ PBS buffer is too high, which causes overheating and melting of the gel.

20. Intermittent cooling is required to prevent melting of the gel.

21. All sample solutions and gel and electrophoresis solutions should contain 2 mM $MgCl_2$. We also find that longer running times are necessary because both the DNA and conjugate migrate much slower under these conditions.

22. Use following settings for the image acquisition: Channel 1 (CPT): excitation: Trans-UV, emission: Cy2; Channel 2 (free Cy3-DNA): excitation: Cy3, emission: Cy3.

23. The preformed DNA-CPT nanostructure is prepared without additional magnesium salt since the DNase I buffer already contains divalent ions (2.5 mM $MgCl_2$ and 0.5 mM $CaCl_2$).

## References

1. Giljohann DA, Seferos DS, Mirkin CA et al (2007) Oligonucleotide loading determines cellular uptake of DNA-modified gold nanoparticles. Nano Lett 7:3818–3821

2. Cutler JI, Zhang K, Mirkin CA et al (2011) Polyvalent nucleic acid nanostructures. J Am Chem Soc 133:9254–9257

3. Cutler JI, Auyeung E, Mirkin CA (2012) Spherical nucleic acids. J Am Chem Soc 134:1376–1391

4. Cheetham AG, Zhang P, Cui H et al (2013) Supramolecular nanostructures formed by anticancer drug assembly. J Am Chem Soc 135:2907–2910

5. Chan JM, Zhang L, Farokhzad OC et al (2010) Spatiotemporal controlled delivery of nanoparticles to injured vasculature. Proc Natl Acad Sci U S A 107:2213–2218

6. Wang Y, Gao S, Yang YY et al (2006) Co-delivery of drugs and DNA from cationic core–shell nanoparticles self-assembled from a biodegradable copolymer. Nat Mater 5:791–796

7. Ankala VS, Fenteany G (2002) Selective deprotection of either alkyl or aryl silyl ethers from aryl, alkyl bis-silyl ethers. Tetrahedron Lett 43:4729–4732

8. Haba K, Popkov M, Shabat D et al (2005) Single-triggered trimeric prodrugs. Angew Chem Int Ed Engl 44:716–720

# Chapter 15

# Enzyme-Responsive Nanoparticles for the Treatment of Disease

## Cassandra E. Callmann and Nathan C. Gianneschi

## Abstract

Nanomedicine for cancer therapy seeks to treat malignancies through the selective accumulation of therapeutics in diseased tissue. Nanoparticles offer the convenience of high drug loading capacities and can be readily decorated with targeting moieties, drugs, and/or diagnostics. Our lab has pioneered a new tissue targeting strategy where enhanced accumulation of nanomaterials occurs as a result of morphology changes to the material in response to overexpressed enzymes in diseased tissues. Herein, we describe the general strategy for the preparation of these enzyme-responsive nanoparticles (ER-NPs) for therapeutic applications.

Key words Nanoparticles, Enzyme-responsive nanoparticles, Self-assembly, Nanomedicine

## 1 Introduction

A central objective of nanomedicine is to expand the therapeutic window of small molecule drugs by packaging them in nanoscale carriers that shield them from the body until they reach their intended target, where the payload is then released. Overall, the goal is to increase the efficacy of the therapeutic, while decreasing its toxicity associated with off-target accumulation. Toward this end, vast arrays of nanoparticle systems have been developed, based on both passive accumulation methods, such as the enhanced permeability and retention (EPR) effect in tumors [1–3], and active accumulation mechanisms, such as receptor-ligand association [4–6], pH gradients [7–9], redox processes [10–12], and exogenous stimuli [13, 14].

The Gianneschi lab has developed a new type of nanoparticle system that is designed to respond to enzymatic signals endogenous to diseased tissue [15–20]. The system utilizes an active accumulation mechanism that is distinctly different from conventional methods. Instead of relying on the inherent $K_D$ of receptor-ligand binding to accumulate material at disease sites, we instead take

Sarah Hurst Petrosko and Emily S. Day (eds.), *Biomedical Nanotechnology: Methods and Protocols*, Methods in Molecular Biology, vol. 1570, DOI 10.1007/978-1-4939-6840-4_15, © Springer Science+Business Media LLC 2017

advantage of the kinetics of enzymatic action and catalytic amplification to retain materials. Specifically, the materials are designed to respond to matrix metalloproteinases (MMPs), which are proteases upregulated in the progression of many inflammatory diseases, including certain cancers [21–24] and post-myocardial infarction [25–28]. Our materials, herein referred to as enzyme-responsive nanoparticles (ER-NPs), are functionalized with MMP-responsive peptides on the nanoparticle shell. When exposed to MMPs, the peptides are cleaved at their recognition sequences. This cleavage disrupts the nanoparticle structure and induces a nano- to micro-scale morphology change in the ER-NPs. For in vivo applications, this translates to long retention times (on the order of days to weeks) in the tissue of interest, making it possible to deliver therapeutic cargo [15] or imaging agents [16, 17, 20].

A key feature of our materials is that they can be prepared with a multitude of functionalities in a well-controlled manner in high fidelity. This is achieved, in part, by preparing the polymer building blocks for particle assembly via Ring Opening Metathesis Polymerization (ROMP). ROMP is a living polymerization technique [29–34], where initiators are capable of producing well-defined, highly reproducible block copolymers with exceptionally low dispersities. Because these materials are generated with ROMP, one can precisely control block lengths and can incorporate multiple functional moieties in a single system. Figure 1 shows a general scheme for how this is accomplished. With this methodology one can polymerize peptides, hydrophobic drugs, and diagnostic agents, all of which are key components of the ER-NP systems. The utility of the approach makes it easy to envision expanding this technology to include other high-value molecular cargo.

Irrespective of application, the enzyme responsive nanoparticles (ER-NPs) follow a common synthetic scheme in their formation and are programmed to respond following exposure to MMPs (Fig. 1). In general, amphiphilic block copolymers (Fig. 1a) containing the appropriate cargo (drug or diagnostic agent) and MMP-responsive peptides are generated via ROMP, which assemble into nanoparticles upon dialysis into aqueous medium from a suitable organic solvent (Fig. 1b). Once assembled, the nanoparticles are characterized via DLS and TEM and assayed for susceptibility to MMPs (Fig. 1c). Finally, the nanoparticles are investigated in their intended applications. Herein, we present a general strategy for the preparation of ER-NPs for therapeutic applications.

## 2   Materials

### 2.1   Peptides and Peptide-based ROMP Monomers

1. Fmoc-protected amino acids (L- and D-versions) for solid phase peptide synthesis (see **Note 1**).

2. Peptide synthesis resins (*see* **Note 2**).

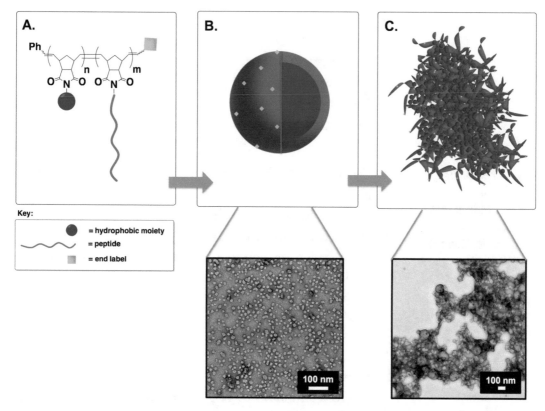

**Fig. 1** General Functional Scheme of ER-NPs. (**a**) Generalized ROMP peptide polymer amphiphile structure, where the hydrophobic moiety may be a drug or inert material, the peptide is MMP-responsive, and the end-label may be a diagnostic agent. (**b**) Upon dialysis, the peptide polymer amphiphiles assemble into nanoparticles, with the hydrophobic moiety in the core and peptide (and diagnostic agent, if used) coating the outside. Shown is a representative TEM image of ER-NPs synthesized by our lab. (c) When exposed to MMPs, the ER-NPs undergo a drastic morphology change. Shown is a representative TEM image depicting such a change. TEM images reproduced with permission from [15]

3. 20% methyl piperidine in DMF for Fmoc deprotection.

4. 0.4 M HBTU in DMF.

5. 1 M DIPEA in DMF.

6. *N*-(Glycine)-*cis*-5-norbornene-*exo*-2,3-dicarboximide [35].

7. Resin-cleavage cocktail solution (*see* **Note 2**).

8. Cold diethyl ether.

*2.2 Ring Opening Metathesis Polymerization: Direct Method (see Note 3)*

1. Catalyst—(IMesH$_2$)(C$_5$H$_5$N$_2$)(Cl)$_2$Ru=CHPh [33].

2. Peptide monomer.

3. Hydrophobic drug monomer [15].

4. Dye-labeled termination agents [36].

5. Dry DMF.

6. Ethyl vinyl ether.

7. Inert atmosphere.

8. Air-free reaction vessel.

9. Magnetic stir bar and stir plate.

10. Diethyl ether.

11. Size Exclusion Chromatography with Multi-Angle Light Scattering (SEC-MALS) for polymer analysis.

**2.3 Ring Opening Metathesis Polymerization: Post-Polymerization Modification Method (see Note 3)**

1. Catalyst—$(IMesH_2)(C_5H_5N_2)(Cl)_2Ru=CHPh$ [33].

2. NHS-norbornenyl ester [18].

3. Enzyme-responsive peptide (*see* **Note 1**).

4. Hydrophobic drug monomer [15].

5. Dye-based chain transfer agents [36] for dye-labeled polymer ends.

6. NHS-modified dye and amine monomer for post-polymerization incorporation of dyes [18].

7. Dry DMF.

8. Ethyl vinyl ether.

9. Inert ($N_2$) atmosphere.

10. Air-free reaction vessel (or glove box with inert atmosphere).

11. Magnetic stir bar and stir plate.

12. Diethyl ether.

13. SEC-MALS for polymer analysis.

14. Centrifuge.

**2.4 Dialysis Reagents**

1. ROMP polymer.

2. Dimethyl Sulfoxide (DMSO).

3. DMF.

4. Dulbecco's Phosphate Buffered Saline (DPBS): 2.67 mM KCl, 1.47 mM $KH_2PO_4$, 138 mM NaCl, 8.09 mM $Na_2HPO_4$, pH = 7.4 (*see* **Note 4**).

5. 3500 MWCO dialysis tubing and/or cups.

6. Dynamic Light Scattering (DLS) instrument for hydrodynamic radius determination.

**2.5 Matrix Metalloproteinase Degradation of ER-NPs**

1. ER-NPs.

2. MMP-9 (Calbiochem, USA).

3. 24 mM *p*-aminophenyl mercuric acetate in 0.1 M NaOH.

4. Heating Block.

5. RP-HPLC for analysis.

**2.6 Transmission Electron Microscopy**

1. ER-NPs.
2. TEM carbon grids (Ted Pella Inc.).
3. Emitech K350 glow discharge unit.
4. 1% uranyl acetate stain.
5. FEI Tecnai G2 Sphera Microscope.

# 3 Methods

**3.1 Preparation of Peptide Substrates**

1. Determine peptide sequence to be synthesized (*see* **Note 1** and **Note 5**).
2. If using a peptide synthesizer, program sequence into unit and initiate synthesis based on the instrument being used. In this case, skip **steps 3–8**.
3. If synthesizing by hand, weigh out the appropriate amounts of amino acids and resin, based on desired coupling conditions (*see* **Note 2**).
4. Swell the resin of choice (*see* **Note 2**) for 45 min with DMF in a peptide synthesis vessel.
5. Add first amino acid to vessel.
6. Add HBTU and DIPEA to vessel (resin/amino acid/HBTU/DIPEA 1:3:3:4) and shake for 45 min.
7. Drain vessel; add 20% 4-methylpiperidine in DMF and shake for 20 min for Fmoc deprotection.
8. Add next amino acid to vessel, repeat **steps 6–7**. Repeat until the entire sequence is synthesized.
9. If synthesizing a peptide to be used directly as a monomer for ROMP (*see* **Note 2**), couple *N*-(Glycine)-*cis*-5-norbornene-*exo*-2,3-dicarboximide [35] at the last step, using the same coupling conditions as in **steps 6–7**.
10. Cleave the peptide from the resin, using the correct cleavage cocktail for resin (*see* **Note 2**).

**3.2 Ring Opening Metathesis Polymerization: Direct Polymerization Method (see Note 3 and Fig. 2)**

1. Determine desired block lengths (*see* **Note 6**), and calculate amounts of all monomers and catalyst needed.
2. In a flask that has been flushed with $N_2$ (or in a glove box), dissolve hydrophobic drug monomer in dry DMF at a concentration between 0.1 and 0.8 M.
3. In a separate vessel, dissolve catalyst in dry DMF at a concentration between 0.01 and 0.05 M.
4. Quickly add correct volume of catalyst to hydrophobic drug monomer to initiate polymerization.

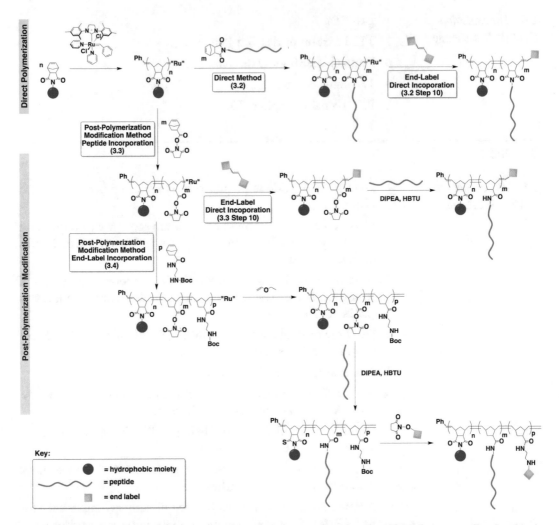

**Fig. 2 General Synthetic Scheme for Generation of Enzyme-Responsive ROMP Polymers**. The first block of our peptide polymer amphiphiles is polymerized in the same manner, regardless of peptide addition method (direct or post-polymerization modification). After the completion of the first block, however, two different directions may be taken—direct polymerization of the functional material/peptide (proceeding to the right of the first block in the scheme), or polymerization of a functionalizable norbornenyl derivative for post-polymerization modification (proceeding below the first block in the scheme)

5. Stir reaction with magnetic stir bar for 2 h to ensure complete polymerization of drug monomer.

6. Remove a 30 μL aliquot for SEC-MALS analysis of the homo-polymer (the first block of the copolymer). Quench aliquot with ~1 μL ethyl vinyl ether for 20 min before analyzing (Fig. 3a).

7. Dissolve peptide monomer in dry DMF at a concentration of ~0.05 M.

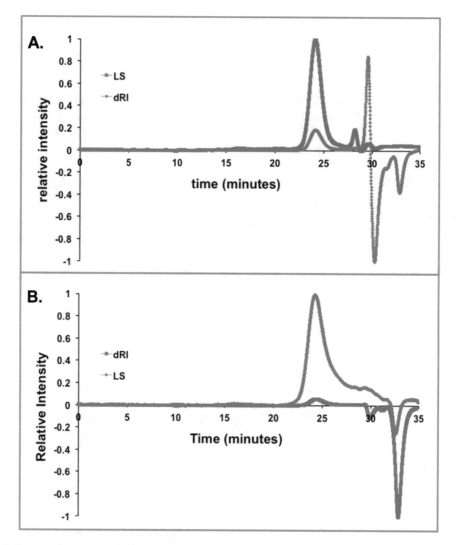

**Fig. 3** SEC-MALS Traces of ROMP polymers. To determine molecular weight, materials are analyzed via SEC-MALS after A) polymerization of the first (hydrophobic) block and B) polymerization of the second (hydrophilic) block. Polymer sequence: (Paclitaxel)$_{10}$-(NorG-GPLGLAGGERDG)$_3$. Source: Reproduced with permission from [15]

8. Add appropriate volume of peptide monomer to vessel containing the remaining solution. Stir for 3 h to ensure complete polymerization (*see* **Note 7**).

9. Remove a 30 μL aliquot for SEC-MALS analysis of the diblock copolymer. Quench aliquot with ~1 μL ethyl vinyl ether for 20 min before analyzing (Fig. 3b).

10. If incorporating a dye as an end-label, add 1.2 equivalents of the desired dye-based chain transfer agent and stir for an additional 2 h.

11. Quench catalyst by addition of ethyl vinyl ether; stir 25 min to ensure complete quenching.

12. Transfer polymer solution to a 50 mL centrifuge tube and reduce DMF volume to <2 mL.

13. Precipitate polymer from solution by adding ~40 mL of cold diethyl ether to centrifuge tube. Centrifuge at 4000 rpm in a microcentrifuge for 7 min. Decant ether, and dry polymer cake *in vacuo* overnight.

### 3.3 Ring Opening Metathesis Polymerization: Post-Polymerization Modification for Peptide Incorporation

1. Determine desired block lengths (*see* **Note 6**), and calculate amounts of all monomers and catalyst needed.

2. In a flask that has been flushed with $N_2$ (or, in a glove box), dissolve hydrophobic drug monomer in dry DMF at a concentration between 0.1 and 0.8 M.

3. In a separate vessel, dissolve catalyst in dry DMF at a concentration between 0.01 and 0.05 M.

4. Quickly add correct volume of catalyst to hydrophobic drug monomer to initiate polymerization.

5. Stir reaction with magnetic stir bar for 2 h to ensure complete polymerization of drug monomer.

6. Remove a 30 μL aliquot for SEC-MALS analysis of the homopolymer (the first block of the copolymer). Quench aliquot with ~1 μL ethyl vinyl ether for 20 min before analyzing.

7. Dissolve NHS-norbornenyl ester in dry DMF at a concentration of 0.1–0.7 M.

8. Add appropriate volume of NHS-norbornenyl ester to vessel containing the remaining solution. Stir for 40 min to ensure complete polymerization.

9. Remove a 30 μL aliquot for SEC-MALS analysis of the diblock copolymer. Quench aliquot with ~1 μL ethyl vinyl ether for 20 min before analyzing.

10. If incorporating a dye as an end-label, add 1.2 equivalents of the desired dye-based chain transfer agent and stir for an additional 2 h.

11. Quench catalyst by addition of ethyl vinyl ether; stir 25 min to ensure complete quenching.

12. Transfer polymer solution to a 50 mL centrifuge tube and reduce DMF volume to <2 mL.

13. Precipitate polymer from solution by adding ~40 mL of cold diethyl ether to centrifuge tube. Centrifuge at 4000 rpm for 7 min. Decant ether, and dry polymer cake *in vacuo* overnight.

14. Dissolve dried polymer in 4:1 DMF:DMSO.

15. Add desired number of equivalents of peptide (*see* **Note 6**) to flask.

16. Add excess (4–16 equivalents) DIPEA to flask.

17. Stir at room temperature overnight.

18. Precipitate polymer with cold diethyl ether. Centrifuge.

19. Decant ether and wash precipitate with cold methanol twice.

20. Dry newly synthesized peptide-polymer amphiphile (PPA) *in vacuo* overnight.

21. Analyze degree of peptide conjugation by SEC-MALS against unconjugated polymer.

**3.4 Ring Opening Metathesis Polymerization: Post-Polymerization Modification for Dye Incorporation**

1. If following direct polymerization method, follow **steps 1–9** in Subheading 3.2. If following post-polymerization modification method, follow **steps 1–9** in Subheading 3.3.

2. Dissolve amine monomer in dry DMF at a concentration of ~0.06 M.

3. Add appropriate volume of amine monomer to vessel containing the remaining solution. Stir for 40 min to ensure complete polymerization.

4. Remove a 30 μL aliquot for SEC-MALS analysis of the diblock copolymer. Quench aliquot with ~1 μL ethyl vinyl ether for 20 min before analyzing.

5. Quench catalyst by addition of ethyl vinyl ether; stir 25 min to ensure complete quenching.

6. Transfer polymer solution to a 50 mL centrifuge tube and reduce DMF volume to <2 mL.

7. Precipitate polymer from solution by adding ~40 mL of cold diethyl ether to centrifuge tube. Centrifuge at 4000 rpm for 7 min. Decant ether, and dry polymer cake *in vacuo* overnight.

8. If following direct polymerization method, skip to **step 16**.

9. If following post-polymerization modification method, dissolve dried polymer in 4:1 DMF:DMSO.

10. Add desired number of equivalents of peptide (*see* **Note 6**) to flask.

11. Add excess (4–16 equivalents) DIPEA to flask.

12. Stir at room temperature overnight.

13. Precipitate polymer with cold diethyl ether. Centrifuge at 4000 rpm in a microcentrifuge for 7 min.

14. Decant ether and wash precipitate with cold methanol twice.

15. Dry newly synthesized peptide-polymer amphiphile (PPA) *in vacuo* overnight.

16. Dissolve polymer in DMF.

17. Add 1.2 equivalents of desired NHS-modified dye to the polymer solution.

18. Add 1.2 equivalents of DIPEA to the polymer solution.

19. Stir overnight at room temperature.

20. Precipitate polymer with cold diethyl ether. Centrifuge at 4000 rpm in a microcentrifuge for 7 min.

21. Decant ether and dry polymer *in vacuo* overnight.

**3.5  ER-NP Formation**

1. Dissolve polymer in DMSO or DMF at a concentration of 0.1–10.0 mg/mL (*see* **Note 8**).

2. In a drop-wise fashion, slowly add 1.0–1.5 volume equivalents of DPBS (*see* **Note 4** and **Note 9**) to facilitate ER-NP formation.

3. Transfer the ER-NP solution to dialysis tubing (or dialysis cup) and dialyze into 1.0 L of DPBS.

4. Change DPBS twice daily for three total water changes.

5. Remove ER-NPs from dialysis tubing/cups and analyze size distributions by DLS (Fig. 4a).

6. Store in vials at 4 °C.

**3.6  Analysis of ER-NPs by TEM (Fig. 4b)**

1. Glow discharge and plasma-clean carbon grid.

2. Place small (4 μL) aliquot of ER-NP on grid. Allow to adhere for 5 min.

3. Rinse grid with distilled water.

4. Add 1% uranyl acetate solution (stain) to grid. Rinse immediately with water.

5. Allow sample to dry for 5 min.

6. Analyze with a microscope.

**3.7  Enzymatic Degradation of ER-NPs with MMPs**

1. Activate MMP-9 by adding 0.4 μL of a 24 mM *p*-aminophenyl mercuric acetate in 0.1 M NaOH to 5 μL enzyme solution, and heating for 2 h at 37 °C (*see* **Note 10**).

2. In one vial, treat 100–120 μM ER-NP (concentration with respect to peptide) with MMP-9 (100 μU, 1.25 μL) for 24 h at 37 °C.

3. As a positive control, treat 100–120 μM of authentic peptide with MMP-9 (100 μU, 1.25 μL) for 24 h at 37 °C.

4. As a negative control, heat 100–120 μM of authentic peptide for 24 h at 37 °C.

5. Inactivate MMP-9 by heating solutions to 65 °C for 20 min.

6. Analyze all samples with RP-HPLC, collecting the peaks corresponding to peptide fragments for ESI analysis.

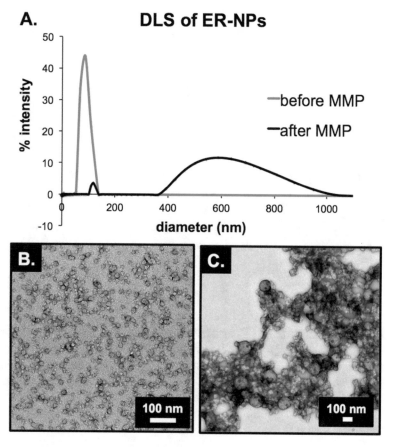

**Fig. 4** Characterization of ER-NPs. (**a**) DLS trace of ER-NPs before treatment with MMP (red trace) and after treatment with MMP (black trace). (**b**) Negative stain, dry-state TEM image of ER-NPs before treatment with MMP. (**c**) Negative stain, dry-state TEM image of ER-NPs after exposure to MMP. Polymer sequence: (Paclitaxel)$_{10}$-(NorG-GPLGLAGGERDG)$_3$. Source: Reproduced with permission from [15]

The cleavage efficiency can be calculated from the fraction of the area under the peak, relative to the area of the negative control (Fig. 5).

7. Analyze aggregation of ER-NPs by DLS and TEM (Fig. 4b, c).

# 4    Notes

1. In our experience, the peptide sequence to be used for forming ER-NPs does not have to be exactly the same in every application—however, all peptide sequences we use contain the recognition sequence **GPL<u>GL</u>AG**. MMP-9 and -12 cleave

# HPLC Analysis of MMP Cleavage

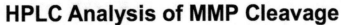

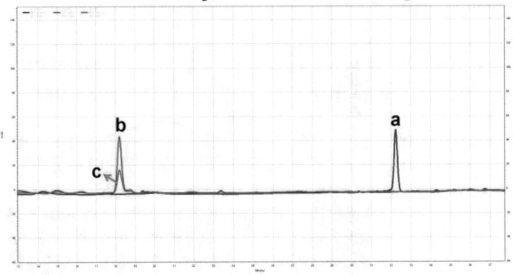

**Fig. 5** HPLC Analysis of ER-NPs for MMP cleavage efficiency. Peak "a" corresponds to an authentic sample of the intact peptide, prior to MMP degradation. Peak "b" corresponds to the degradation product of the authentic peptide sample, after exposure to MMP. Peak "c" corresponds to the degradation product of the peptide on the ER-NP, after exposure to MMP. Efficiency of peptide cleavage on the ER-NP by MMP is calculated from the ratio of the area under peak "c" to that of peak "b." Polymer sequence: $(Phenyl)_{21}$-[(Nor-COOH)$_3$ -(Nor-GPLGLAGGERDG)$_3$], where "Nor" represents norbornenyl derivative. Source: Reproduced with permission from [17]

this sequence at the residues underlined. Additional amino acids on the C-terminal end of the recognition sequence may be added, based on the needs of the researcher. Things to consider on this front include the overall charge of the peptide, the hydrophilicity of the peptide, and whether or not the peptide will be directly incorporated into a ROMP monomer [37]. The sequences that we have had success with in our lab include:

GPLGLAGGWGERDGS [16, 17, 20].
GPLGLAGGERDG [15, 19].
GPLGLAGKWAAAAKAAAAK [18]

2. The choice of resin for solid phase peptide synthesis (SPPS) will depend upon the peptide sequence to be made. In our publications, we use several different resins (Rink Amide, MBHA, and Wang). Thus, the cleavage cocktails used will also depend on the choice of resin and amino acid sequence, the protocols for which can be readily found in the literature.

3. One of the benefits of ER-NPs is that their substituent ROMP polymers can be generated in different ways, yet still have the same resulting function as nanoparticles. Indeed, our lab has

successfully utilized several different polymerization schemes toward the formation of ER-NPs, which suggests that this technique is robust. Figure 1 outlines the polymerization routes we have utilized, which should serve as a useful guide in determining which synthetic route the researcher should utilize.

4. When dialyzing from organic solvent to aqueous medium, it may be necessary to use a different buffer than what is indicated in this text. We have found that, when working with certain peptide sequences, PBS solutions that contain calcium and magnesium lead to aggregation of polymers, rather than nanoparticle formation. Alternative buffers that have worked for our group include pure deionized water; Tris(hydroxymethyl)aminomethane (Tris) buffer: 50 mM, pH=8.5; and phosphate buffered saline (PBS): 40 mM $Na_2PO_4$, pH=8.0.

5. As a negative control for ER-NP function, we synthesize D-amino acid versions (D-peptides) of all peptides and generate polymers with them. To do this, split the polymer solutions in half before adding the second block in the direct polymerization method, and add L-peptide monomers to one portion and D-peptide monomers to the second.

6. In our work, we have successfully generated ER-NPs with hydrophobic:hydrophilic ratios of 21:8 [17, 18], 10:3 [15, 19], and 20:2 [20]. However, the actual lengths of each block (hydrophobic and hydrophilic) may need to be adjusted by the researcher to generate well-defined nanoparticles. If using the direct polymerization method for the peptide component, it is important to note that the degree of polymerization of the peptide will impact its susceptibility to proteolysis by MMPs [38], with longer block lengths showing marked decrease in susceptibility.

7. The time required for complete polymerization of the peptide block will depend upon concentration and sequence identity [37]. Therefore, when working with a new peptide monomer, it is best to monitor the polymerization by NMR to determine the length of time needed to fully polymerize the peptide. This is achieved by monitoring the disappearance of the monomer's olefin peak at ~6.30 ppm and the appearance of the polymer's cis/trans olefin peaks at 5.73 and 5.50 ppm (Fig. 6).

8. It may be necessary to use a different organic solvent for dissolving the polymer, depending on the polymer components. DMSO and DMF work well in general, but other solvents that are miscible with water may be needed to form well-defined spherical nanoparticles. In addition, the initial polymer concentration may need to be adjusted so that aggregation does not occur. It is therefore recommended to complete a small solvent and concentration screen, to determine the ideal parameters for each specific system.

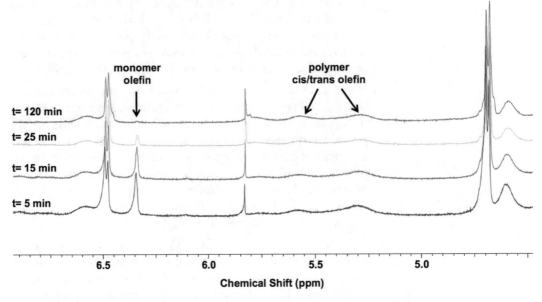

**Fig. 6** Monitoring Polymerization by ¹H–NMR. It is possible to monitor polymerization progression by NMR by observing the disappearance of the monomer olefin peak (~6.30 ppm) and appearance of the polymer cis/trans olefin peaks (~5.7 and ~5.2 ppm) as a function of time. In this representative example, some of the monomer is already consumed at t=5 min (blue trace), as evidenced by the appearance of the cis/trans peaks. The polymerization proceeds until t=120 min (magenta trace), when the monomer olefin peak has completely disappeared

9. Slow addition of aqueous buffer to the polymer solution is important—generally, a slower addition leads to better-defined nanomaterials and reduces the chances of unwanted aggregation. It may be useful to use a syringe pump to control the addition of buffer [20], and extra care should be taken until the critical water concentration (~30–50% by volume) is reached. This parameter, as with the choice of buffers, may need to be adjusted based on the specific polymer system.

10. Though this protocol uses MMP-9, we have also had success with MMP-2 and MMP-12.

## References

1. Maeda H, Nakamura H, Fang J (2013) The EPR effect for macromolecular drug delivery to solid tumors: improvement of tumor uptake, lowering of systemic toxicity, and distinct tumor imaging in vivo. Adv Drug Deliv Rev 65(1):71–79. doi:10.1016/j.addr.2012.10.002

2. Kobayashi H, Watanabe R, Choyke PL (2014) Improving conventional enhanced permeability and retention (EPR) effects; what is the appropriate target? Theranostics 4 (1):81–89. doi:10.7150/thno.7193

3. Torchilin V (2011) Tumor delivery of macromolecular drugs based on the EPR effect. Adv Drug Deliv Rev 63(3):131–135. doi:10.1016/j.addr.2010.03.011

4. Pasqualini R, Koivunen E, Kain R, Lahdenranta J, Sakamoto M, Stryhn A, Ashmun RA,

Shapiro LH, Arap W, Ruoslahti E (2000) Aminopeptidase N is a receptor for tumor-homing peptides and a target for inhibiting angiogenesis. Cancer Res 60(3):722–727

5. Low PS, Antony AC (2004) Folate receptor-targeted drugs for cancer and inflammatory diseases. Adv Drug Deliv Rev 56 (8):1055–1058. doi:10.1016/j.addr.2004.02. 003

6. Arap W, Pasqualini R, Ruoslahti E (1998) Cancer treatment by targeted drug delivery to tumor vasculature in a mouse model. Science 279(5349):377–380. doi:10.1126/science. 279.5349.377

7. Calderón M, Welker P, Licha K, Fichtner I, Graeser R, Haag R, Kratz F (2011) Development of efficient acid cleavable multifunctional prodrugs derived from dendritic polyglycerol with a poly(ethylene glycol) shell. J Control Release 151(3):295–301. doi:10.1016/j. jconrel.2011.01.017

8. Du J-Z, Du X-J, Mao C-Q, Wang J (2011) Tailor-made dual pH-sensitive polymer–doxorubicin nanoparticles for efficient anticancer drug delivery. J Am Chem Soc 133 (44):17560–17563. doi:10.1021/ja207150n

9. Doncom KEB, Hansell CF, Theato P, O'Reilly RK (2012) pH-switchable polymer nanostructures for controlled release. Polym Chem 3 (10):3007–3015. doi:10.1039/C2PY20545A

10. Liu G, Wang X, Hu J, Zhang G, Liu S (2014) Self-immolative polymersomes for high-efficiency triggered release and programmed enzymatic reactions. J Am Chem Soc 136 (20):7492–7497. doi:10.1021/ja5030832

11. Phillips DJ, Patterson JP, O'Reilly RK, Gibson MI (2014) Glutathione-triggered disassembly of isothermally responsive polymer nanoparticles obtained by nanoprecipitation of hydrophilic polymers. Polym Chem 5(1):126–131. doi:10.1039/C3PY00991B

12. Ryu J-H, Chacko RT, Jiwpanich S, Bickerton S, Babu RP, Thayumanavan S (2010) Self-cross-linked polymer nanogels: A versatile nanoscopic drug delivery platform. J Am Chem Soc 132(48):17227–17235. doi:10. 1021/ja1069932

13. Blum AP, Kammeyer JK, Rush AM, Callmann CE, Hahn ME, Gianneschi NC (2015) Stimuli-responsive nanomaterials for biomedical applications. J Am Chem Soc 137 (6):2140–2154. doi:10.1021/ja510147n

14. Torchilin VP (2010) Passive and active drug targeting: Drug delivery to tumors as an example. In: Schäfer-Korting M (ed) Drug Delivery. Springer, Berlin Heidelberg, Berlin,

Heidelberg, pp 3–53. doi:10.1007/978-3-642-00477-3_1

15. Callmann CE, Barback CV, Thompson MP, Hall DJ, Mattrey RF, Gianneschi NC (2015) Therapeutic enzyme-responsive nanoparticles for targeted delivery and accumulation in tumors. Adv Mater 27(31):4611–4615. doi:10.1002/adma.201501803

16. Chien M-P, Carlini AS, Hu D, Barback CV, Rush AM, Hall DJ, Orr G, Gianneschi NC (2013) Enzyme-directed assembly of nanoparticles in tumors monitored by in vivo whole animal imaging and ex vivo super-resolution fluorescence imaging. J Am Chem Soc 135 (50):18710–18713. doi:10.1021/ja408182p

17. Chien M-P, Thompson MP, Barback CV, Ku T-H, Hall DJ, Gianneschi NC (2013) Enzyme-directed assembly of a nanoparticle probe in tumor tissue. Adv Mater 25(26):3599–3604. doi:10.1002/adma.201300823

18. Chien M-P, Thompson MP, Lin EC, Gianneschi NC (2012) Fluorogenic enzyme-responsive micellar nanoparticles. Chem Sci 3 (9):2690–2694. doi:10.1039/C2SC20165H

19. Daniel KB, Callmann CE, Gianneschi NC, Cohen SM (2016) Dual-responsive nanoparticles release cargo upon exposure to matrix metalloproteinase and reactive oxygen species. Chem Commun 52(10):2126–2128. doi:10. 1039/C5CC09164K

20. Nguyen MM, Carlini AS, Chien M-P, Sonnenberg S, Luo C, Braden RL, Osborn KG, Li Y, Gianneschi NC, Christman KL (2015) Enzyme-responsive nanoparticles for targeted accumulation and prolonged retention in heart tissue after myocardial infarction. Adv Mater 27(37):5547–5552. doi:10.1002/adma. 201502003

21. Egeblad M, Werb Z (2002) New functions for the matrix metalloproteinases in cancer progression. Nat Rev Cancer 2(3):161–174 http://www.nature.com/nrc/journal/v2/ n3/suppinfo/nrc745_S1.html

22. Gialeli C, Theocharis AD, Karamanos NK (2011) Roles of matrix metalloproteinases in cancer progression and their pharmacological targeting. FEBS J 278(1):16–27. doi:10. 1111/j.1742-4658.2010.07919.x

23. Kessenbrock K, Plaks V, Werb Z (2010) Matrix Metalloproteinases: Regulators of the Tumor Microenvironment. Cell 141(1):52–67. doi:10.1016/j.cell.2010.03.015

24. Rundhaug JE (2003) Matrix Metalloproteinases, angiogenesis, and cancer: commentary re: A. C. Lockhart et al., Reduction of wound angiogenesis in patients treated with

BMS-275291, a broad spectrum matrix metalloproteinase inhibitor. Clin. Cancer Res., 9: 00–00, 2003. Clinical Cancer Research 9 (2):551–554.

25. Creemers EEJM, Cleutjens JPM, Smits JFM, Daemen MJAP (2001) Matrix metalloproteinase inhibition after myocardial infarction: A new approach to prevent heart failure? Circ Res 89(3):201–210. doi:10.1161/hh1501. 094396

26. Phatharajaree W, Phrommintikul A, Chattipakorn N (2007) Matrix metalloproteinases and myocardial infarction. Can J Cardiol 23 (9):727–733

27. Spinale FG (2007) Myocardial matrix remodeling and the matrix metalloproteinases: Influence on cardiac form and function. Physiol Rev 87(4):1285–1342. doi:10.1152/physrev. 00012.2007

28. Vanhoutte D, Schellings M, Pinto Y, Heymans S (2006) Relevance of matrix metalloproteinases and their inhibitors after myocardial infarction: a temporal and spatial window. Cardiovasc Res 69(3):604–613. doi:10.1016/j. cardiores.2005.10.002

29. Bielawski CW, Grubbs RH (2000) Highly efficient ring-opening metathesis polymerization (ROMP) using new ruthenium catalysts containing N-heterocyclic carbene ligands. Angew Chem Int Ed 39(16):2903–2906. doi:10. 1002/1521-3773(20000818)39:16<2903:: AID-ANIE2903>3.0.CO;2-Q

30. Bielawski CW, Grubbs RH (2007) Living ring-opening metathesis polymerization. Prog Polym Sci 32(1):1–29. doi:10.1016/j. progpolymsci.2006.08.006

31. Leitgeb A, Wappel J, Slugovc C (2010) The ROMP toolbox upgraded. Polymer 51 (14):2927–2946. doi:10.1016/j.polymer. 2010.05.002

32. Sanford MS, Love JA, Grubbs RH (2001) A versatile precursor for the synthesis of new ruthenium olefin metathesis catalysts. Organometallics 20(25):5314–5318. doi:10.1021/ om010599r

33. Sanford MS, Love JA, Grubbs RH (2001) Mechanism and activity of ruthenium olefin metathesis catalysts. J Am Chem Soc 123 (27):6543–6554. doi:10.1021/ja010624k

34. Scholl M, Ding S, Lee CW, Grubbs RH (1999) Synthesis and activity of a new generation of ruthenium-based olefin metathesis catalysts coordinated with 1,3-Dimesityl-4,5-dihydroimidazol-2-ylidene Ligands. Org Lett 1 (6):953–956. doi:10.1021/ol990909q

35. Conrad RM, Grubbs RH (2009) Tunable, temperature-responsive polynorbornenes with side chains based on an elastin peptide sequence. Angew Chem Int Ed 48(44):8328–8330. doi:10.1002/anie.200903888

36. Thompson MP, Randolph LM, James CR, Davalos AN, Hahn ME, Gianneschi NC (2014) Labelling polymers and micellar nanoparticles via initiation, propagation and termination with ROMP. Polym Chem 5 (6):1954–1964. doi:10.1039/C3PY01338C

37. Kammeyer JK, Blum AP, Adamiak L, Hahn ME, Gianneschi NC (2013) Polymerization of protecting-group-free peptides via ROMP. Polym Chem 4(14):3929–3933. doi:10. 1039/C3PY00526G

38. Blum AP, Kammeyer JK, Yin J, Crystal DT, Rush AM, Gilson MK, Gianneschi NC (2014) Peptides displayed as high density brush polymers resist proteolysis and retain bioactivity. J Am Chem Soc 136(43):15422–15437. doi:10. 1021/ja5088216

# Chapter 16

## NanoScript: A Versatile Nanoparticle-Based Synthetic Transcription Factor for Innovative Gene Manipulation

**Kholud Dardir, Christopher Rathnam, and Ki-Bum Lee**

### Abstract

Cellular reprogramming and stem cell-based therapies have shown tremendous potential in the field of regenerative medicine. To that end, developing tools to control stem cell fate is an attractive area of research for replacing damaged and diseased cells and reestablishing functional connections for tissue repair. Transcription factor (TFs) proteins are well known to regulate gene expression and direct stem cell fate. Inspired by natural TFs, NanoScript, a nanoparticle (NP)-based platform, mimics TFs to afford control over gene expression and stem cell fate for regenerative medicine. Here, we describe the construction of the NanoScript platform, which is designed with tunable properties to replicate the structure and function of TFs to bind to specific portions of the genome and regulate gene expression in a way that does not involve viral delivery.

**Key words** NanoScript, Nanoparticle, Transcription factor, Gene manipulation, Nonviral delivery, Gene regulation

## 1 Introduction

Transcription factors (TFs) are proteins that bind to a targeted DNA sequence to regulate gene expression and ultimately determine cell fate by controlling the activation or repression of the specified gene. [1] TFs are composed of three major domains: (1) a DNA-binding domain (DBD), which binds to a specified DNA sequence, (2) an activation/repression domain (AD/RD), which recruits transcriptional factors to initiate transcription for activation or repression of a gene, and (3) a nuclear localization signal (NLS), which penetrates the nuclear membrane to gain entry into the nucleus. NanoScript is composed of the same three major domains all bound by a nanoparticle, which acts as a linker domain, and can be further enhanced by incorporating an epigenetic modulator,

The original version of this chapter was revised. An erratum to the chapter can be found at DOI: 10.1007/978-1-4939-6840-4_23

Sarah Hurst Petrosko and Emily S. Day (eds.), *Biomedical Nanotechnology: Methods and Protocols*, Methods in Molecular Biology, vol. 1570, DOI 10.1007/978-1-4939-6840-4_16, © Springer Science+Business Media LLC 2017

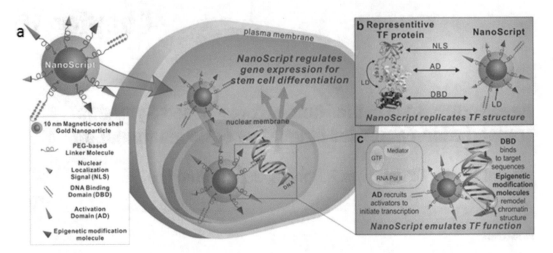

**Fig. 1 Schematic of NanoScript transcription factor (TF) proteins**. (**a**) Small molecules and peptides are assembled by 10 nm magnetic-core gold shell nanoparticle (MCNP) (which can be substituted) to develop the NanoScript platform. (**b**) The four essential TF domains are effectively replicated by NanoScript. The MCNP mimics the linker domain (LD) and serves to tether the small molecules together. (**c**) The DBD binds to complementary DNA sequences, the AD recruits transcriptional machinery components such as RNA polymerase II (RNA Pol II), and the epigenetic modification molecule for modulating the chromatin state [3]. Modified from http:/pubs.acs.org/doi/full/10.1021/nn501589f

which allows for chromatin remodeling and better access to the specified gene sequence [2] (Fig. 1a). In addition, NanoScript can be tailored to target different genes to differentiate stem cells into various lineages, such as myocytes, chondrocytes, and neurons [3–6]. Given the platform's tunable properties, it has immense potential for the regulation of genes responsible for stem cell differentiation as well as uses in other applications.

To construct NanoScript, the first step is to synthesize the nanoparticle linker domain, which serves to connect the AD, NLS, and DBD (Fig. 2a). Several types of nanoparticles can be used as linker domains, including gold NPs (AuNPs), silica NPs, and magnetic NPs. The main criteria for an optimal linker domain are good biocompatibility, facile surface conjugation, and small size. AuNPs are very accessible since they can be purchased or easily synthesized. They have numerous advantages and can be used as imaging agents, due to their strong surface plasmon resonance and ability to scatter light, and therapeutic delivery agents, due to their ability to be facilely bioconjugated [7]. To enhance transfection efficiency, a highly magnetic zinc-doped iron oxide core-gold shell NP (MCNP) can be used as the linker domain, and this type of system will be described in this chapter. This structure allows for noninvasive MRI imaging, nonviral, target-specific delivery, and enhanced cellular uptake using magnetofection (in which a magnetic field is used to attract the particles to the surface of the cell) all while retaining the aforementioned advantages of gold as a shell material [8, 9]. Thus, these combined advantages make MCNPs

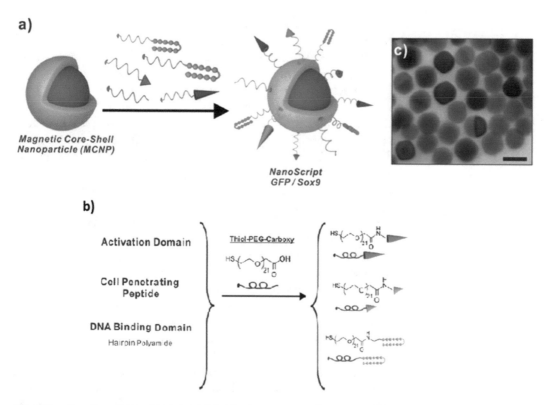

**Fig. 2 Construction of NanoScript.** (a) Small molecules and peptides are assembled on magnetic-core shell gold nanoparticle (MCNP) (which can be substituted) to develop the NanoScript platform. (b) The four essential TF domains: The Activation Domain (AD), The Cell Penetrating Peptide (CPP), and the DNA Binding Domain (DBD) are all conjugated to a thiol-PEG-carboxy and then bound to the nanoparticle. (c) A transition electron micrograph of NanoScript with a scale bar of 20 nm [6]

efficient linker domains for NanoScript. To enhance the effects of the synthetic TF, an epigenetic modulator can be incorporated to modify the chromatin structure, allowing NanoScript greater access to previously inaccessible portions of the genome to more efficiently modulate the expression of genes responsible for inducing stem cell differentiation [10].

The DBD is composed of a hairpin polyamide that is designed to bind to a predefined DNA sequence. The DBD typically targets the promoter region of the desired genes, and it has a high binding affinity [11]. The sequence of the hairpin polyamide consists of the amino acids N-methylpyyrole (Py) and N-methylimidazole (Im). Py binds to adenine and thymine, while Im binds to guanine and cytosine [12]. The sequence of Py and Im can be tailored to target the gene sequence of interest, which is one of the factors that makes NanoScript highly versatile (Fig. 2b).

The AD, or transactivation peptide, works along with the DBD to induce transcription and upregulate gene expression. The AD initiates the transcription process by recruiting the cell's transcriptional machinery to the binding site. Furthermore, the AD can be replaced with an RD to repress and transiently down-regulate gene expression.

The NLS aids cell membrane penetration, and it allows Nano-Script to reach its final destination, nuclear DNA. The NLS grants the particles entry into the nucleus through the nuclear pore complex. Due to the limited diameter of the nuclear pore, the MCNP-based platform is carefully designed to possess a hydrodynamic diameter of approximately 34 nm for efficient nuclear uptake [13] (Fig. 2c). Once NanoScript has successfully reached its target, an epigenetic modulator, N-(4-Chloro-3-(trifluoromethyl) phenyl)-2-ethoxybenzamide (CTB), can modify the chromatin structure by loosening the DNA and allowing for improved access to the binding sites for transcription [14]. This can be achieved by conjugating CTB to the MCNPs along with the aforementioned domains to modify the chromatin structure without altering the DNA sequence [15].

It is important to note that there have not been mechanistic studies of NanoScript with respect to receptor interactions, DNA binding, and the recruitment of transcriptional machinery. However, based on the literature, we can propose a mechanism of action for NanoScript. While the mechanism of cell-penetrating peptides has not been completely elucidated, there have been a significant number of studies devoted to understanding how they shuttle molecules and other entities into the cell [16, 17]. Specifically, studies have shown that the TAT peptide, which natively possesses a NLS granting it and its cargo nuclear access, causes cellular uptake of quantum dots by micropinocytosis [16]. TAT itself has been shown to permeate the cell membrane even after clathrin- and caveolin-mediated pathways, the two major mechanisms of cellular uptake, were knocked down proving that TAT does not need to undergo receptor-mediated endocytosis for efficient cellular delivery [17]. However, some studies arrive at contradictory conclusions regarding what happens to TAT- or NLS-conjugated nanoparticles after they have entered the cell. Some studies claim cytoplasmic and nuclear localization of the nanoparticles while others show nanoparticles trapped in endosomes. It has been proposed that these discrepancies in results are likely due to the use of various cell lines and cargos being delivered to the cell causing the cells to process them in different ways [16]. In the case of NanoScript, nuclear and cytoplasmic localization has been observed in both mesenchymal stem cells and neural stem cells [3].

Once inside the cytosol, the TAT peptide shuttles NanoScript into the nucleus by interacting with the nuclear pore complexes, which act as the gatekeeper for nuclear uptake. These complexes bind NanoScript, translocate into the nucleus, and then release the cargo on the other side [18]. As such, the addition of the nuclear localization signal TAT onto NanoScript allows for its robust nuclear localization and transcriptional activation. Once inside the nucleus, the next step is DNA binding. The hairpin polyamide facilitates binding of NanoScript to the DNA through hydrogen bonding interactions with the minor groove of the DNA double helix. As previously described, the pyrrole-imidazole pairs allow the polyamide to hydrogen bond and form van der Waals interactions with the specified Watson-Crick base pairs [19]. The polyamides were carefully designed to target the promoter of the genes that we are interested in regulating.

Lastly, after binding to the DNA promoter region NanoScript must activate transcription. The aforementioned activation domain that was previously described is able to directly bind or interact with transcription factor II B (TFIIB), which is a general transcription factor that is involved in coordinating the interactions between RNA polymerase to the DNA and other transcription factors, forming the preinitiation complex [20]. This in turn activates transcription of the target gene. Furthermore, as previously mentioned, due to the highly tunable nature of NanoScript, it can also be used to repress genes simply by replacing the activation domain with a repressor peptide. This repressor peptide then acts by recruiting the transcriptional corepressors—the Groucho and the transducin-like enhancer (TLE) family proteins. These proteins prevent the transcription of genes by both preventing the preinitiation complex from forming as well as changing the chromatin landscape, altering the accessibility of the TFs to certain genes and their commensurate expression [21].

This chapter describes a novel biomimetic nanoparticle-based synthetic transcription factor. This platform has the ability to alter gene expression profiles in cells for various applications including inducing cellular differentiation. This allows for a novel nonviral method to alter gene expression that is safe and effective and can potentially be used for the treatment of diseases and disorders.

# 2 Materials

## 2.1 MCNP Synthesis (See Note 1)

1. 10 nm MNP: 1.5174 mmol $Fe(Acac)_3$, 0.4825 mmol $FeCl_2$, 0.3338 mmol $ZnCl_2$, 10 mmol 1,2-hexadecandiol, 6 mmol oleic acid, 6 mmol oleylamine, 20 mL tri-n-octylamine, and chloroform.

2. MCNP: 20 mL of tri-n-octylamaine, 0.3 mmol (100 μL) oleylamine, 60 μL of a 5 mg/300 μL, stock solution of $AuCl_3$, chloroform, trisodium citrate, and tetramethylammonium hydroxide.

3. Glassware and stir bars that have been soaked in aqua regia, rinsed with deionized water and acetone, and then dried and stored in an oven.

4. A temperature controller, a heating mantel, and a heating probe for control the temperature and the heating rate.

5. 10,000 MCFWO filter for purification.

**2.2 Polyamides and Small Molecule Preparation and Conjugation Chemistry**

1. Small vials, such as cryogenic tubes, to mix and shake the small amounts of solutions added.

2. Hairpin polyamides synthesized using conventional solid phase synthesis methods.

3. Custom-ordered AD and NLS.

4. CTB synthesized using established organic methods.

**2.3 Cell Culture and Analysis**

1. Cell culture media and equipment (e.g., tissue culture plates).

2. Standard PCR kit and primers for chondrogenic genes, such as aggrecan, sox9, and collagen II.

3. Standard antibody staining supplies (e.g., bovine serum albumin, normal goat serum, formalin, triton X-100), primary mouse antibody against collagen and primary rabbit antibody against aggrecan, and secondary antibodies.

# 3    Methods

**3.1 MCNP Synthesis [6, 22, 23]**

1. First, synthesize the 10 nm MNP cores using thermal decomposition methods. Add 1.5174 mmol $Fe(Acac)_3$, 0.4825 mmol $FeCl_2$, 0.3338 mmol $ZnCl_2$, 10 mmol 1,2-hexadecandiol, 6 mmol oleic acid, 6 mmol oleylamine, and 20 mL tri-n-octylamine to a three-neck round-bottom flask setup with a small condenser and stir at 150 °C under vacuum for approximately 45 min (*see* **Note 2**).

2. Remove the vacuum and increase the reaction temperature to 200 °C at a rate of 4 °C per min and hold the temperature there for 2 h (*see* **Note 3** and **Note 4**).

3. Increase the temperature to 265 °C at a rate of 4 °C per min and hold the temperature there for 30 min.

4. Cool the reaction mixture to room temperature.

5. Purify the reaction mixture by dispersing it in ethanol.

6. Centrifuge the reaction mixture at $9,600 \times g$ in a microcentrifuge several times to produce a dry-looking pellet.

7. Disperse the particles and store them in chloroform.

8. Characterize the particles by dynamic light scattering (DLS) and transition electron microscopy (TEM).

9. To reduce gold salts onto the MNP cores, add 5 mg of the 10 nm magnetic cores to a 50 mL three-neck round-bottom flask and mix in 20 mL of tri-n-octylamaine.

10. Heat the mixture to 60 °C under vacuum for 10 min to evaporate the chloroform.

11. Cool the mixture to room temperature, add 100 µL (0.3 mmol) oleylamine and 60 µL of a 5 mg/300 µL stock solution of $AuCl_3$, and heat the mixture to 70 °C under vacuum to evaporate the solvent.

12. Increase the temperature to 150 °C at a rate of 10 °C per min and hold the temperature there for 4 h.

13. Cool to room temperature and centrifuge at 9,600 × $g$ to collect the particles, which should then be purified in chloroform and magnetically decanted several times.

14. Store the resulting particles in a minimal amount of chloroform.

15. Render the particles water soluble by carrying out a ligand exchange in a trisodium citrate buffer prepared in 1 M tetramethylammonium hydroxide using ultrasonication and then purify them in deionized water and magnetically decant them several times.

16. Store the resulting particles in a known amount of deionized water.

17. Characterize the particles using DLS and TEM (Fig. 2c) (*see* **Note 5**).

### 3.2 Polyamide and Small Molecule Preparation and Conjugation Chemistry

1. Conjugate the DBD, AD, NLS, and CTB. First, prepare the PEG ester linker using EDC/NHS coupling (*see* **Note 6**).

2. Prepare a 50 mM PEG solution in DMF.

3. Prepare a 1 M NHS solution in DMF.

4. Prepare a 1 M EDC solution, initially dissolved in a minimal amount of deionized water, and by bringing it to volume with DMF.

5. Combine 25 µL of PEG, 6.25 µL of NHS, and 6.25 µL of EDC and shake or stir for 30 min–2 h.

6. Prepare the small molecules by combining them with the PEG ester linker molecule (SH-PEG-COOH) and then conjugating them to the MCNPs.

7. First, weigh out the DBD, AD, NLS, and CTB in separate vials and dissolve them in DMF to obtain a 10 mM solution (*see* **Note 7**).

8. Combine 12.5 µL of each small molecule separately with 3.75 µL of the PEG ester and shake for 2 h. You should have four separate mixtures from this step termed, SH-PEG-COOH-DBD, SH-PEG-COOH-AD, SH-PEG-COOH-NLS, and SH-PEG-COOH-CTB.

9. Based on the desired percentages of each component of the particle surface, combine the appropriate amounts of each

mixture from Subheading 3.2, **step 8** then slowly add 1 mL of MCNPs, and shake or stir for 2 h (*see* **Note 8**).

10. Purify the particles using a 10,000 MCFWO filter to remove any unreacted excess material (*see* **Note 9**).

11. Analyze the particles using a Zetasizer to obtain their zeta potential and hydrodynamic size and image them using a TEM (*see* **Note 10**).

*3.3 Delivery of NanoScript*

1. Once NanoScript has been synthesized, deliver it to cells (*see* **Note 11**) by adding it to culture media (*see* **Note 12**). Supplement with serum-free media to aid in cell uptake.

2. Pipette the nanoparticle solution directly into cell culture media and swirl gently to disperse the particles.

3. Place the plate containing the cells and NanoScript on a sterile magnet in a cell culture incubator for 30 min to increase the uptake of NanoScript into cells (*see* **Note 13**).

4. After 4 h, detransfect the cells by removing the culture media and replacing it with fresh differentiation media (*see* **Note 14**).

5. After transfection, care for the cells as required for that cell type.

*3.4 Evaluating NanoScript*

1. Measure size of NanoScript using transmission electron microscopy and dynamic light scattering (*see* **Note 15**).

2. Quantify the robust nuclear localization of NanoScript and cellular uptake using inductively coupled plasma atomic emission spectroscopy (ICP-OES), dye-labeled nanoparticles, and cross-sectional TEM (Fig. 3c, d, h, i).

3. Measure the efficacy of NanoScript to induce transcriptional activation using PCR (utilized primers for chondrogenic genes, such as aggrecan, sox9, and collagen II) and immunostaining (utilizing primary mouse antibody against collagen and primary rabbit antibody against aggrecan) (Fig. 3a, b, e) (*see* **Note 16**).

4. Measure cell function using functional tests. For chondrogenesis, functional tests were not performed; however, for neuronal differentiation, calcium imaging and patch clamp recording showed that induced neurons behaved similarly to natural neurons (Fig. 3f, g).

# 4   Notes

1. If MCNPs are not accessible, they can easily be substituted with several other types of NPs, such as AuNPs.

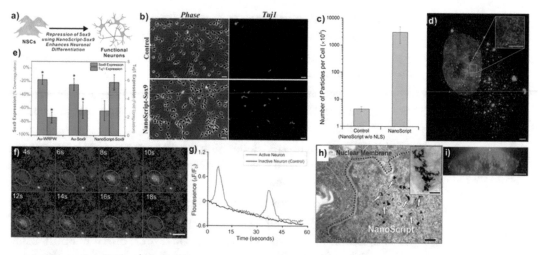

Fig. 3 **Evaluating NanoScript**. (a) Schematic guiding neural stem cells to becoming functional mature neurons. (b) Tuj1 immunofluorescence images showing efficient conversion of NSCs to functional neurons. (c, d) Number of NanoScript particles that enter the cell, dye labeled articles for fluorescence imaging. (e) PCR showing NanoScript control of transcription in the cell. (f) Calcium imaging for testing functionality of induced neurons. (h, j) Cross-sectional TEM and 3D illumination microscopy for determining nuclear localization of NanoScript [3, 6]. Modified from http:/pubs.acs.org/doi/full/10.1021/nn501589f

2. Since the particles are magnetic, the stir bar, which is magnetic, should be constantly stirring at a high setting throughout the reaction time. When removing the particles from the reaction vessel, use a magnet to remove the stir bar from the solution as quickly as possible to avoid particles sticking to the magnetic stir bar.

3. To keep the reaction at an elevated temperature, use insulation, such as steel wool.

4. When stirring under atmosphere at elevated temperatures, it is recommended to have a small vent to release excess pressure.

5. The MCNPS should appear a pink/red color throughout the whole synthesis. If the solution turns purple or colorless, that is an indication that the particles have aggregated.

6. An examples of AD is {GLY}{D-SER}{D-ASP}{D-ALA}{D-LEU}{D-ASP}{D-ASP}{D-PHE}{D-ASP}{D-LEU}{D-ASP}{D-MET}{D-LEU}{GLY}{D-SER}{D-ASP}{D-ALA}{D-LEU}{D-ASPD-ASP}{D-PHE}{D-ASP}{D-LEU}{D-ASP}{D-MET}{D-LEU}{GLY}{D-SER}. An example of NLS is CALN-NAGRKKRRQRRR. An example of DBD is NGN2.

7. Weigh out the minimal amount of peptide and small molecules needed since the solutions have a short shelf life.

8. To attain the percentages calculate the surface area of the MCNPs and determine the ratio of each component. In the past we have added 25% CTB, 20% TAT, 20% DBD, 25% AD, and 10% PEG ester.

9. Aggregation can occur if peptides are added at too high a concentration.

10. NanoScript can be stored at 4 °C for up to two weeks while maintaining efficacy.

11. NanoScript has been tested in adipose-derived mesenchymal stem cells, human neuro progenitor cells, and otic progenitor cells.

12. Dosing may vary depending on the application; however, we have seen that a dose of 2 nM is effective and cytotoxic effects are not observed.

13. Magnetofection has been shown to double the amount of NanoScript uptaken compared to regular transfection [2].

14. Studies have shown that by Day 7 of differentiation using NanoScript, 95% of NanoScript is removed from the cells. Avoid harsh pipetting as it may damage the cells.

15. Our studies have shown the size to be 34 nm.

16. We have studied chondrogenesis, myogenesis, and neurogenesis.

## References

1. Reik W (2007) Stability and flexibility of epigenetic gene regulation in mammalian development. Nature 447:425–432

2. Spitz F, Furlong EEM (2012) Transcription factors: from enhancer binding to developmental control. Nat Rev Genet 13:613–626

3. Patel S, Jung D, Yin PT, Carlton P, Yamamoto M, Bando T, Sugiyama H, Lee K-B (2014) NanoScript: a nanoparticle-based artificial transcription factor for effective gene regulation. ACS Nano 9:8959–8967

4. Patel S, Yin PT, Sugiyama H, Lee K-B (2015) Inducing stem cell myogenesis using NanoScript. ACS Nano 9:6909–6917

5. Patel S, Pongkulapa T, Yin PT, Pandian G, Rathnam C, Bando T, Vaijayanthi T, Sugiyama H, Lee K-B (2015) Integrating epigenetic modulators into NanoScript for enhanced chondrogenesis of stem cells. J Am Chem Soc 137:4598–4601

6. Patel S, Chueng STD, Yin PT, Dardir K, Song Z, Pasquale N, Kwan K, Sugiyama H, Lee K-B (2015) Induction of stem-cell-derived functional neurons by NanoScript-based gene repression. Angew Chem Int Ed Engl 54:11983–11988

7. Giljohann DA, Seferos DS, Daniel WL, Massich MD, Patel PC, Mirkin CA (2010) Gold nanoparticles for biology and medicine. Angew Chem Int Ed Engl 49:3280–3294

8. Colombo M, Carregal-Romero S, Casula MF, Gutiérrez L, Morales MP, Böhm IB, Heverhagen JT, Prosperi D, Parak WJ (2012) Biological applications of magnetic nanoparticles. Chem Soc Rev 41:4306–4334

9. Lim J, Majetich SA (2013) Composite magnetic–plasmonic nanoparticles for biomedicine: manipulation and imaging. Nano Today 8:98–113

10. Robertson KD (2002) DNA methylation and chromatin: unraveling the tangled web. Oncogene 21:5361–5379

11. Dervan PB, Edelson BS (2003) Recognition of the DNA minor groove by pyrrole-imidazole polyamides. Curr Opin Struct Biol 13:284–299

12. Melander C, Burnett R, Gottesfeld JM (2004) Regulation of gene expression with pyrrole_imidazole polyamides. J Biotechnol 112:195–220

13. Hoelz A, Debler EW, Blobel G (2011) The structure of the nuclear pore complex. Annu Rev Biochem 80:613–643

14. Zaret KS, Carroll JS (2011) Pioneer transcription factors: establishing competence for gene expression. Genes Dev 25:2227–2241

15. Portela A, Esteller M (2010) Epigenetic modifications and human disease. Nat Biotechnol 28:1057–1068

16. Lévy R, Shaheen U, Cesbron Y, Sée V (2010) Gold nanoparticles delivery in mammalian live cells: a critical review. Nano Rev 1. doi:10. 3402/nano.v1i0.4889

17. Ter-Avetisyan G et al (2009) Cell entry of arginine-rich peptides is independent of endocytosis. J Biol Chem 284:3370–3378

18. Newmeyer DD, Forbes DJ (1988) Nuclear import can be separated into distinct steps in vitro: nuclear pore binding and translocation. Cell 52:641–653

19. Pilch DS, Poklar N, Baird EE, Dervan PB, Breslauer KJ (1999) The thermodynamics of polyamide–DNA recognition: hairpin polyamide binding in the Minor Groove of duplex DNA. Biochemistry 38:2143–2151

20. Nyanguile O, Uesugi M, Austin DJ, Verdine GL (1997) A nonnatural transcriptional coactivator. Proc Natl Acad Sci 94:13402–13406

21. Chen G, Courey AJ (2000) Groucho/TLE family proteins and transcriptional repression. Gene 249:1–16

22. Shah BP, Pasquale N, De G, Tan T, Ma J, Lee K-B (2014) Core–shell nanoparticle-based peptide therapeutics and combined hyperthermia for enhanced cancer cell apoptosis. ACS Nano 8:9379–9387

23. Sun S, Zeng H, Robinson DB, Raoux S, Rice PM, Wang SX, Li G (2004) Monodisperse MFe2O4 (M = Fe, Co, Mn) nanoparticles. J Am Chem Soc 126:273–279

# Chapter 17

# Glucose-Responsive Insulin Delivery by Microneedle-Array Patches Loaded with Hypoxia-Sensitive Vesicles

## Jicheng Yu, Yuqi Zhang, and Zhen Gu

## Abstract

In this chapter, we describe the preparation of glucose-responsive vesicles (GRVs) and the fabrication of GRV-loaded microneedle-array patches for insulin delivery. The GRVs were formed of hypoxia-sensitive hyaluronic acid (HS-HA), the synthesis of which is presented in detail. We also describe the procedure to evaluate the in vivo efficacy of this smart patch in a mouse model of chemically induced type 1 diabetes through transcutaneous administration.

**Key words** Diabetes, Drug delivery, Glucose-responsive, Hypoxia-sensitive, Microneedle

## 1 Introduction

Diabetes, which currently affects 415 million people worldwide, is one of the most common chronic metabolic diseases of children and adults [1, 2]. It is essential for diabetic patients to self-monitor their blood glucose levels and inject themselves with the correct dose of insulin [3]. However, this traditional route of care is painful and imprecise, which can cause serious complications such as limb amputation, blindness, and kidney failure [2, 4]. In addition, hypoglycemia caused by excessive insulin dosing may lead to behavioral and cognitive disturbance, seizure, brain damage, and even death [5]. "Smart" glucose-responsive insulin systems that mimic the function of pancreatic $\beta$-cells and "secrete" insulin in response to elevated blood glucose levels are desirable to improve glycemic control and quality of life for diabetic patients [1, 6]. One example is the wearable closed-loop electronic insulin pump that combines a continuous glucose-monitoring sensor and an external insulin infusion pump [2]. However, several challenges such as the lag in blood glucose equilibration with the interstitium and biofouling still remain today. In addition, chemically controlled closed-loop formulations have also been widely explored [1, 2, 6]. In this

Sarah Hurst Petrosko and Emily S. Day (eds.), *Biomedical Nanotechnology: Methods and Protocols*, Methods in Molecular Biology, vol. 1570, DOI 10.1007/978-1-4939-6840-4_17, © Springer Science+Business Media LLC 2017

strategy, an insulin-loaded matrix with glucose-responsive elements can undergo structural changes regulated by glucose concentration changes, resulting in glucose-dependent insulin release; typical glucose-responsive moieties include glucose oxidase (GOx), glucose-binding proteins (GPB), and phenylboronic acid (PBA) [7–11]. Despite these promising platforms, challenges still persist in demonstrating a system that could combine fast responsiveness, ease of administration, and biocompatibility without long-term side effects.

We have recently developed a microneedle (MN)-patch device consisting of glucose-responsive vesicles (GRVs) for insulin delivery [12]. Instead of using pH-sensitive material as reported before [7–9], these vesicles were formed by hypoxia-sensitive hyaluronic acid (HS-HA) and encapsulated insulin and GOx, which can convert glucose to gluconic acid, consuming oxygen to generate a hypoxic environment (Fig. 1a). The HS-HA was synthesized by conjugating amine-functionalized 2-nitroimidazole (NI) with hyaluronic acid (HA) through an amide bond. The hydrophobic NI group, as the hypoxia-sensitive component, can be reduced into hydrophilic 2-aminoimidazoles *via* a single-electron reaction catalyzed by nitroreductases with bioreducing agents in a hypoxic environment [13, 14]. The reduced product is water-soluble, which facilitates the dissociation of GRVs. The GRVs were further incorporated with a MN-array patch for ease of administration [15, 16]. When the patch is applied onto the skin, the MNs are exposed to the high interstitial fluid glucose in the vascular and lymph capillary networks [17]. As the blood glucose level rises, a localized hypoxic environment is generated within the GRVs due to the enzymatic activity of GOx, which promotes the release of insulin (Fig. 1b). We demonstrated that this "smart insulin patch" with a

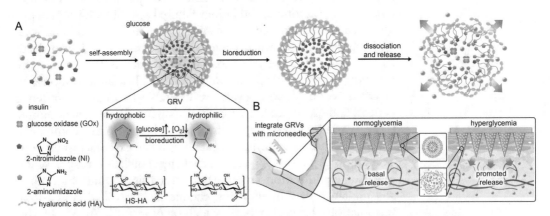

**Fig. 1** Schematic of the glucose-responsive insulin delivery system using hypoxia-sensitive vesicle-loading MN-array patches. (**a**) Formation and mechanism of GRVs composed of HS-HA. (**b**) Schematic of the GRV-containing MN-array patch (smart insulin patch) for in vivo insulin delivery triggered by a hyperglycemic state to release more insulin. Reproduced from ref. 12 with permission

novel trigger mechanism provided a tight glucose regulation with rapid responsiveness and reliable avoidance of hypoglycemia in a mouse model of chemical-induced type 1 diabetes. This study demonstrates the potential benefit of an on-demand MN-array patch that can be activated and self-regulated in response to physiological signals.

In this chapter, we describe in detail the synthesis and preparation of this glucose-responsive insulin device. The methods provided are appropriate for the evaluation of the smart insulin patch in a mouse model of chemical-induced type 1 diabetes.

## 2   Materials

All the solutions are prepared using ultrapure, deionized water (18 M$\Omega$·cm, Millipore), and analytical grade reagents. All reagents were used as received.

### 2.1   Synthesis of HS-HA and m-HA

1. 2-Nitroimidazole (NI).
2. Potassium carbonate ($K_2CO_3$).
3. $N,N$-dimethylformamide (DMF).
4. 6-(Boc-amino) hexyl bromide.
5. Methanol.
6. Ethyl acetate.
7. Hydrochloric acid (HCl).
8. Hyaluronic acid (HA) (300 KDa).
9. 1-ethyl-3-(3-dimethylaminopropyl)carbodiimide (EDC).
10. $N$-Hydroxysuccinimide (NHS).
11. Methacrylic anhydride.
12. Sodium hydroxide (NaOH).
13. Ethanol.
14. Human recombinant insulin (27.5 IU/mg of Zn salt, Life Technology).
15. Glucose oxidase (GOx).
16. Dialysis membrane.

### 2.2   Preparation of MN-Array Patches

1. Silicone mold (Blueacre Technology Ltd).
2. Irgacure 2959.
3. $N,N'$-methylenebisacrylamide.

### 2.3   Animal Experiments

1. Streptozotocin (STZ)-induced adult diabetic mice (male C57B6; Jackson Laboratory).
2. Clarity GL2Plus glucose meter (Clarity Diagnostics).

## 3    Methods

### 3.1  Synthesis of Hypoxia-Sensitive Polymer, HS-HA

HS-HA was synthesized by chemical conjugation with the 6-(2-nitroimidazole) hexylamine through amide formation (Fig. 2).

#### 3.1.1  Synthesis of 6-(2-Nitroimidazole) Hexylamine

1. Dissolve NI (0.15 g, 1.3 mmol) in DMF and add $K_2CO_3$ (0.28 g, 2.0 mmol) under stirring conditions.

2. Into this solution, add 6-(Boc-amino) hexyl bromide (0.39 g, 1.4 mmol) dropwise. Stir the reaction mixture at 80 °C for 4 h.

3. Remove solid impurities from the reaction mixture using a filter (*see* **Note 1**), and wash with methanol. Collect the residual solution, and evaporate the solvent using a rotary evaporator to obtain the crude product (*see* **Note 2**).

4. Suspend the crude mixture into 10 mL of water, and extract with 10 mL of ethyl acetate. Collect the organic layer and dry over sodium sulfate, then concentrate it.

5. Redissolve the product in 10 mL of methanol on ice, and add 5 mL of 1.25 M HCl in methanol while stirring. Stir the reaction mixture for 2 h.

6. Remove organic solvent using rotary evaporator.

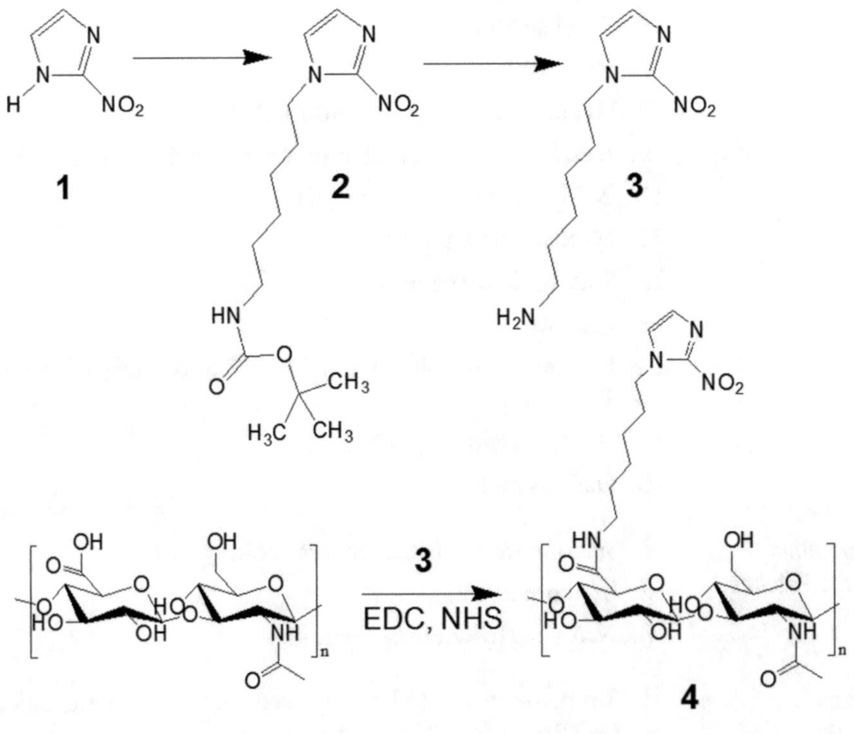

**Fig. 2** Synthesis scheme of hypoxia-sensitive hyaluronic acid (HS-HA). Reproduced from ref. 12 with permission

*3.1.2  Synthesis of*
*Hypoxia-Sensitive HA*

1. Activate HA (0.24 g) in water with EDC (0.56 g, 3.4 mmol) and NHS (0.39 g, 3.4 mmol) for 15 min.

2. Dissolve 0.18 g of 6-(2-nitroimidazole) hexylamine in 5 mL of DMF.

3. Add the 6-(2-nitroimidazole) hexylamine solution into the activated HA solution dropwise while stirring. Stir overnight at room temperature.

4. The following day, dialyze the reaction solution (MWCO 8 KDa) against a 1:1 mixture of DI water and methanol for 1 day, and then against DI water for 2 days.

5. Freeze-dry the dialysate to obtain a fluffy yellow solid.

### 3.2  Preparation of Glucose-Responsive Vesicle Formulation

1. Dissolve 20 mg of HS-HA in 2 mL of water/methanol (1:1 vol/vol).

2. Dissolve 10 mg of human insulin and 1 mg of GOx in 1 mL of water (*see* **Note 3**).

3. Add the HS-HA solution into the insulin and GOx solution dropwise while stirring. Stir for 2 h at 4 °C.

4. Dialyze the mixture (MWCO 3 KDa) against DI water for 1 day to remove the methanol.

5. Adjust the pH value of the resulting GRV suspension to 5.3 (the pI of insulin) with 0.1 M HCl solution. Centrifuge at $6200 \times g$ for 10 min in a microcentrifuge.

6. Filter the supernatant using a centrifugal filter (100,000 Da MWCO, Millipore).

7. Adjust the pH value of the resulting suspension to 7.4, and store at 4 °C. The resulting vesicle encapsulating insulin and enzyme is designated as GRV (E + I).

8. In order to prepare vesicles without glucose-specific enzymes (GRV (I)), dissolve 10 mg of human insulin in 1 mL of water as in Subheading 3.2, **step 2**. Then, follow **steps 3–7** in Subheading 3.2 to prepare GRV (I).

9. For vesicles with half doses of enzymes (GRV (1/2E + I)), dissolve 10 mg of human insulin and 0.5 mg of GOx in 1 mL of water as in Subheading 3.2, **step 2**. Then, follow **steps 3–7** in Subheading 3.2 to prepare GRV (1/2E + I).

10. Perform transmission electron microscopy (TEM) analysis to verify the formation of GRVs. Figure 3a shows a representative TEM image of the GRVs.

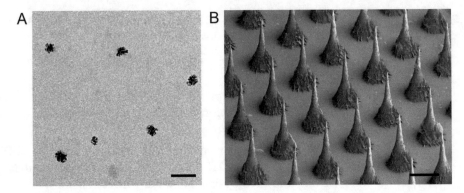

**Fig. 3** (**a**) TEM images of GRVs encapsulating insulin and enzyme. Scale bar is 200 nm. (**b**) SEM image of an MN array. Scale bar is 200 μm. Reproduced from ref. 12 with permission

*3.3 Synthesis of Methacrylated HA (m-HA) (for Microneedle Fabrication)*

1. Dissolve 1.0 g of HA in 50 mL of water at 4 °C, and add 0.8 mL of methacrylic anhydride dropwise while stirring.

2. Adjust the reaction solution to pH 8–9 with 5 M NaOH solution and stir for 24 h at 4 °C.

3. The following day, add the reaction solution into 500 mL of acetone dropwise to precipitate out the crude product. Collect the solid by centrifugation and wash with ethanol three times.

4. Suspend the crude product in DI water and purify by dialysis (MWCO 8 KDa) against DI water for 2 days.

5. Freeze-dry the dialysate to obtain a fluffy white solid.

*3.4 Preparation of GRV-Loaded MNs*

1. Suspend the prepared GRVs in DI water (100 mg/mL).

2. Deposit 35 μL of the prepared GRV solution onto the MN mold surface using a pipette (*see* **Note 4**). Place the mold under vacuum for 5 min, follow by centrifugation using a Hettich Universal 32R centrifuge for 20 min at 440 × *g*. Repeat this process three times until the GRV solution layer is completely dry.

3. Apply a piece of 4 cm × 9 cm silver adhesive tape around the 2 cm × 2 cm MN mold baseplate.

4. Dissolve Irgacure 2959 (0.1 g) into 50 mL of *m*-HA solution (2 g) and *N,N'*-methylenebisacrylamide (1 g) in water (*see* **Note 5**).

5. Add 3 mL of the prepared solution into the MN mold reservoir, followed by centrifugation for 20 min at 440 × *g*. Place the mold under a vacuum desiccator at 25 °C until the *m*-HA layer is completely dry.

6. Separate the resulting MN-array patch from the mold (*see* **Note 6**), and place it under a UV irradiation lamp (365 nm) for 10 s to initiate the crosslinking polymerization (*see* **Note 7**).

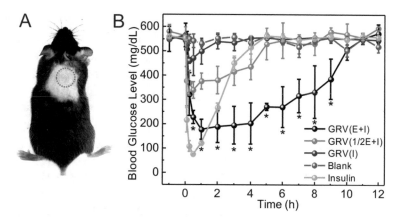

**Fig. 4** (**a**) Mouse dorsum and relevant skin (the area within *red dashed line*) transcutaneously treated with an MN-array patch. (**b**) Blood glucose levels in STZ-induced diabetic mice after treatment with blank MNs containing only crosslinked HA, MNs loaded with human recombinant insulin, MNs loaded with GRVs containing insulin and enzyme (GRV(E + I)), MNs loaded with GRVs containing insulin and half amount of enzyme (GRV(1/2E + I)), or MNs loaded with GRVs containing insulin (GRV(I)). Reproduced from ref. 12 with permission

7. Store the prepared MN-array patch at 4 °C.

8. Perform scanning electron microscopy (SEM) analysis to verify the fabrication of the MNs. Figure 3b shows a representative SEM image of a GRV-loaded MN-array patch.

**3.5 Assessment of In Vivo Efficacy in STZ-Induced Type 1 Diabetic Mice**

1. Randomly divide the mice into experimental groups (5 mice/ group), and subject them to different MN-array patches: blank MNs containing only cross-linked HA, MNs loaded with human recombinant insulin, MNs loaded with GRV (E + I), MNs loaded with GRV (1/2E + I), or MNs loaded with GRV (I) (insulin dose: 10 mg/kg for each mouse). Remove the dorsal hair with hair remover under anesthesia.

2. Administer the MN-array patches onto the mice dorsal skin and fix with a topical skin adhesive (Fig. 4a) (*see* **Note 8**).

3. Monitor the glucose levels of each mouse by measuring the tail vein blood sample ($\sim$3 μL) with a glucose meter (at 5, 15, 30, and 60 min, and once per hour afterward) until a return to stable hyperglycemia is observed (Fig. 4b).

# 4 Notes

1. Excess $K_2CO_3$, which is insoluble in DMF, is removed by filteration to obtain the crude product.

2. The vacuum condition in the rotary evaporator does not induce the reduction of the NI group, since the reductive

reaction is catalyzed by nitroreductases with bioreducing agents [13, 14].

3. Insulin is insoluble in water at pH 7.4. To prepare the insulin solution, dissolve insulin in water by adjusting the pH to 3.0 with a 0.1 M HCl solution. After completely dissolving the insulin, adjust the pH to 7.4 with a 0.1 M NaOH solution.

4. To prepare blank MNs containing only cross-linked HA, directly follow **steps 3–7** (Subheading 3.2). To prepare MNs loaded with human recombinant insulin, MNs loaded with GRV (E + I), MNs loaded with GRV (1/2E + I), and MNs loaded with GRV (I), deposit insulin, GRV (E + I), GRV (1/2E + I), or GRV (I) solution onto the MN mold, respectively.

5. After dissolving Irgacure 2959, the resulting solution should be protected from light to avoid undesirable polymerization.

6. Medical tape can help to remove MNs from the mold. The patch base can be further tailored to fit the injection.

7. For your safety, UV-blocking eye protection goggles should be used when performing the UV-initiated polymerization.

8. Make sure that the MNs are inserted perpendicular to the skin to avoid breakage.

## Acknowledgment

This work was supported by the grants from the American Diabetes Association (ADA) to Z.G. (1-14-JF-29 and 1-15-ACE-21) and a grant from NC TraCS, NIH's Clinical and Translational Science Awards (CTSA, NIH grant 1UL1TR001111) at UNC-CH.

## References

1. Mo R et al (2014) Emerging micro-and nano-technology based synthetic approaches for insulin delivery. Chem Soc Rev 43 (10):3595–3629

2. Veiseh O et al (2015) Managing diabetes with nanomedicine: challenges and opportunities. Nat Rev Drug Discov 14(1):45–57

3. Owens DR, Zinman B, Bolli GB (2001) Insulins today and beyond. Lancet 358 (9283):739–746

4. Bratlie KM et al (2012) Materials for diabetes therapeutics. Adv Healthc Mater 1 (3):267–284

5. Ohkubo Y et al (1995) Intensive insulin therapy prevents the progression of diabetic microvascular complications in Japanese patients with non-insulin-dependent diabetes mellitus: a randomized prospective 6-year study. Diabetes Res Clin Pract 28(2):103–117

6. Ravaine V, Ancla C, Catargi B (2008) Chemically controlled closed-loop insulin delivery. J Control Release 132(1):2–11

7. Gu Z et al (2013) Injectable nano-network for glucose-mediated insulin delivery. ACS Nano 7 (5):4194–4201

8. Gu Z et al (2013) Glucose-responsive microgels integrated with enzyme nanocapsules for closed-loop insulin delivery. ACS Nano 7 (8):6758–6766

9. Tai W et al (2014) Bio-inspired synthetic nanovesicles for glucose-responsive release of insulin. Biomacromolecules 15(10):3495–3502

10. Chou DH-C et al (2015) Glucose-responsive insulin activity by covalent modification with

aliphatic phenylboronic acid conjugates. Proc Natl Acad Sci 112(8):2401–2406

11. Kim SW et al (1990) Self-regulated glycosylated insulin delivery. J Control Release 11 (1):193–201

12. Yu J et al (2015) Microneedle-array patches loaded with hypoxia-sensitive vesicles provide fast glucose-responsive insulin delivery. Proc Natl Acad Sci 112(27):8260–8265

13. Nunn A, Linder K, Strauss HW (1995) Nitroimidazoles and imaging hypoxia. Eur J Nucl Med 22(3):265–280

14. Krohn KA, Link JM, Mason RP (2008) Molecular imaging of hypoxia. J Nucl Med 49(Suppl 2):129S–148S

15. Prausnitz MR (2004) Microneedles for transdermal drug delivery. Adv Drug Deliv Rev 56 (5):581–587

16. Yang SY et al (2013) A bio-inspired swellable microneedle adhesive for mechanical interlocking with tissue. Nat Commun 4:1702

17. Heo YJ et al (2011) Long-term in vivo glucose monitoring using fluorescent hydrogel fibers. Proc Natl Acad Sci 108(33):13399–13403

# Chapter 18

# Electrospun Nanofiber Scaffolds and Their Hydrogel Composites for the Engineering and Regeneration of Soft Tissues

## Ohan S. Manoukian, Rita Matta, Justin Letendre, Paige Collins, Augustus D. Mazzocca, and Sangamesh G. Kumbar

## Abstract

Electrospinning has emerged as a simple, elegant, and scalable technique that can be used to fabricate polymeric nanofibers. Pure polymers as well as blends and composites of both natural and synthetic ones have been successfully electrospun into nanofiber matrices for many biomedical applications. Tissue-engineered medical implants, such as polymeric nanofiber scaffolds, are potential alternatives to autografts and allografts, which are short in supply and carry risks of disease transmission. These scaffolds have been used to engineer various soft tissues, including connective tissues, such as skin, ligament, and tendon, as well as nonconnective ones, such as vascular, muscle, and neural tissue. Electrospun nanofiber matrices show morphological similarities to the natural extracellular matrix (ECM), characterized by ultrafine continuous fibers, high surface-to-volume ratios, high porosities, and variable pore-size distributions. The physiochemical properties of nanofiber matrices can be controlled by manipulating electrospinning parameters so that they meet the requirements of a specific application.

Nanostructured implants show improved biological performance over bulk materials in aspects of cellular infiltration and in vivo integration, taking advantage of unique quantum, physical, and atomic properties. Furthermore, the topographies of such scaffolds has been shown to dictate cellular attachment, migration, proliferation, and differentiation, which are critical in engineering complex functional tissues with improved biocompatibility and functional performance. This chapter discusses the use of the electrospinning technique in the fabrication of polymer nanofiber scaffolds utilized for the regeneration of soft tissues. Selected scaffolds will be seeded with human mesenchymal stem cells (hMSCs), imaged using scanning electron and confocal microscopy, and then evaluated for their mechanical properties as well as their abilities to promote cell adhesion, proliferation, migration, and differentiation.

**Key words** Nanofiber scaffolds, Electrospinning, Tissue engineering, Regenerative medicine, Soft tissue, Biodegradable polymers, Hydrogels, Skin, Tendon, Nerve

## 1 Introduction

More than two decades ago the field of tissue engineering began to rapidly grow as an interdisciplinary field, applying principles of engineering and the life sciences toward the development of

Sarah Hurst Petrosko and Emily S. Day (eds.), *Biomedical Nanotechnology: Methods and Protocols*, Methods in Molecular Biology, vol. 1570, DOI 10.1007/978-1-4939-6840-4_18, © Springer Science+Business Media LLC 2017

biological substitutes that restore, maintain, or improve tissue function [1]. Tissue engineering approaches generally involve the fabrication and use of three-dimensional (3D) scaffolds composed of biocompatible materials. These biocompatible materials, hereafter referred to as biomaterials, can support cell in-growth and proliferation, whilst eliciting no adverse immune effects [2].

Autografts and allografts remain the "gold standard" treatment for many orthopedic illnesses and injuries. Biological grafts, despite their many merits, have several limitations including limited availability, donor site morbidity, immunogenicity, and possible disease transmission [3, 4]. Tissue-engineered implants, such as biodegradable 3D porous scaffolds, have emerged as a viable alternative to these biological autografts and allografts to repair/regenerate damaged tissues and restore functionality. As such, scaffolds of both natural and/or synthetic origin are being designed to mimic the structures and functions of native tissues [5]. The ideal scaffold for tissue engineering must satisfy a number of often conflicting criteria: (1) the pores must be at an appropriate density and size to allow for optimal cell migration; (2) the surface area must be sufficient with chemistries that encourage cell adhesion, proliferation, migration, and differentiation; and (3) the degradation rate must closely match the regeneration rate of the target natural tissue [6].

In recent years, nanomaterials have emerged as useful components in varying scientific and medical applications. The field of nanoscience and technology is concerned with natural and artificial structures with at least one dimension on the nanometer scale (i.e., in the range from 1 µm down to 1 nm) [7]. Nanomaterials have unusual properties as compared to conventional macro-sized and bulk materials due to their nanoscale dimensions. These nanoscale material properties are based on the "quantum effect." Nanostructures have extraordinarily high surface area-to-volume ratios, and they can display tunable optical emissions and super paramagnetic behavior, which can be successfully exploited for a variety of health care applications ranging from drug delivery to biosensing [7]. Biomaterial scaffolds often have exceptionally high surface areas and porosities because they are comprised of fibers with nanoscale diameters; this property enables them to encourage superior cell infiltration and adhesion. Their nanoscale size also grants them superior mechanical and degradation properties compared to larger, micro-sized porous materials [7].

Polymer fibers with sizes in the micrometer regime (ranging from 10 to 100 µm) can be produced by conventional melt, dry, or wet spinning processes. Nanofibers refer to fibers with diameters less than 1 µm [7]. Nanofibers are fabricated using a variety of techniques, including drawing, template synthesis, phase separation, molecular self-assembly, and electrospinning [7] (Table 1). Electrospinning is a broadly useful, and relatively recently developed, technology that relies on the application of an electrostatic

**Table 1**
**Comparison of different nanofiber fabrication techniques: advantages and disadvantages**

| Fabrication technique | Advantages | Disadvantages |
|---|---|---|
| Drawing | • Simple equipment | • Discontinuous process<br>• Not scalable<br>• No control on fiber dimensions |
| Template synthesis | • Continuous process<br>• Fiber dimensions can be varied using different templates | • Not scalable |
| Temperature-induced phase separation | • Simple equipment<br>• Convenient fabrication process<br>• Mechanical properties of fiber matrices can be varied by changing polymer composition | • Not scalable<br>• No control on fiber dimensions<br>• Limited to specific polymers |
| Molecular self-assembly | • Produces only smaller nanofibers of few nanometers in diameter and few micrometers in length<br>• Complex functional structures | • Not scalable<br>• No control on fiber dimensions<br>• Complex process |
| Electrospinning | • Simple instrumentation<br>• Continuous process<br>• Cost-effective<br>• Scalable<br>• Ability to fabricate fiber diameters ranging from few nanometers to several micrometers<br>• Aligned and random-oriented fibers | • Jet instability<br>• Toxic solvents<br>• Large, bulky housing for instruments<br>• High voltage danger |

force to drive fiber formation. A high electric potential, often between 5 and 30 kV, is applied to a pendant droplet of polymer solution from a syringe. The polymer jet ejected from the surface of the charged solution overcomes surface tensions and is attracted to a grounded or oppositely charged collector target (stationary or moving).

The electrospinning process is simple, elegant, reproducible, continuous, and scalable. It is possible to fabricate fibers ranging from ~ 3 nm to 6 μm in diameter and several meters in length. Both natural and synthetic polymer, as well as hybrid polymers, can be electrospun into nanofiber matrices and then used in a variety of biomedical applications [7–11]. Several processing parameters affect the formation of electrospun fibers, including the needle diameter, solution flow rate, applied electric potential, polymer solution concentration, polymer molecular weight, and the distance to the target. The effects of polymer solution concentration on both fiber diameter and quality can be seen in Fig. 1. The size, shape, orientation, and dynamics of the target collector also plays

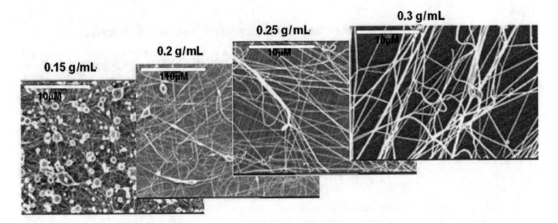

**Fig. 1** SEM micrographs of Poly(lactic-co-glycolic acid) (PLAGA) fiber matrices electrospun at various polymer solution concentrations while using constant spinning parameters at an applied voltage of 20 kV/cm, a flow rate of 2 mL/h, and ambient parameters. Increases in the polymer concentration (from 0.15 to 0.3 g/mL) leads to increases in polymer viscosity and decreases in bead density. A concentration of 0.25 g/mL resulted in bead-free nanofibers and the fiber diameters increased when higher concentrations were used. Reprinted with permission from © 2004 Wiley Periodicals, Inc. [12]

an important role in determining the outcome of the electrospinning process. The use of a stationary collector results in random nanofiber deposition, often with shorter fiber lengths, while the use of a rotating collector results in more oriented nanofibers, which are typically longer in length. Nanofibers may also be deposited in alignment using special collector configurations, which subject the charged fibers to local electric fields, forcing them to align in a particular orientation.

Although electrospun polymeric nanofibers can be fabricated for a variety of biomedical applications, this chapter focuses only on electrospun nanofibrous scaffolds and hydrogel composites and their related soft tissue engineering and regeneration applications. Particular focus will be given to their application in the regeneration of soft connective tissues (skin and tendons) as well as soft nonconnective tissues (nerves).

## 2   Materials

All chemicals used in the synthesis of the sodium alginate (SA)-coated polycaprolactone (PCL) scaffolds were used without further modification.

### 2.1   Electrospinning

1. Organic solvents: methylene chloride and ethanol (200 proof, lab grade).

2. Inorganic solvents: hydrogen peroxide, phosphate buffered saline (PBS), hydrochloric acid, sodium hydroxide, calcium chloride, deionized water.

3. Synthetic polymer components: polycaprolactone (PCL) (MW 80 kDa), alginic acid.

4. Glass vials (28 × 95 mm) with screw threads.

5. Parafilm.

6. Vortex mixer.

7. Syringe: 10 mL Luer-Lok tip.

8. Dispensing needle: 18-gauge, 1.0″ length, blunt end, stainless steel, threaded cap.

9. DC power: HV power supply.

10. Pump: Aladdin-1000 syringe.

11. Lab jack.

12. Grounded target: Fisherbrand Aluminum Foil.

13. Servomotor.

14. Wires and alligator clamps.

15. Centrifuge.

16. Thermo Scientific Smart Fourier Transform Infrared (FTIR) Spectrometer.

17. Cork-borer #5 (area = 1.96 cm$^2$).

18. Desiccator.

19. Universal test machine.

*2.2 Scanning Electron Microscopy*

1. Double-sided carbon tape (8 × 20 mm).

2. Aluminum SEM specimen mount stubs.

3. Sharp blade or scalpel.

4. Sputter coater with gold foil.

*2.3 Confocal Microscopy of Live and Dead Cells/ Immuno histochmeistry/Cell Proliferation*

1. Dulbecco's PBS 1×.

2. Live/Dead Viability/Cytotoxicity kit for mammalian cells.

3. Fine-tip tweezers.

4. Lab-Tek two-well glass chamber slide.

5. Human mesenchymal stem cells (hMSCs).

6. Quant-iT™ PicoGreen dsDNA assay (P7589).

7. Triton X-100.

8. BioTek plate reader.

9. Calcein-AM.

10. Ethidium homodimer-1 (EthD-1).

11. Formaldehyde.

12. Bovine serum albumin (BSA).

13. Tween 80 (TBST).

14. Rabbit-antiS-100 primary antibody.

15. AlexaFluor 488 secondary antibody.

## 3   Methods

### 3.1   Fabrication of Nanofiber Matrices

Electrospinning is based on the principle of inducing a static electrical charge on the molecules of a solution, which causes the liquid to stretch into a fiber [13]. This technique has advantages over other fabrication techniques due to its simplicity and elegance. In addition, fabrication using this method is both continuous and reproducible, unlike with other techniques [7]. Electrospinning can be used to produce both aligned and random fiber lattice matrices following established protocols [14–17].

Parallel electrodes may be used to align fibers in an oriented fashion (Fig. 2). The orientation of the collector has a direct impact on fiber deposition. Therefore, the parallel set up creates an electric field profile, which forces the charged nanofiber to span the gap area [13].

An alternative electrospinning setup, seen in Fig. 3, uses a grounded rotating plate to draw the charged fibers from the needle, coating the plate. When the force of the electric field overcomes the surface tension of the PCL solution in the syringe, the polymer is drawn towards the grounded plate. The rotation of the plate controls the orientation of the fibers—when the plate is stationary, the

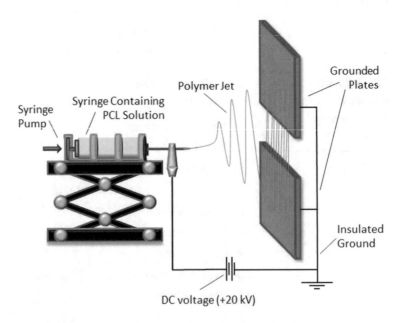

**Fig. 2** Diagram of a parallel-electrode collector system for the electrospinning of aligned fibers

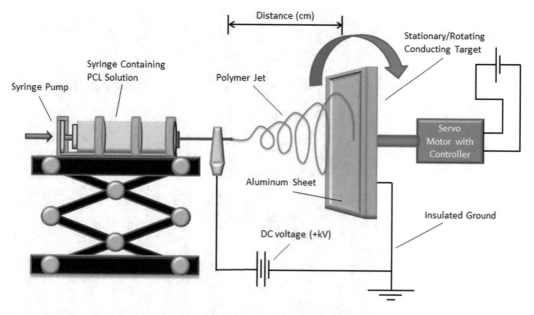

**Fig. 3** Schematic of the electrospinning set-up. As seen here, the syringe is placed atop the lab jack within the Aladdin-1000 syringe pump. The positive lead is attached to the 18-gauge needle via an alligator clamp, and the polymer jet is drawn towards the grounded aluminum sheet mounted on the plate/servo motor apparatus. The plate is either made to be stationary or rotate to produce aligned or randomly oriented fiber matrices, respectively, by powering the motor

fibers align in an ordered fashion, whereas if the plate is rotating the fibers take on a random configuration.

Well-established laboratory techniques can be used to fabricate both randomly oriented and aligned fiber lattice matrices (Fig. 4) (*see* **Notes 1–6**).

1. Prepare a 12.5 wt% solution of PCL by mixing methylene chloride and ethanol in an 85:15 ratio and adding the appropriate amount of PCL.

2. Draw the solution into the syringe and screw the 18-gauge needle onto the tip.

3. Fasten the syringe into the apparatus containing a mechanism to press the syringe pump, and release the polymer at a flow rate of 1 mL/h with a 20-kV applied voltage and a 15–20 cm working distance.

4. Place aluminum foil to cover the plate attached to the servo motor to ensure easy removal of the fibers following the electrospinning procedure [18].

5. As shown in Fig. 2, connect a grounded cable to the aluminum foil sheet and a positive lead to the 18-gauge needle via alligator clamps. For parameter optimization, refer to **Note 4**.

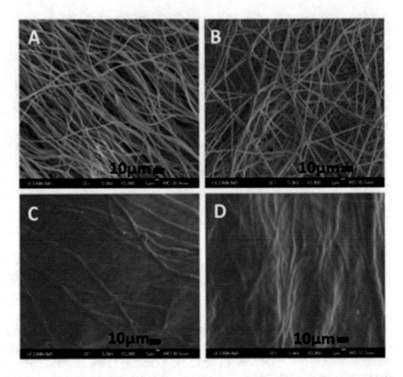

**Fig. 4** Scanning electron microscope images of (**a**) aligned PCL fibers, (**b**) randomly distributed PCL fibers, (**c**) aligned PCL fibers coated with LMW SA, and (**d**) randomly distributed PCL fibers coated with LMW SA.

*3.2 Preparation of Sodium Alginate*

PCL fiber meshes lack the optimal hydrophilic features that promote cell adhesion; coating them with sodium alginate (SA) can increase their hydrophilicity and impart them with desirable tensile properties/stiffness. Many forms of SA, including gels, beads, and sponges, can be utilized in tissue engineering and drug delivery [19–21].

Recently, efforts have been made to convert high-molecular weight sodium alginate to low molecular weight sodium alginate. This can be done using chemical treatments (such as hydrogen peroxide) without altering the chemical structure and bioactivity of the SA. Depolymerization can be used to alter the molecular weight; the molecular weights obtained depend on the temperature of the reaction, the exposure time, and the concentration of the depolymerizing chemical [22].

1. In order to obtain a low-molecular weight (LMW) sodium alginate (SA) (*see* **Note 7**), an oxidative degradation with hydrogen peroxide at specified temperatures is conducted to initiate a depolymerization process [22]. Add three milliliters of hydrogen peroxide dropwise to a 100-mL 1.0 wt% SA solution, stirring constantly. Ensure the reaction pH and

temperature are maintained at 6 °C and 40 °C, respectively, throughout the process (*see* **Notes 5** and **6**).

2. Carry out this reaction for two time points, 30 and 60 min, to obtain two low molecular weight SAs. Visually it should be apparent that SA depolymerization caused significant lowering of the original SA solution viscosity.

3. To recover the alginate in the reactive solution, precipitate out the SA using 95% ethanol and separate it by centrifugation. Wash the polymer samples several times with deionized water, lyophilize them, and keep them desiccated until further use.

**3.3 Characterization of Sodium Alginate**

1. Record the Fourier transform infrared (FTIR) spectra on the prepared SA samples using a Thermo Scientific Smart FTIR at room temperature in attenuated total reflectance (ATR) mode.

2. Keep the scaffolds in direct contact with the crystal probe and compressed for FTIR measurement. Spectra should be scanned in the range of 3500–500 cm$^{-1}$ at a resolution of 4 cm$^{-1}$, completing 64 scans [18] (*see* **Note 8**).

**3.4 Fabricating Sodium Alginate-Coated Nanofibers**

Although PCL has ideal in vivo mechanical and biocompatible properties, it has a very long degradation time and its lack of hydrophilicity undermines host cell adhesion to the graft. As biomedical engineering moves more towards the self-regeneration of tissues with the aid of a resorbable scaffold, this nonbiodegradability becomes an issue. One way to combat the issue of nonbiodegradability is to form a loosely connected matrix of the body fiber, in this case PCL, and surround this scaffold with a biodegradable substance.

1. SA was chosen as the degradable substance for this particular study. After preparation of low molecular weight sodium alginate (LMW SA) (*see* Subheadings 3.2 and 3.3), cut the nanofiber matrices into circular regions of approximately 1.96 cm$^2$ using a cork-borer #5. Soak these sections in a DI water solution containing 3 wt% SA.

2. After 10 min in the SA bath, desiccate the discs under vacuum pressure for 1 h to ensure uniform SA coating.

3. To stabilize the coating, transfer the discs to a 0.1 M calcium chloride solution in order to induce cross-linking.

4. For the scaffolds used in cell studies, use a SA precursor solution containing 0.004 wt% laminin (40 μg/mL), a protein found in the basal membrane of epithelial cells and known to promote cell adhesion [13].

### 3.5 Degradation Studies of SA-Coated Scaffolds

In order to determine the rate of SA erosion from the composite matrices, the weights of the scaffold in both a hydrated and dry state are calculated and compared. Observing the weight loss pattern gives insight into how the structure degrades. SA erosion affects how cells adhere to and penetrate into the scaffold. The nanotopographical features of the matrix influence cell behavior significantly, and such features can be engineered to tune the scaffolds properties [7].

The ability for cells to penetrate through a scaffold plays a crucial role in how the ECM will be secreted and form. How cells can penetrate deep into a scaffold influences how ECM will be secreted by cells as well. This is of great importance because electrospun nanofiber matrices have morphological similarities to the natural ECM. Therefore, these nanofiber matrices represent dynamic systems that provide surfaces for cell attachment, especially for nanofiber skin grafts [7].

1. Weigh circular SA-coated dry samples and incubate them in capped vials containing 20 mL of PBS at a specific pH of 7.4 and temperature of 37 °C. Change the PBS every 24 h.

2. Following a 48-h incubation period, weight both the dry and wet samples (considered the 0-day weight or the hydrated 0-day weight, respectively).

3. Isolate the samples at further time points in order to measure the wet and dry weights.

4. Present the weight loss as a function of time, with the PCL fiber matrix serving as a control and allowing for analysis of weight loss over time [13]. The weight loss is calculated by subtracting the 0-day weight for either the wet or dry samples, respectively.

### 3.6 Mechanical Testing

Electrospun nanofibrous scaffolds and hydrogel composites for soft tissue engineering and regeneration applications must have appropriate mechanical properties in order to be effective. Due to the high surface area and tiny pores present in nanofiber matrices, mechanical properties vary significantly when compared to the original bulk material. The chemical composition of the scaffold influences its mechanical integrity (Fig. 6). Studies have shown that scaffolds composed of co-spun poly-lactide-co-glycolide polymer (PLAGA) and chitosan have improved mechanical properties compared to those of the single polymers alone [14, 23] (Fig. 5). The fabrication procedure and fiber alignment also impacts the scaffold properties [24]; poly(ε-caprolactone) nanofibers have shown increased tensile moduli when mandrel speed was increased [15, 25]. Therefore, multiple parameters affecting mechanical properties can be toggled when designing a nanofiber mesh for a given application [7].

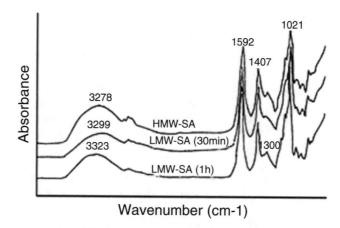

**Fig. 5** FTIR-ATR spectra of SA polymers before depolymerization (HMW-SA), after 30 min of depolymerization (LMW-SA 30 min) and after 1 h of depolymerization (LMW-SA 1 h) in a hydrogen peroxide solution. The decreasing intensity at wavelengths 1400, 1300, and 1070 $cm^{-1}$ indicate that depolymerization occurs at the 1,4-glycosidic linkages, decreasing the monomer chain length and overall molecular weight of individual SA polymers

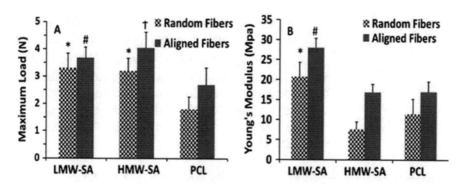

**Fig. 6** Tensile properties of the fiber-hydrogel composite fiber matrices. (**a**) Maximum load and (**b**) Young's modulus for both wet and dry conditions where LMW SA shows superior tensile properties within a uniform fiber matrix

1. Determine the tensile properties of both aligned and random fiber matrices, as well as SA-coated composites, using Instron test equipment.

2. Cut the composite samples into dog-bone shapes with a 20 × 10 mm test section.

3. Stretch the specimens at a constant speed of 10 mm/min to failure. Perform tests according to the ASTM standards [16].

**3.7 Cell Culture Studies**

Tissue engineering involves understanding the interactions between the scaffold and cells. As previously noted, electrospun nanofiber scaffolds have a high surface area-to-volume ratio, providing a large area for cell attachment. It is critical to know how cells preferentially adhere to a scaffold when utilizing it to create a natural wound healing response. In the case of nerve regeneration, electrospun nanofiber nerve grafts must allow for both neurite outgrowth as well as neural stem cell (NSCs) differentiation [7, 26].

1. Use human mesenchymal stem cells (hMSCs) in supplemented media after five passages in order to conduct cell culture experiments.

2. Prior to seeding cells on the scaffolds, scaffolds must be sterilized by immersing them in 70% ethanol for 30 min, exposing them to UV light for 45 min on each side, and washing with sterile PBS (*see* **Note 9**).

3. Incubate the scaffolds in a cell suspension of 100,000 cells/scaffold in a sterile tube. The seeding concentration on the scaffold is found following standard protocol for cell count of cellular stock solution.

4. Rotating the tube on a benchtop rocker for 4 h to ensure uniform cell attachment in this dynamic environment [13].

5. After 24 h, transfer the cell-seeded scaffolds to a 48-well plate and switch the growth media for neural induction media. Neural induction media is supplemented media that is used in order to accelerate neurite differentiation. (For more information on stem cells and scaffold grafts, *see* **Notes 10** and **11**, respectively).

**3.8 Cell Proliferation**

Cell proliferation on the fiber-hydrogel composite was quantified by measuring present cellular DNA [18].

1. Set-up the Quant-iT™ PicoGreen dsDNA assay (P7589).

2. For each measurement, collect three samples from each cell culture time (3, 7, 14, and 21 days).

3. Wash cellular constructs twice with PBS.

4. Transfer the cells to a new well plate.

5. Add 0.3 mL of 1% Triton X-100 solution to lyse the cells.

6. Conduct three freeze-thaw cycles with the well plates.

7. Mix contents thoroughly with a pipette for cell lysate extraction.

8. Transfer 20 μL of sample DNA and 80 mL (component B) and 100-mL (component A) kit reagents into the new well plate.

9. Cover well plates with aluminum foil to exclude from light exposure.

10. Incubate well plates for 5 min.

11. Set-up BioTek (energy HT) plate reader.

12. Measure fluorescence at ex/em 485/535 nm.

13. Use a standard curve to convert the optical readings to DNA concentration.

14. Repeat twice for analysis in duplicate for each well.

**3.9 Live/Dead Assay-Cell Viability**

In tissue engineering, challenges in wound healing can only be solved if scaffolds, cells, and molecules are able to interact with one another. This requires cells to proliferate at the wound site, and scaffolds and molecules must support and promote healthy, viable tissue. Polymer selection for scaffold development yields different levels of cell proliferation as well as viability across the cellular construct [27, 28]. Following a report on a collection of biomaterials for these regenerative applications [19], the SA-coated nanofibers developed required further assessment of cell viability. Both hMSCs and iPSCs on the composite fiber-hydrogel matrices were measured. Live cells can be identified via imaging with this live/dead cell viability kit (Fig. 7). Such cells are attributed with widespread intracellular esterase activity. This activity converts nonfluorescent cell-permeant calcein AM to calcein, which is brightly fluorescent and retained within a live cell. EthD-1 is blocked by live cells' intact plasma membranes but enters the dead cells whose membranes are damaged, binding to nucleic acids and enhancing fluorescence 40-fold. As a result, the dead cells can be identified by intense red fluorescence.

1. Wash polymer grafts with PBS twice.

2. Incubate them with 2 μM of calcein-AM and 4 μM of ethidium homodimer-1 (EthD-1).

3. Identify live and dead cells by their respective fluorescence.

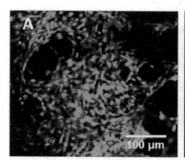 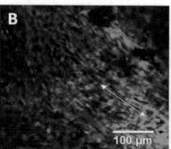 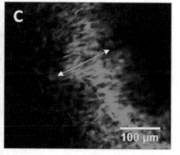

**Fig. 7** Live-dead assay confocal microscopic images (10× magnification) for low molecular weight sodium alginate. Images (**a**), (**b**), and (**c**) are PCL fiber matrices coated at 3, 10, and 14 days, respectively

### 3.10 Scanning Electron Microscopy (SEM)

To characterize the sample surface topography, fiber lattice, and composite structure, scanning electron microscopy (SEM) should be conducted for cell-seeded membranes. SEM is beneficial particularly because of its large depth of field, allowing for thorough characterization of the surface morphology of engineered tissues.

1. Collect the cell-seeded membranes and attach them using 3% glutaraldehyde/paraformaldehyde.

2. Allow membranes proper fixation overnight.

3. Wash samples with distilled water, dry under vacuum, and keep desiccated.

4. Sputter-coat the scaffold surfaces with Au/Pd using a Hummer V sputtering system. Sputter coating provides the sample with a thin conductive surface layer that prevents the sample from interference and becoming charged by the electron beam emitted by the SEM during imaging.

5. Image the samples for characterization using a JEOL 6335F FESEM, or similar.

6. Qualitatively characterize the surface topography of the composite structure and the fiber lattice, highlighting the presence of cells.

### 3.11 Immuno histochemistry

Immunohistochemistry (IHC) was conducted to identify protein expression in cells. This versatile methodology utilizes the interaction of antigens and antibodies to localize specific antigens [29].

1. Collect the cellular grafts from culture times including 7, 14, and 21 days.

2. Wash grafts in PBS and fix with 4% formaldehyde in PBS for 40 min.

3. Wash grafts in PBS for 40 min three times.

4. Mount the constructs into optimal cutting temperature medium (OCT) and freeze at $-80\ ^{\circ}C$.

5. Obtain slices 40 μm thick from the samples and mount on them on Superfrost glass slides.

6. Perform the blocking step with 5% bovine serum albumin (BSA) in Tris buffered saline with 1% Tween 80 (TBST).

7. Incubate the cells in the samples overnight at $4\ ^{\circ}C$ in primary antibody (rabbit-antiS-100, 1:500) in blocking buffer (5% BSa in TBST).

8. Incubate the cell constructs in secondary antibody AlexaFluor 488 (1:500) in blocking buffer for 1.5 h.

9. Mount the samples in Fluoroshield with DAPI and cure for 10 min.

10. Image the samples using a Zeiss LSM Pascal5 confocal microscope, or similar.

## 4   Notes

1. Prior to beginning the electrospinning process, ensure that the needle is filled with solution and is free of air bubbles—these will cause the process to halt and will require a manual reset. Simply push the solvent through the needle by applying a small amount of pressure to the syringe plunger until a small amount of the solution begins to drip through the other side. Wipe away the excess solution with a Kimwipe, and push the plunger gently to form a small head of solution at the tip of the needle.

2. In case any adjustments need to be made to the setup before, during, or after the electrospinning procedure, the apparatus should be set up in a plastic casing with a transparent sliding plastic door for ease of access on one side. The apparatus cannot be left in an open air environment both to prevent harmful exposure of chemical fumes evaporating out of the syringe as well as to prevent foreign material from mixing with the fibers.

3. Immediately after beginning the electrospinning process, watch the flow of the polymer jet moving towards the grounded plate—the jet should be moving toward the plate when fully charged. The spherical head present after pushing the air out of the needle originally should be a conical shape, known as a Taylor cone [30].

4. This electrospinning time and procedure was optimized to produce a ~ 100-$\mu$m-thick fiber lattice consisting of sparsely spaced fibers with diameters ranging from 600 to 900 nm. The images were analyzed to determine the distribution in fiber diameter using ImageJ (NIH) software [31]. To obtain specifically oriented fibers, a different setup should be used as seen in Fig. 2.

5. When handling all chemicals be sure to use the proper laboratory safety procedures and equipment. Latex gloves and safety goggles should be worn at all times to mitigate risk of injury while performing lab work. Work with all chemicals in the hood.

6. Prepare all of the equipment to be used prior to beginning any experiments. Wash all vials, spatulas, weighing dishes, etc., first and store them in a clean area for use during experimentation so that procedures can be carried out without hesitation; some of the solvents, such as ethanol and methylene chloride, have a very high vapor pressure and will evaporate quickly. To ensure this loss of solvent is kept to a minimum, it is best to be prepared prior to beginning any procedures involving these chemicals.

7. This study used LMW SA obtained at 60 min in the scaffold design and for further characterization [18]. LMW SA-coated nanofibers, as compared to HMW SA-coated nanofibers, display a more uniform infiltration of the nanofiber matrix and nanofiber coating. Due to this uniformity, it is more difficult for defect formation and propagation, ultimately leading to a higher tensile strength in the LMW SA-coated fibers.

8. The depolymerization of the polymer backbone occurred at the 1,4-glycosidic linkage [18]. Band intensities before and after depolymerization SA were compared to confirm the backbone structure and depolymerization, which can be seen in Fig. 5.

9. Sterility is of upmost importance in order to avoid bacterial or microbe contamination. Any potential contaminants may have a detrimental impact on the results obtained as well as compromise the reliability of the results. When the scaffold is subject to ethanol, it may shrink. This can be accommodated by making the graft thicker and of a larger diameter.

10. Induced pluripotent stem cells (iPSCs) have also been studied in terms of nerve graft cell studies. iPSCs can be seeded onto a scaffold in order to determine neural differentiation using neural basal medium supplemented with growth factors. These growth factors include BDNF, GDNF, and IGF-1.

11. Skeletal muscle grafts have been created using nonwoven PGA fiber meshes seeded with myoblasts. This electrospun scaffold allowed for the viability of myoblasts; markers of skeletal muscle differentiation, including alpha sacromeric actin and desmin, can also be stained [18]. More recent studies have shown adhesion and proliferation of a myoblast cell line on electrospun DegraPol with no apparent toxicity [31]. These cells have also been seeded on a gelatin or fibronectin nonwoven electrospun PLLA fiber mesh. Myotubes on the aligned mesh were able to successfully mimic myotubes on native muscular tissue [2, 19].

12. As with all studies, the variation and interpretation of results is dependent upon the data in aggregate and the relation between respective data points within the set. It allows for appropriate conclusions to be drawn and quantifies the significance of differences within the data. Here, the value defining significance is $p < 0.05$. Quantitative results were expressed as mean ± standard deviation and analyzed using the Student's t-test or ANOVA as well as a post hoc test. To better understand what can be drawn from each test, Student's t-test aims to compare the means of two different treatment groups, while ANOVA compares means for more than two different treatment groups. In the latter case, the treatment groups are typically distributed with a common variance.

# Acknowledgements

The authors acknowledge funding from the Connecticut Regenerative Medicine Research Fund (Grant Number: 15-RMB-UCHC-08), the Department of Defense (OR120140), and the National Science Foundation Award Number IIP-1311907, IIP-1355327, and EFRI-1332329. University of Connecticut Health Research Excellence Program Convergence Grant No. 401473-10300-20.

# References

1. Heineken F, Skalak R (1991) Tissue engineering: a brief overview. J Biomech Eng 113 (2):111–112

2. Yost MJ, Price RL, Simpson DG, Yan W, Terracio L (2008) Cardiac patch engineering. Encyclopedia of biomaterials and biomedical engineering, 2nd edn. p 542–551

3. Whitlock EL, Tuffaha SH, Luciano JP, Yan Y, Hunter DA, Magill CK, Moore AM, Tong AY, Mackinnon SE, Borschel GH (2009) Processed allografts and type I collagen conduits for repair of peripheral nerve gaps. Muscle Nerve 39(6):787–799

4. Ray WZ, Mackinnon SE (2010) Management of nerve gaps: autografts, allografts, nerve transfers, and end-to-side neurorrhaphy. Exp Neurol 223(1):77–85. doi:10.1016/j.expneurol.2009.03.031

5. James R, Kesturu G, Balian G, Chhabra AB (2008) Tendon: biology, biomechanics, repair, growth factors, and evolving treatment options. J Hand Surg Am 33(1):102–112

6. Lannutti J, Reneker D, Ma T, Tomasko D, Farson D (2007) Electrospinning for tissue engineering scaffolds. Mater Sci Eng C 27 (3):504–509

7. Kumbar S, James R, Nukavarapu S, Laurencin C (2008) Electrospun nanofiber scaffolds: engineering soft tissues. Biomed Mater 3 (3):034002

8. Kumbar SG, Nair LS, Bhattacharyya S, Laurencin CT (2006) Polymeric nanofibers as novel carriers for the delivery of therapeutic molecules. J Nanosci Nanotechnol 6 (9–10):2591–2607

9. Kumbar SG, Bhattacharyya S, Sethuraman S, Laurencin CT (2007) A preliminary report on a novel electrospray technique for nanoparticle based biomedical implants coating: precision electrospraying. J Biomed Mater Res B Appl Biomater 81(1):91–103

10. Kumbar SG, Nukavarapu SP, James R, Nair LS, Laurencin CT (2008) Electrospun poly (lactic acid-co-glycolic acid) scaffolds for skin tissue engineering. Biomaterials 29(30):4100–4107

11. Li W-J, Jiang YJ, Tuan RS (2006) Chondrocyte phenotype in engineered fibrous matrix is regulated by fiber size. Tissue Eng 12 (7):1775–1785

12. Katti DS, Robinson KW, Ko FK, Laurencin CT (2004) Bioresorbable nanofiber-based systems for wound healing and drug delivery: optimization of fabrication parameters. J Biomed Mater Res B Appl Biomater 70(2):286–296

13. Teo WE, Inai R, Ramakrishna S (2016) Technological advances in electrospinning of nanofibers. Sci Technol Adv Mater 12:013002

14. Guadalupe E, Ramos D, Shelke NB, James R, Gibney C, Kumbar SG (2015) Bioactive polymeric nanofiber matrices for skin regeneration. J Appl Polym Sci 132. doi:10.1002/app.41879

15. Cheng Y, Ramos D, Lee P, Liang D, Yu X, Kumbar SG (2014) Collagen functionalized bioactive nanofiber matrices for osteogenic differentiation of mesenchymal stem cells: bone tissue engineering. J Biomed Nanotechnol 10 (2):287–298

16. Choi JS, Lee SJ, Christ GJ, Atala A, Yoo JJ (2008) The influence of electrospun aligned poly($\varepsilon$-Caprolactone)/collagen nanofiber meshes on the formation of self-aligned skeletal muscle myotubes. Biomaterials 29 (19):2899–2906

17. Lee P, Manoukian OS, Zhou G, Wang Y, Chang W, Yu X, Kumbar SG (2016) Osteochondral scaffold combined with aligned nanofibrous scaffolds for cartilage regeneration. RSC Adv 6(76):72246–72255

18. Shelke NB, Lee P, Anderson M, Mistry N, Nagarale RK, Ma X, Yi X, Kumbar SG (2016) Neural tissue engineering: nanofiber-hydrogel based composite scaffolds. Polym Adv Technol 27(1):42–51. doi:10.1002/pat.3594

19. Shelke NB, James R, Laurencin CT, Kumbar SG (2014) Polysaccharide biomaterials for

drug delivery and regenerative engineering. Polym Adv Technol 25(5):448–460

20. Jen AC, Wake MC, Mikos AG (1996) Review: hydrogels for cell immobilization. Biotechnol Bioeng 50(4):357–364

21. Kumbar SG, Dave AM, Aminabhavi TM (2003) Release kinetics and diffusion coefficients of solid and liquid pesticides through interpenetrating polymer network beads of polyacrylamide-g-guar gum with sodium alginate. J Appl Polym Sci 90(2): 451–457

22. Mao S, Zhang T, Sun W, Ren X (2012) The depolymerization of sodium alginate by oxidative degradation. Pharm Dev Technol 17 (6):763–769

23. Duan B, Wu L, Yuan X, Hu Z, Li X, Zhang Y, Yao K, Wang M (2007) Hybrid nanofibrous membranes of PLGA/Chitosan fabricated via an electrospinning array. J Biomed Mater Res A 83:868–878

24. Jaiswal D, Roshan J, Shelke NB, Harmon MD, Brown JL, Hussain F, Kumbar SG (2015) Gelatin nanofiber matrices derived from Schiff base derivative for tissue engineering applications. J Biomed Nanotechnol 11(11):2067–2080. doi:10.1166/jbn.2015.2100

25. Li WJ, Mauck RL, Cooper JA, Yuan X, Tuan RS (2007) Engineering controllable anisotropy in electrospun biodegradable nanofibrous scaffolds for musculoskeletal tissue engineering. J Biomech 40:1686–1693

26. Anderson M, Shelke NB, Manoukian OS, Yu X, McCullough LD, Kumbar SG (2015) Peripheral nerve regeneration strategies: electrically stimulating polymer based nerve growth conduits. Crit Rev Biomed Eng 43(2–3)

27. Ahmed I, Ponery AS, Nur-E-Kamal A, Kamal J, Meshel AS, Sheetz MP, Schindler M, Meiners S (2007) Morphology, cytoskeletal organization, and myosin dynamics of mouse embryonic fibroblasts cultured on nanofibrillar surfaces. Mol Cell Biochem 301:241–249

28. Sahoo S, Ouyang H, Goh JC, Tay TE, Toh SL (2006) Characterization of a novel polymeric scaffold for potential application in tendon/ligament tissue engineering. Tissue Eng 12:91–99

29. Schacht V, Kern JS (2015) Basics of Immunohistochemistry. J Invest Dermatol 135(3):1–4. doi:10.1038/jid.2014.541

30. Yarin AL, Koombhongse S, Reneker DH (2001a) Taylor cone and jetting from liquid droplets in electrospinning of nanofibers. J Appl Phys 90:4836–4846

31. James R, Kumbar SG, Laurencin CT, Balian G, Chhabra AB (2011) Tendon tissue-engineering: adipose 1 derived stem cell and GDF-5 mediated regeneration using electrospun matrix systems. Biomed Mater 6(2):025011. doi:10.1088/1748-6041/6/2/025011

# Chapter 19

# Application of Hydrogel Template Strategy in Ocular Drug Delivery

## Crystal S. Shin, Daniela C. Marcano, Kinam Park, and Ghanashyam Acharya

## Abstract

The hydrogel template strategy was previously developed to fabricate homogeneous polymeric microparticles. Here, we demonstrate the versatility of the hydrogel template strategy for the development of nanowafer-based ocular drug delivery systems. We describe the fabrication of dexamethasone-loaded nanowafers using polyvinyl alcohol and the instillation of a nanowafer on a mouse eye. The nanowafer, a small circular disk, is placed on the ocular surface, and it releases a drug as it slowly dissolves over time, thus increasing ocular bioavailability and enhancing efficiency to treat eye injuries.

**Key words** Drug delivery, Nanowafer, Hydrogel, Dexamethasone, Eye, Cornea

## 1 Introduction

Drug delivery to the eye, although seemingly simple, is a challenging task [1, 2]. The wet ocular surface is exposed to the external environment, which inadvertently increases the risk of injuries and infections [3]. Ocular surface diseases in the cornea may seem benign, but can compromise normal vision and eventually lead to vision loss if not properly treated.

Topical drug therapy is the most common treatment for ocular surface diseases including inflammation, injuries, and dry eye [4, 5]. Eye drop formulations account for 90% of available ophthalmic formulations, and this treatment method is preferred by patients due to its easy installation [2]. However, eye drop therapy is not highly efficient providing low ocular bioavailability with rapid fluctuations of drug concentration since less than 5% of the applied dose reaches the targeted ocular tissue [6]. Most of the topical ophthalmic formulations are in solution, and their effectiveness is affected by the solubility and stability of the active pharmaceutical ingredients. The physiology of the ocular surface presents barriers

Sarah Hurst Petrosko and Emily S. Day (eds.), *Biomedical Nanotechnology: Methods and Protocols*, Methods in Molecular Biology, vol. 1570, DOI 10.1007/978-1-4939-6840-4_19, © Springer Science+Business Media LLC 2017

for the drug molecules to entering the epithelium due to reflex tearing, tight epithelial junctions, and nasolacrimal drainage, which shortens the contact time on the ocular surface [7]. In addition, drug absorption on the ocular surface is hindered by the systemic clearance by blood capillaries in the conjunctiva [8].

Recently, several nanoparticle suspensions have been developed for ophthalmic application to improve the drug bioavailability in the eye [9, 10]. Nanotechnology-based drug delivery systems, such as polymeric micelles, nanoparticles, liposomes, and dendrimers, have been developed to enhance the efficiency of the ocular drug delivery systems [11–14]. In addition, drug-loaded contact lenses have been shown to increase drug bioavailability in comparison to eye drops [15–18]. Although these advances are promising, there is still a need for a drug delivery system that is safe, efficacious, noninvasive, and patient compliant.

Previously, we developed the hydrogel template strategy to fabricate polymer microparticles with homogeneous size and shape that can function as multifunctional drug delivery vehicles [19–21]. We showed that a hydrogel-forming biopolymer (i.e., gelatin) can serve as a template to imprint predefined microstructures.

In this report, we demonstrate the versatility of the hydrogel template strategy for the development of nanowafer-based ocular drug delivery systems [22, 23]. The nanowafer is a small circular disk fabricated with biopolymers via the modified hydrogel template strategy. The drug-loaded nanowafer is placed on the ocular surface, and it releases the drug as it slowly dissolves over time thus increasing ocular bioavailability and enhancing efficiency.

There are numerous biocompatible polymers that have been explored in biomedical and pharmaceutical applications [24]. Among these, a water-soluble, polyvinyl alcohol was selected to serve as a template, which contains nano-reservoirs. Polyvinyl alcohol (PVA) is a mucoadhesive polymer that is used clinically as artificial tear eye drops [25]. The PVA nanowafer functions both as a drug delivery vehicle and a lubricant. During the course of the drug release, the nanowafer slowly dissolves thus lubricating the ocular surface and then it eventually disappears.

## 2    Materials

### 2.1    Fabrication of a Polydimethylsiloxane (PDMS) Template

1. Silicon mask containing array of nano-wells (i.e., 500 nm length × 500 nm width × 500 nm depth).

2. Sylgard 184 Silicone elastomer kit, including curing agent and siloxane (Dow Corning).

3. Oven or hot plate.

### 2.2 Fabrication of Nanowafers

1. PDMS template.
2. Poly(vinyl alcohol), Mw 146,000–186,000 Da, 87–89% hydrolyzed.
3. Dexamethasone sodium phosphate (*see* **Notes 1** and **2**).
4. Fluorescein solution for imaging.
5. Ethanol.
6. Nanopure water.

### 2.3 Instillation of Nanowafers on Mouse Corneas

1. Nanowafers.
2. Fluorescein eye drops.
3. Balanced salt solution.
4. Micropipettes.
5. Ketamine and xylazine for anesthesia.
6. Forceps.
7. Stereomicroscope and/or epifluorescence microscope.

## 3 Methods

The process of fabricating nanowafers consists of two steps: the fabrication of a PDMS template containing nano-posts and the fabrication of nanowafers containing nano-reservoirs filled with drugs (Figs. 1 and 2). Upon fabrication of the drug-loaded nano-wafers, the instillation of nanowafers on mouse corneas is illustrated. Alternatively, fluorescein-loaded nanowafers and fluorescein eye drops are instilled on mouse corneas. The fluorescence intensities can be quantified and compared to demonstrate that the nanowafer

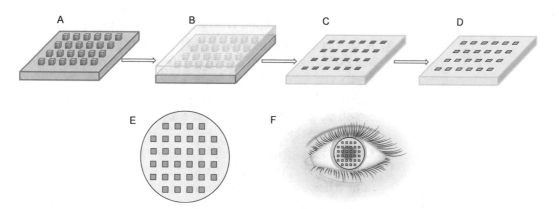

**Fig. 1** Schematic of nanowafer fabrication. (**a**) PDMS template containing nano-posts. (**b**) PVA solution is transferred into the PDMS template. (**c**) PVA film now contains arrays of nano-reservoirs. (**d**) Nano-reservoirs filled with drug (**e, f**) An illustration of a circular nanowafer and instillation on the cornea

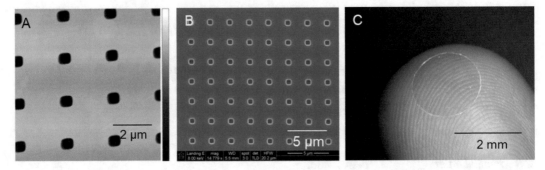

**Fig. 2** Images of a nanowafer: Atomic force microscopic image (**a**) and a scanning electron microscopic image (**b**) showing well-defined nano-reservoirs on PVA film. (**c**) A nanowafer placed on a fingertip demonstrating into transparency

increases the drug residence time on the ocular surface and drug diffusion into the corneal epithelium (*see* **Note 3**).

*3.1 Fabrication of PDMS Template*

1. Prepare a PDMS solution by mixing curing agent and siloxane in a 1:10 ratio.

2. Pour the polydimethylsiloxane (PDMS) solution onto a silicon mask containing nano-reservoirs (500 nm length × 500 nm width, 500 nm depth) and cure it at 60 °C for 24 h (*see* **Notes 4 and 5**).

3. Once cured, the PDMS template is carefully separated from the silicone mask and now contains a 3″ × 3″ pattern of nano-posts (500 nm length × 500 nm width, 500 nm height).

*3.2 Fabrication of Drug-Loaded Nanowafers*

1. Prepare a polyvinyl alcohol (PVA) solution (4%, w/v) with three parts ethanol and two parts water (*see* **Note 6**).

2. Prepare a dexamethasone solution by dissolving 1 mg of dexamethasone sodium phosphate in 1 mL of 4% PVA solution. Alternatively, 10 μL of fluorescein solution can be added to 1 mL of 4% PVA solution for fluorescence imaging (*see* **Note 7**).

3. Pipet 5 mL of PVA solution onto the PDMS template and keep it at 60 °C for 30 min or until the PVA forms a clear film wafer, whichever comes first (*see* **Note 8**).

4. Peel the PVA wafer containing nano-reservoirs from the PDMS template, and then place it on a flat surface with the nano-reservoirs facing up.

5. Using a micropipette, transfer the dexamethasone solution onto the PVA wafer, then swiftly swipe the solution across the wafer using a razor blade to fill the nano-reservoirs (*see* **Note 9**).

6. Punch the drug-filled nanowafer into small circular nanowafers, 2 mm in diameter (*see* **Note 10**).

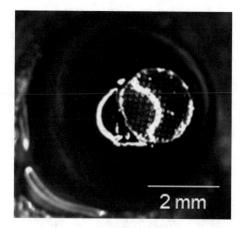

**Fig. 3** Bright field image of a nanowafer instilled on a mouse cornea

**3.3 Instillation of Nanowafers on Mouse Corneas**

1. Anesthetize a mouse via intraperitoneal injection of ketamine (100 mg/kg) and xylazine (10 mg/kg) (*see* **Notes 11** and **12**).

2. Place the nanowafer on the cornea using forceps (Fig. 3). To hydrate the ocular surface, instill 2 μL of balanced salt solution (BSS) (*see* **Note 13**).

3. To demonstrate the enhanced drug residence time, instill a fluorescein-loaded nanowafer and 2 μL of fluorescein eye drops on the corneas of two different mice. Obtain fluorescence images at specific time intervals and compare them to evaluate the fluorescein's molecular diffusion into the cornea using each instillation methodology (*see* **Note 14**).

# 4 Notes

1. Dexamethasone sodium phosphate is a water-soluble, corticosteroid drug with potent anti-inflammatory properties. It is commonly used in ophthalmic solution to treat ocular inflammation and dry eye-related corneal diseases [26, 27].

2. Any other water-soluble drugs also can be used to fabricate nanowafers.

3. Fluorescein is a fluorescent compound that can be easily observed using a fluorescence microscope. Since dexamethasone is not fluorescent, fluorescein is used to fabricate nanowafers to visualize the nano-reservoirs. It is also easy to trace on the ocular surface once instilled on mouse corneas.

4. It is important to cure the PDMS on a flat surface to fabricate the PDMS template with uniform thickness.

5. Before curing, air bubbles from the PDMS solution should be removed. One way to remove air bubbles is to place it under vacuum.

6. A polyvinyl alcohol solution can be prepared on a heat/stirring plate at 50 °C with a magnetic stirring bar. In this case, the solution needs to be prepared in a capped Pyrex bottle.

7. A fluorescein solution should be placed in a light protected tube or covered in foil. Once prepared, the solution should be refrigerated.

8. When transferring the solution, it is ideal to minimize air bubbles, which can affect the surface of the PVA film. In addition, the PDMS template should be kept on a flat surface to obtain even thickness of the film.

9. When swiping with the blade, evenly apply pressure to avoid breaking the PVA wafer.

10. Prior to punching, the drug-filled wafer should be dried.

11. Proper training for animal care and use should be approved by your institution before animal studies are performed.

12. Any other methods that are approved by institutional guidelines can be used to anesthetize animals.

13. Instillation will be easier if the nanowafer is placed while observing it under a stereomicroscope.

14. The same concentration of fluorescein is used to fabricate fluorescein-loaded nanowafer and prepare fluorescein eye drops.

# References

1. Urtti A (2006) Challenges and obstacles of ocular pharmacokinetics and drug delivery. Adv Drug Deliv Rev 58:1131–1135

2. Novack GD (2009) Ophthalmic drug delivery: development and regulatory considerations. Clin Pharmacol Ther 85:539–543

3. Kim YC, Chiang B, Wu X et al (2014) Ocular delivery of macromolecules. J Control Release 190:172–181

4. Gaudana R, Ananthula HK, Parenky A et al (2010) Ocular drug delivery. Am Assoc Pharm Sci J 12:348–360

5. Ranta VP, Urtti A (2006) Transscleral drug delivery to the posterior eye: prospects of pharmacokinetic modeling. Adv Drug Deliv Rev 58:1164–1181

6. Mannermaa E, Vellonen KS, Urtti A (2006) Drug transport in corneal epithelium and blood_retina barrier: emerging role of transporters in ocular pharmacokinetics. Adv Drug Deliv Rev 58:1136–1163

7. Jarvinen K, Jarvinen T, Urtti A (1995) Ocular absorption following topical delivery. Adv Drug Deliv Rev 16:3–19

8. Salminen L (1990) Review: systemic absorption of topically applied ocular drugs in humans. J Ocul Pharmacol 6:243–249

9. Diebold Y, Calonge M (2010) Applications of nanoparticles in ophthalmology. Prog Retina Eye Res 29:596–609

10. Gershkovich P, Wasan KM, Barta CA (2008) A review of the application of lipid-based systems in systemic, dermal, transdermal, and ocular drug delivery. Crit Rev Ther Drug 25:545–584

11. Choy YB, Park JH, McCarey BE et al (2008) Mucoadhesive microdiscs engineered for

ophthalmic drug delivery: effect of particle geometry and formulation on preocular residence time. Invest Ophthalmol Vis Sci 49:4808–4815

12. Chang E, McClellan AJ, Farley WJ et al (2011) Biodegradable PLGA-based drug delivery systems for modulating ocular surface disease under experimental murine dry eye. J Clin Exp Ophthalmol 2:191. doi:10.4172/2155-9570.1000191

13. Aksungur P, Demirbilek M, Denkbas EB et al (2011) Development and characterization of cyclosporine a loaded nanoparticles for ocular drug delivery: cellular toxicity, uptake, and kinetic studies. J Control Release 151:286–294

14. Shah M, Edman MC, Janga SR et al (2013) A rapamycin-binding protein polymer nanoparticle shows potent therapeutic activity in suppressing autoimmune dacryoadenitis in a mouse model of Sjogren's syndrome. J Control Release 171:269–279

15. Gulsen D, Chauhan A (2004) Ophthalmic drug delivery through contact lenses. Invest Ophthalmol Vis Sci 45:2342–2347

16. Carvalho IM, Marques CS, Oliveira RS et al (2015) Sustained drug release by contact lenses for glaucoma treatment–a review. J Control Release 202:76–82

17. Garhwal R, Shady SF, Ellis EJ et al (2012) Sustained ocular delivery of ciprofloxacin using nanospheres and conventional contact lens materials. Invest Ophthalmol Vis Sci 53:1341–1352

18. Singh K, Nair AB, Kumar A et al (2011) Novel approaches in formulation and drug delivery using contact lenses. J Basic Clin Pharm 2:87–101

19. Acharya G, Shin CS, McDermott M et al (2010) The hydrogel template method for fabrication of homogeneous nano/micro particles. J Control Release 141:314–319

20. Acharya G, Shin CS, Vedantham K et al (2010) A study of drug release from homogeneous PLGA microstructures. J Control Release 146:201–206

21. Acharya G, McDermott M, Shin SJ et al (2011) Hydrogel templates for the fabrication of homogeneous polymer microparticles. Methods Mol Biol 726:179–185

22. Coursey TG, Henriksson JT, Marcano DC et al (2015) Dexamethasone nanowafer as an effective therapy for dry eye disease. J Control Release 213:168–174

23. Yuan X, Marcano DC, Shin CS et al (2015) Ocular drug delivery nanowafer with enhanced therapeutic efficacy. ACS Nano 9:1749–1758

24. Ludwig A (2005) The use of mucoadhesive polymers in ocular drug delivery. Adv Drug Deliv Rev 57:1595–1639

25. Moshirfar M, Pierson K, Hanamaikai K et al (2014) Artificial tears potpourri: a literature review. Clin Ophthalmol 8:1419–1433

26. Nagelhout TJ, Gamache DA, Roberts L et al (2005) Preservation of tear film integrity and inhibition of corneal injury by dexamethasone in a rabbit model of lacrimal gland inflammation-induced dry eye. J Ocul Pharmacol Ther 21:139–148

27. Patane MA, Cohen A, From S et al (2011) Ocular iontophoresis of EGP-437 (dexamethasone phosphate) in dry eye patients: results of a randomized clinical trial. Clin. Ophthalmol. 5:633–643

# Chapter 20

# High-Accuracy Determination of Cytotoxic Responses from Graphene Oxide Exposure Using Imaging Flow Cytometry

## Sandra Vranic and Kostas Kostarelos

## Abstract

Graphene and other 2D materials have received increased attention in the biomedical field due to their unique properties and potential use as carriers for targeted drug delivery or in regenerative medicine. Before the exploitation of graphene-based materials in biomedicine becomes a reality, it is necessary to establish the full toxicological profile and better understand how the material interacts with cells and tissues. Because specific properties, such as flake size and surface chemistry, might determine whether graphene can achieve therapeutic efficacy without causing toxicity, it is important to develop highly accurate and reliable screening techniques to accurately assess the biocompatibility of different types of graphene-based materials. In this protocol, we describe a method to achieve accurate determination of the cytotoxic response following in vitro exposure to large graphene oxide (L-GO) sheets using annexin V/propidium iodide staining and the Imagestream® platform. The proposed protocol is especially suitable for the toxicity assessment of carbonaceous materials that form aggregates in cell culture media, which is a common occurrence. We describe how to best gate out any interfering signals coming from the material by visual inspection and by using powerful software, thus performing the analysis of cellular death on a selected population of cells with higher accuracy and statistical relevance compared to conventional flow cytometry.

**Key words** Graphene oxide, Imagestream, Imaging flow cytometry, Toxicity, Annexin V/propidium iodide, Cell viability assay, 2D material, Material agglomeration, Apoptosis, Necrosis

## 1  Introduction

Graphene-based nanomaterials started receiving considerable attention due to their unique chemical and physical properties, such as enhanced electron mobility and thermal conductivity, mechanical strength, and distinctive optical characteristics, which can be exploited in biomedicine [1]. Graphene and other 2D materials can be used as carriers for targeted drug delivery, cancer treatment agents *via* photo-thermal therapy, or scaffolds for nerve regeneration [2, 3]. However, in vitro and in vivo knowledge regarding the safety and biocompatibility of graphene-based materials is still being gathered. Such studies are critically important to provide better insight into the interactions of 2D materials with

Sarah Hurst Petrosko and Emily S. Day (eds.), *Biomedical Nanotechnology: Methods and Protocols*, Methods in Molecular Biology, vol. 1570, DOI 10.1007/978-1-4939-6840-4_20, © Springer Science+Business Media LLC 2017

cells and tissues, thus helping us to further understand their biological safety profile.

The assays used to determine cytotoxicity of graphene-based nanomaterials in vitro are similar to those developed for other carbon-based nanostructures (e.g., carbon nanotubes) and for other types of nanomaterials in general. The most commonly used cytotoxicity tests are colorimetric assays, such as the "modified" lactate dehydrogenase (LDH) and water-soluble tetrazolium salts (WST-8) assays; these are considered to be the most reliable because potential interferences of nanomaterials with the components of the assay might be avoided [4–6]. Carbon-based materials have repeatedly been found to interfere with the MTT assay, which is also widely used to determine cytotoxicity induced by nanomaterial exposure [6, 7]. More sophisticated techniques, such as flow cytometry, can offer higher accuracy, but often with a lower throughput [8]. The advantage of flow cytometry is that the sampled cells can be sorted based on different parameters such as their size, granularity, or fluorescence, all of which might be influenced by the interactions between cells and the nanomaterial. A popular flow cytometry-based assay used to study the cytotoxicity of carbonaceous nanomaterials involves annexin V/propidium iodide staining, which we described previously [9]. Using this technique, unstained events on flow cytometry bivariate plots are distinguished from the stained events. The unstained events are considered to be live cells, annexin V-positive events belong to early apoptotic cells, propidium iodide-positive events indicate necrotic cells, while both annexin V and propidium iodide-positive events represent either late apoptotic or necrotic cells.

A potential issue in assessing the cytotoxic responses of cells to carbon-based materials in vitro using flow cytometry comes from the fact that the material commonly tends to agglomerate when dispersed in cell culture media or after interacting with molecules secreted by the cells [10, 11]. The size of such agglomerates can be similar to the size of a cell and therefore can appear as an "unstained event" on the annexin V/propidium iodide bivariate plot. Subsequently, this results in an overestimate of the number of live cells in the sample, which can lead to inaccurate conclusions regarding the cytotoxicity of the material [9]. Gating systems provided in classic flow cytometry software offer the possibility of excluding such interferences; however, the accuracy of the gating is limited as the events being gated in or out cannot be visually inspected.

More advanced flow cytometry techniques and instrumentation, such as Imagestream®, are combining flow cytometry with high-resolution imaging, therefore providing both statistical power over the acquired data in conjunction with the possibility of viewing each individual acquired event. Using this technique, it is possible to distinguish whether each acquired event in the bivariate plot is the result of an aggregated material or a cell. Moreover, using further software analysis, it is possible to gate out the material based on its contrast properties (the contrast properties of aggregated materials and cells differ). The combination of those capabilities offers superior accuracy in comparison to classic flow cytometry and colorimetric assays. Imagestream® has so far been used to accurately assess the uptake of different nanomaterials, including carbon nanotubes in vitro [11–14] as well as to determine the extent of cell death induced by different pharmacological agents [15, 16].

In this protocol, we describe a method that can be used to accurately determine the cytotoxic responses of mammalian cells following exposure to large (>2 μm and <20 μm in lateral dimension), but thin (1–2 layers, 0.6 nm thickness of one layer) graphene oxide (GO) sheets using annexin V/propidium iodide staining and Imagestream®. We provide an explanation on how best to gate out any interfering signals from the material itself by visual inspection and by applying the software features provided with the instrument, and in this way, to perform analyses of cellular death on a selected population of cells with high accuracy and statistical relevance. The proposed protocol is suitable for the assessment of the toxicity of carbonaceous materials that form aggregates in the cell culture media; however, it can be further optimized and used for any type of carbon- or non-carbon-based material.

## 2  Materials

### 2.1  L-GO Preparation

1. Large GO (L-GO) material dispersed in sterile, endotoxin-free water (2.4 mg/mL).

2. RPMI 1640 cell culture medium with L-glutamine and sodium bicarbonate (R8758, Sigma-Aldrich, Merck, UK).

3. Fetal Bovine Serum (FBS, Sigma-Aldrich, Merck, UK).

4. 15 mL sterile, plastic tubes (Corning, Costar, Sigma-Aldrich, Merck, UK).

5. Vortex.

### 2.2  Cell Culture

1. Adherent immortalized lung epithelial cell line Beas-2B (CRL-9609, ATCC).

2. Cell culture medium appropriate for the cell line studied. For the Beas-2B cell line, the RPMI 1640 cell culture medium with 20 mM glutamine (R8758, Sigma-Aldrich, Merck, UK) and supplemented with 10% FBS (Gibco, Thermo Scientific, UK), 50 U/mL penicillin, and 50 μg/mL streptomycin (all from Sigma-Aldrich, Merck, UK) was used.

3. 0.05% trypsin with 0.53 mM ethylenediaminetetraacetic acid (EDTA) tetra-sodium salt (T3924, Sigma-Aldrich, Merck, UK).

4. Six-well flat-bottom plates (Corning, Costar, Sigma-Aldrich, Merck, UK).

5. T-75 sterile flasks (Corning, Costar, Sigma-Aldrich, Merck, UK).

6. Incubator set at 37 °C and 5% $CO_2$.

7. 1.5 mL micro centrifuge tubes.

8. 5, 10, and 25 mL serological pipettes (VWR, UK).

9. 10 μL, 200 μL, and 1 mL pipette tips (Starlab, UK).

10. Centrifuge ($210 \times g$ for 5 min) for pelleting cells.

11. 15 mL sterile, plastic tubes (Corning, Costar, Sigma-Aldrich, Merck, UK).

12. Annexin V, Alexa Fluor® 488 conjugate (A13201, Thermo Fisher Scientific, UK).

13. Annexin Binding Buffer (V13246, Thermo Fisher Scientific, UK).

14. Propidium iodide (P4864, Sigma Aldrich, UK).

15. 5% dimethyl sulfoxide (DMSO) (D2650, >99.7%, sterile, filtered, Sigma-Aldrich, Merck, UK).

16. Dulbecco's Phosphate Buffered Saline (PBS), with $MgCl_2$ and $CaCl_2$ (D8662, Sigma-Aldrich, Merck, UK).

17. Trypan blue (T8154, 0.4% solution, Sigma-Aldrich, Merck, UK).

# 3    Methods

This protocol allows determination of cytotoxic responses from the exposure to L-GO material incubated with Beas-2B cells for 24 h. The protocol is especially suitable for materials that contain one structural dimension at the micron scale or smaller-sized nanomaterials that form aggregates of sizes similar to that of a cell, which biases quantitative assessment of toxicity. Time points and concentrations of treatment as well as the cell type or the type of the material and its surface functionalization can be modified (*see* **Note 1**).

## 3.1    Preparation of L-GO Dispersions

1. Synthesize L-GO sheets from graphite powder (Sigma-Aldrich) according to a previously described modified Hummers method and purification protocols [8, 17]. The lateral dimensions of the L-GO flakes are between 2 and 20 μm, with a thickness ranging between 1 and 2 layers. Disperse L-GO material in complete cell culture medium (RPMI1640 cell culture medium + 10% FBS) to obtain a concentration of 0.05 mg/mL, which is the highest concentration of treatment

for the cells (*see* **Note 2**). Prepare the dispersion shortly before the treatment in a 15 mL sterile, plastic tube. Vortex thoroughly immediately after the preparation and again before treating the cells.

2. In order to determine the concentration of the L-GO that is inducing a significant decrease in cellular viability compared to the untreated cells, perform a dose escalation study. Prepare successive dilutions of the material (0.025 and 0.0125 mg/mL) by diluting a concentrated solution of the material (0.05 mg/mL) with complete cell culture medium in 15 mL sterile, plastic tubes.

### 3.2 Cell Culture Treatment and Preparation for Data Collection

1. Grow the cells in T-75 flasks in complete cell culture medium until they reach 80% confluence, and then passage them. In order to detach cells from the support, rinse them first with 1 mL of trypsin-EDTA at 37 °C. Incubate the cells with 3 mL of trypsin-EDTA at 37 °C for no longer than 5 min.

2. Detach cells by up and down pipetting, then place them in a 15 mL sterile tube, and add 10% FBS (300 μL) to stop the action of trypsin-EDTA.

3. Count cells and determine the number of live cells per mL using a trypan blue dye exclusion assay.

4. Seed 20,000 cells/cm² in six-well plates, using 2 mL of complete cell culture medium per well, and incubate them for 48 h to allow the cells to reach 80% confluence (*see* **Note 3**, *see* Fig. 1 for a schematic of the cell preparation and treatment protocol).

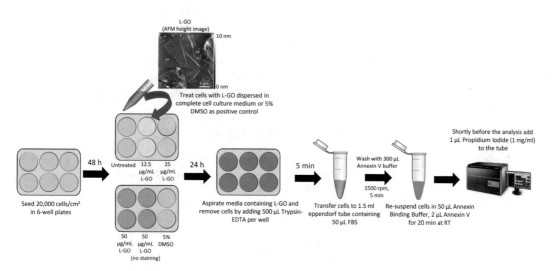

**Fig. 1** Schematic of the cell treatment protocol and staining procedure. Atomic Force Microscopy (AFM) height image shows lateral dimensions and thickness of L-GO flakes used for the treatment

5. Prepare the following controls, which will ensure proper setup of the instrument laser and successful subsequent data analysis: (1) untreated (complete cell culture medium will be added) and unstained cells (the background autofluorescence control), (2) untreated (complete cell culture medium will be added) cells, stained with annexin V/propidium iodide (the negative control for the dose escalation study), (3) cells treated with the highest concentration of the material (0.05 mg/mL) (*see* Subheading 3.1, **step 2**), but left unstained (necessary to determine if, upon interaction with the cells, the material emits signal when excited with the same laser used to excite the annexin V/propidium iodide dyes), (4) cells treated with 5% DMSO and stained with annexin V and cells treated with 5% DMSO and stained with propidium iodide (the single-stained positive controls, necessary to create a compensation matrix and remove spectral overlap), and (5) cells treated with 5% DMSO and stained with both annexin V and propidium iodide (a double-stained positive control, necessary for the setup of the excitation laser of the instrument and as a positive control for the dose escalation study).

6. Prepare the samples. Treat the cells with 3 mL of the L-GO material dispersed in complete cell culture medium (*see* Subheading 3.1, **step 2**). These will be stained with both annexin V and propidium iodide to carry out a dose escalation study.

7. After treatment, incubate the cells at 37 °C in a 5% $CO_2$ humidified atmosphere for 24 h.

8. After the incubation period is finished, aspirate the media from all of the samples and controls (*see* **Note 4**).

9. Remove the adherent cells by adding 500 μL trypsin-EDTA to each well and incubate the cells at 37 °C for 5 min in a humidified atmosphere.

10. Detach the cells from the support by up and down pipetting and transfer the cells from one well to a 1.5 mL Eppendorf tube containing 10% FBS (50 μL) to stop the action of trypsin-EDTA (*see* **Note 5**).

11. Centrifuge cells at 210 × *g* for 5 min.

12. Carefully remove the supernatant and gently resuspend the cells in 300 μL of 1× Annexin Binding Buffer to wash them (*see* **Notes 6** and **7**).

13. Centrifuge the cells at 210 × *g* for 5 min.

14. Remove the supernatant and resuspend the cells in 50 μL of Annexin Binding Buffer (*see* **Note 8**).

15. For those samples and controls that require annexin V staining, add 2 μL of annexin V-Alexa 488 to each tube.

16. Incubate the cells in the dark at room temperature for 20 min.

17. Place cell suspensions on ice until analysis.

18. Shortly before the analysis, for those samples and controls that also require propidium iodide staining, add 1 μL of propidium iodide (1 mg/mL) to the tubes.

19. Acquire the data.

**3.3  Data Collection (INSPIRE Software)**

The Imagestream® platform processes the data in two steps: data are first acquired using the Amnis INSPIRE™ application provided with the Imagestream® instrument. Next, the IDEAS software, which can be freely downloaded, processes and analyzes the data. This software contains the algorithms and tools required to analyze the images acquired using the INSPIRE application in the first step. Compensation for the spectral crosstalk needs to be calculated from the control single-stained files and applied to all of the experimental files.

1. Turn lasers on according to the excitation/emission spectra of the dyes used. For this protocol, the lasers turned on are: 488 nm (for the excitation of annexin V and propidium iodide) and 785 nm (the side-scatter and bright-field laser).

2. Start the acquisition by running the brightest sample first. In this protocol, we first run the positive control for the cell death (i.e., cells treated with 5% DMSO and subsequently stained with both annexin V and propidium iodide). This step is critical to establish settings of the excitation laser power and to avoid saturation of the fluorescent signal. For the excitation of annexin V and propidium iodide, 488 nm laser power was set at 60 mW, while 785 nm laser was set at 0.02 mW. To ensure the accuracy of the results, the same laser power settings must be used for all of the samples.

3. Select 60× magnification and acquire images with a normal depth of field.

4. Turn on the appropriate fluorescence emission channels. Channel 01 is used for the bright field, Channel 02 for annexin V and Channel 04 for propidium iodide.

5. Create a bivariate plot to gate the cells. This plot should have the "Area_M01" feature on the *x*-axes and "Aspect Ratio Intensity_M01_Brightfield" feature on the *y*-axes. This enables the population of single events to be gated in the analysis and eliminates doublets or signals from debris (Fig. 2). Before starting the acquisition, make sure that at least 5000 events will be acquired.

6. After running the positive control, run the untreated cells (i.e., the untreated and stained cells as a negative control for the dose escalation study and the untreated and unstained cells for the

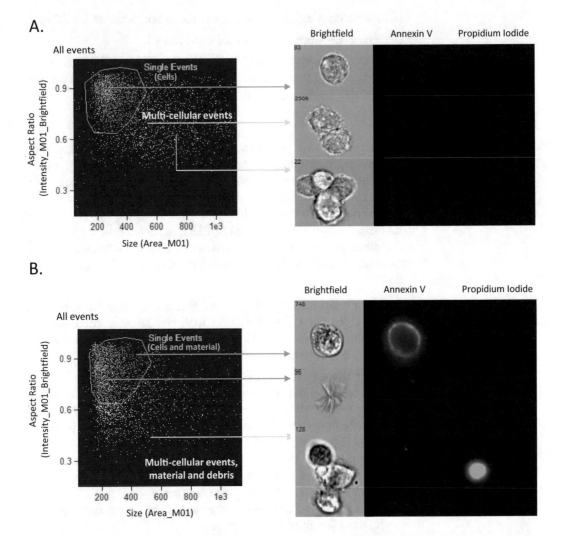

**Fig. 2** Bivariate plots and gating applied to distinguish single events in the population of all acquired events of untreated (**a**) and cells treated with 50 μg/mL L-GO (**b**). The aim of this step is to exclude doublets, multicellular events, and debris from the analysis. Bivariate plot is distinguishing events based on their size using "Area_M01" and "Aspect Ratio Intensity_M01_Brightfield" features provided in the IDEAS software. This separation is based on a Brightfield image using a mask that encompassed whole cell (M01). "Area_M01" is the size of the event in a Brightfield image expressed in square microns and "Aspect ratio" is a measure of the circularity of the event. Aspect ratio of 1 corresponds to a perfect circle (such as rounded single cell), while doublets have Aspect ratio of 0.5. Note that in the population of cells treated with 50 μg/mL L-GO (**b**) gated single events will include not only single cells, but also the material. This can be seen in Imaging Gallery after clicking on a corresponding event on the bivariate plot

autofluorescence check), as well as the cells treated with the highest concentration of the material but left unstained. Finally, run the cells treated with escalating doses of the L-GO material (the samples).

7. The single-stained positive controls should be run last to record files for the compensation matrix (*see* Subheading 3.4, **step 1**). For this purpose, follow the instructions in the "Compensation" tab. Briefly, 500 events need to be acquired with all the channels turned on (except the bright-field and dark-field channels).

**3.4 Data Analysis (IDEAS Software)**

1. Start the data analysis by creating a compensation matrix by following the instructions in the "Compensation" tab. When clicking the "Create new matrix" tab, it will be required to insert the files acquired using the single stains only. The compensation matrix will be automatically generated by IDEAS software. Save it to apply it to all other acquired data files.

2. Create bivariate plots to gate the cells. The first plot should have the "Area_M01" feature on the *x*-axes and "Aspect Ratio Intensity_M01_Brightfield" feature on the *y*-axes. This enables the user to gate the population of single events in the analysis and eliminate doublets or signals from debris (Fig. 2).

3. Create a second bivariate plot based on the single events, selected in Subheading 3.4, **step 2**. This plot will gate the events that are in focus and distinguish them from the events that are not in focus, including the material under study. This step is crucial and a prerequisite for subsequent image-based analysis using IDEAS software. The *x*-axes are labeled "Gradient RMS_M01_Brightfield" and the *y*-axes are labeled "Contrast_M01_Brightfield." Selected events should have high values of gradient and contrast features and should be inspected in the Imaging Gallery before including or excluding them from the gate. All the events with high values of the contrast and gradient will be gated as "Focused events." In order to confirm that selected events include only cells in focus and not the material, inspect all events included in the gate in the preview option in Imaging Gallery (*see* **Note 9** and **10**). Readjust the gate if necessary (Fig. 3).

4. After events involving cells have been selected and separated from those involving nanomaterials, create third bivariate plots using "Focused cells" with the "Intensity_MC_Channel_02" (annexin V) on the *x*-axes and "Intensity_MC_Channel_04" (propidium iodide) on the *y*-axes. Draw the gates using the "Untreated cells" file and create four gates: AV-/PI- (alive cells), AV+ (early apoptotic cells), PI+ (necrotic cells), and AV+/PI+ (late apoptotic and/or necrotic cells). Click the symbol "Σ" in the upper right corner of the bivariate plot. The number of cells and percentages in each gate will appear (Fig. 4).

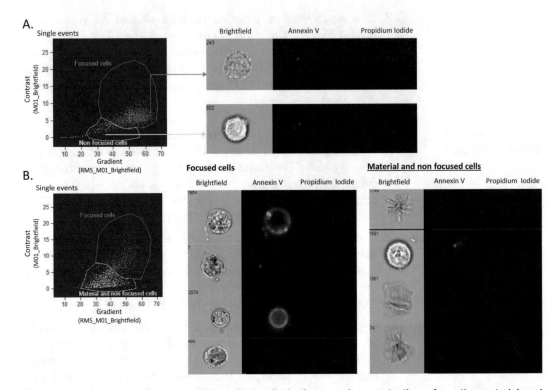

**Fig. 3** Bivariate plot and gating used to select cells in focus and separate them from the material and nonfocused events in untreated (**a**) and cells treated with 50 µg/mL L-GO (**b**). Features for the bivariate dot plot are calculated based on a Brightfield image and using the mask that covers whole cell (M01_Brightfield extension in the name of a feature). As already described, "Contrast_M01_Brightfield" and "Gradient RMS_M01_Brightfield" values for a cell change after interaction with carbon-based materials [13]. Low Gradient RMS (root mean square) and Contrast feature values characterize events that are unfocused, which is the case for some cells and all the material in the analysis. Gating of the events with high values of these features enables to select cells that will be included in the analysis of the cellular death and separate them from the material. Successful separation of the cells from the material needs to be verified in Imaging Gallery by observing all gated cells

5. Once all three plots are created for one experimental condition, create a "Statistic report" template with parameters including percentage of double negative, annexin V single positive, propidium iodide single positive, and annexin V/propidium iodide double positive cells. This sheet can be saved as a template and then applied to all the other control samples using the "Batch Data Files" option in the "Tools" tab.

6. Before exporting the values of all files and plotting them in graphs, make sure that gates are set properly in each of the files. Once the gating is readjusted, export the values and create a graph (Fig. 5, *see* **Note 11**).

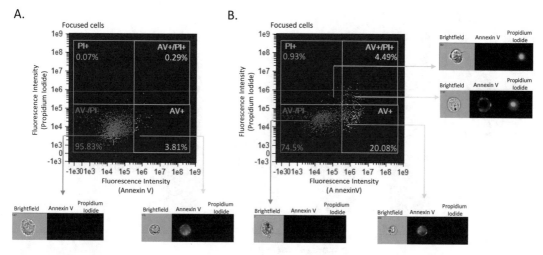

**Fig. 4** Analysis of cellular death of untreated (**a**) and cells treated with 50 μg/mL L-GO (**b**) using annexin V/ propidium iodide staining based on a population of single, focused cells. Bivariate plot includes intensity of fluorescence collected using channel for annexin V (Intensity_MC_Annexin V) and propidium iodide (Intensity_MC_Propidium Iodide). Gates for double negative, single and double positive cells are designed based on a population of "untreated and stained" cells and then applied on all other treatment conditions. Images of cells corresponding to different gates can be previewed in Imaging Gallery (*insets*). Percentage of cells in each gate are calculated by the software and can be found after clicking on a "Σ" symbol in the upper right corner of a plot

## 4    Notes

1. This cell line is relevant to study cytotoxicity in in vitro models representing the exposure to the material by inhalation. Other adherent and nonadherent cell lines (such as A549, MCF-7, MH-S, THP-1, etc.) can be used as well. Cells should be removed from the plate before analysis using trypsin.

2. The highest dose of the L-GO material used for this experiment was 0.05 mg/mL. Higher doses tend to stick to the surface of the cells and quench the fluorescence of the dyes, thus indicating that the material could be less toxic than it really is.

3. It is important to grow and treat cells on six-well plates or larger surfaces to collect enough cells for the analysis. It is required to have at least $10^6$ cells per sample. Treat the cells when they have reached 80% confluence if six-well plates are used; otherwise, the number of collected cells might not be sufficient.

4. After treatment with the material, GO in this case, it is important to remove the supernatant before collecting the cells to prevent the quenching of the fluorescence of the dyes due to the interference of the material with the fluorochrome.

5. The cells can be stained and fixed with paraformaldehyde if the analysis cannot be performed immediately following the

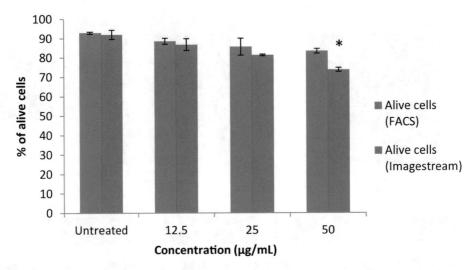

**Fig. 5** Comparison of cellular viability assessed using FACS and Imagestream®. Cells were treated with increasing concentrations of L-GO material dispersed in the complete cell culture medium for 24 h, collected and stained using described annexin V/propidium iodide protocol. Cellular viability was assessed either using flow cytometry (FACS Verse instrument) or imaging flow cytometry (Imagestream®). Higher accuracy of the cytotoxicity assessment was achieved using imaging flow cytometry compared to flow cytometry, especially after treatment with the highest concentration of L-GO material due to a possibility of observing the events included in the analysis and of excluding aggregated material from the analysis. Data are represented as means $\pm$SD ($n = 6$) and were statistically analyzed with IBM SPSS software (version 22) using analysis of variance (one-way ANOVA) with $p < 0.05$ considered significant

treatment. Store the fixed cells at 4 °C. However, because the washing and centrifugation steps required for fixation may introduce further cellular damage, the analysis of nonfixed samples is preferred.

6. If necessary, cells can be carefully washed using Phosphate Buffered Saline (PBS) with $MgCl_2$ and $CaCl_2$; however, during the washing and depending on the cell type, the cells might detach from the support, decreasing the population of analyzed cells. Using this protocol, cells can be analyzed without washing as the material removed from the surface is efficiently excluded from the analysis.

7. Carefully remove the supernatant after the centrifugation step in order not to disturb the pellet and lose cells.

8. Resuspend the cells in a maximum of 60 μL of Annexin Binding Buffer; otherwise, cells will be too diluted to analyze.

9. During the gating of single events and when focusing to gate out the material and nonfocused cells, always observe the cells on the borders of a gate to make sure that the highest accuracy is achieved.

10. With the increasing concentration of the material used for the treatment, the contrast and focus properties of the cells might

change. It is thus allowed to readjust gating in the bivariate plots, by looking at the Imaging Gallery, aiming to exclude the material and nonfocused cells from the analysis.

11. Higher accuracy of the cytotoxicity assessment was achieved using imaging flow cytometry compared to the flow cytometry (FACS Verse instrument) under the same conditions and with the same sample preparation procedures, especially after the treatment with the highest concentration of the L-GO material due to the possibility to observe the events included in the analysis and exclude aggregated material from the analysis (Fig. 5).

## Acknowledgments

This work was supported by grants: the EU FP7-ICT-2013-FET-F GRAPHENE Flagship (no. 604391) and the "RADDEL" Marie Curie Initial Training Network (ITN) grant under the EU's FP7 PEOPLE program. The authors would like to acknowledge Dr. Neus Lozano Valdes for the production of the L-GO material and Dr. Gareth Howell and Antonia Banyard for technical help and fruitful discussions as well as the Medical Research Council (MRC) funded Single Cell Genomics Facility as the funding source for the Imagestream® and the Manchester Collaborative Centre for Inflammation Research (MCCIR) for the FACSVerse instrument. The authors would like to acknowledge Dr. Irene de Lazaro, who read the manuscript and provided critical comments.

## References

1. Kostarelos K, Novoselov KS (2014) Graphene devices for life. Nat Nanotechnol 9 (10):744–745. doi:10.1038/nnano.2014.224 PubMed PMID:25286265

2. Kim J, Jeong C, Kim WJ (2016) Synergistic nanomedicine by combined gene and photothermal therapy. Adv Drug Deliv Rev 98:99–112. doi:10.1016/j.addr.2015.12.018 PubMed PMID:WOS:000370895000009

3. Zhang BM, Wang Y, Zhai GX (2016) Biomedical applications of the graphene-based materials. Mater Sci Eng C Mater Biol Appl 61:953–964. doi:10.1016/j.msec.2015.12.073 PubMed PMID: WOS:000370303600109

4. Ali-Boucetta H, Bitounis D, Raveendran-Nair R, Servant A, Van den Bossche J, Kostarelos K (2013) Purified graphene oxide dispersions lack in vitro cytotoxicity and in vivo pathogenicity. Adv Healthc Mater 2(3):433–441.

doi:10.1002/adhm.201200248    PubMed PMID:23184580

5. Bitounis D, Ali-Boucetta H, Hong BH, Min DH, Kostarelos K (2013) Prospects and challenges of graphene in biomedical applications. Adv Mater 25(16):2258–2268. doi:10.1002/adma.201203700 PubMed PMID:23494834

6. Liao KH, Lin YS, Macosko CW, Haynes CL (2011) Cytotoxicity of graphene oxide and graphene in human erythrocytes and skin fibroblasts. ACS Appl Mater Interfaces 3 (7):2607–2615. doi:10.1021/am200428v PubMed PMID: 21650218

7. Wang X, Mansukhani ND, Guiney LM, Ji Z, Chang CH, Wang M, Liao YP, Song TB, Sun B, Li R, Xia T, Hersam MC, Nel AE (2015) Differences in the toxicological potential of 2D versus aggregated molybdenum disulfide in the lung. Small 11(38):5079–5087. doi:10.1002/

smll.201500906 PubMed PMID:26237579; PMCID:PMC4600460

8. Orecchioni M, Jasim DA, Pescatori M, Manetti R, Fozza C, Sgarrella F, Bedognetti D, Bianco A, Kostarelos K, Delogu LG (2016) Molecular and genomic impact of large and small lateral dimension graphene oxide sheets on human immune cells from healthy donors. Adv Healthc Mater 5(2):276–287. doi:10.1002/adhm.201500606 PubMed PMID:26687729

9. Al-Jamal KT, Kostarelos K (2010) Assessment of cellular uptake and cytotoxicity of carbon nanotubes using flow cytometry. Methods Mol Biol 625:123–134. doi:10.1007/978-1-60761-579-8_11 PubMed PMID:20422386

10. Gosens I, Post JA, de la Fonteyne LJ, Jansen EH, Geus JW, Cassee FR, de Jong WH (2010) Impact of agglomeration state of nano- and submicron sized gold particles on pulmonary inflammation. Part Fibre Toxicol 7(1):37. doi:10.1186/1743-8977-7-37 PubMed PMID:21126342; PMCID:PMC3014867

11. Vranic S, Gosens I, Jacobsen NR, Jensen KA, Bokkers B, Kermanizadeh A, Stone V, Baeza-Squiban A, Cassee FR, Tran L, Boland S (2016) Impact of serum as a dispersion agent for in vitro and in vivo toxicological assessments of TiO2 nanoparticles. Arch Toxicol. doi:10.1007/s00204-016-1673-3 PubMed PMID:26872950

12. Marangon I, Boggetto N, Ménard-Moyon C, Venturelli E, Béoutis ML, Péchoux C, Luciani N, Wilhelm C, Bianco A, Gazeau F (2012) Intercellular carbon nanotube translocation assessed by flow cytometry imaging. Nano Lett 12(9):4830–4837. doi:10.1021/nl302273p PubMed PMID:22928721

13. Marangon I, Boggetto N, Ménard-Moyon C, Luciani N, Wilhelm C, Bianco A, Gazeau F (2013) Localization and relative quantification of carbon nanotubes in cells with multispectral imaging flow cytometry. J Vis Exp 12(82): e50566. doi:10.3791/50566 PubMed PMID:24378540; PMCID:PMC4048057

14. Vranic S, Boggetto N, Contremoulins V, Mornet S, Reinhardt N, Marano F, Baeza-Squiban A, Boland S (2013) Deciphering the mechanisms of cellular uptake of engineered nanoparticles by accurate evaluation of internalization using imaging flow cytometry. Part Fibre Toxicol 10:2. doi:10.1186/1743-8977-10-2 PubMed PMID:23388071; PMCID: PMC3599262

15. George TC, Basiji DA, Hall BE, Lynch DH, Ortyn WE, Perry DJ, Seo MJ, Zimmerman CA, Morrissey PJ (2004) Distinguishing modes of cell death using the ImageStream multispectral imaging flow cytometer. Cytometry A 59(2):237–245. doi:10.1002/cyto.a.20048 PubMed PMID:15170603

16. Rieger AM, Nelson KL, Konowalchuk JD, Barreda DR (2011) Modified annexin V/propidium iodide apoptosis assay for accurate assessment of cell death. J Vis Exp 24(50). doi:10.3791/2597 PubMed PMID:21540825; PMCID:PMC3169266

17. Rauti R, Lozano N, León V, Scaini D, Musto M, Rago I, Ulloa Severino FP, Fabbro A, Casalis L, Vázquez E, Kostarelos K, Prato M, Ballerini L (2016) Graphene oxide nanosheets reshape synaptic function in cultured brain networks. ACS Nano 10(4):4459–4471. doi:10.1021/acsnano.6b00130 PubMed PMID:27030936

# Chapter 21

# Air–Liquid Interface Cell Exposures to Nanoparticle Aerosols

Nastassja A. Lewinski, Nathan J. Liu, Akrivi Asimakopoulou, Eleni Papaioannou, Athanasios Konstandopoulos, and Michael Riediker

## Abstract

The field of nanomedicine is steadily growing and several nanomedicines are currently approved for clinical use with even more in the pipeline. Yet, while the use of nanotechnology to improve targeted drug delivery to the lungs has received some attention, the use of nanoparticles for inhalation drug delivery has not yet resulted in successful translation to market as compared to intravenous drug delivery. The reasons behind the lack of inhaled nanomedicines approved for clinical use or under preclinical development are unclear, but challenges related to safety are likely to contribute. Although inhalation toxicology studies often begin using animal models, there has been an increase in the development and use of in vitro air–liquid interface (ALI) exposure systems for toxicity testing of engineered nanoparticle aerosols, which will be useful for rapid testing of candidate substances and formulations. This chapter describes an ALI cell exposure assay for measuring toxicological effects, specifically cell viability and oxidative stress, resulting from exposure to aerosols containing nanoparticles.

**Key words** Air-interfaced culture, SPIONs, Iron oxide, Nanoparticles, Aerosol

## 1  Introduction

Traditionally, the toxicity of inhaled substances, including engineered nanoparticles, is determined with animal experiments, usually following guidelines proposed by the Organization for Economic Co-operation and Development (OECD). The testing strategies used in animals include subacute to chronic inhalation testing conducted over a month to several months, and acute inhalation testing that involves short-term, 4 h exposures. Acute inhalation testing following OECD test guidelines 403 and 436 starts with a maximum obtainable concentration of 5 mg/L for aerosols, which is a very unrealistic exposure scenario, and reveals toxicity data including gross response (behavioral changes and/or mortality), histopathology, and lethal concentration estimates ($LC_{50}$) [1, 2]. For nanoparticle aerosols, it is difficult to achieve

Sarah Hurst Petrosko and Emily S. Day (eds.), *Biomedical Nanotechnology: Methods and Protocols*, Methods in Molecular Biology, vol. 1570, DOI 10.1007/978-1-4939-6840-4_21, © Springer Science+Business Media LLC 2017

such extreme concentrations, and several published studies do not report acute inhalation toxicity at maximum obtainable concentrations [3–6]. Other limitations include the fact that, when conducting these tests, animals could experience distress or pain, and little is revealed about the pharmacokinetics or mechanism(s) of action of the materials. In addition, it is generally recognized that, despite their necessity before human testing, responses measured in animal models may not accurately predict the responses in humans. Taken together, it is arguable that in vitro testing could be used to verify if maximum obtainable concentrations result in acute toxicity. Animal testing can then commence directly at realistic exposure concentrations to assess the more human relevant subacute (28 days) inhalation toxicity. In this chapter, we discuss recent progress in the field in using air–liquid interface (ALI) exposure systems to assess the in vitro toxicity of aerosolized nanoparticles, and we also provide step-by-step instructions to measure cell viability and oxidative stress following cellular exposure to aerosolized nanoparticles using an ALI exposure system.

Aerosols are tested in vitro using air–liquid interface (ALI) exposure, which involves directly introducing aerosols to cells cultured on permeable membrane supports. This allows for cells to receive nutrients from culture medium touching the basolateral side while exposing the apical side to air as shown in Fig. 1. The rationale behind moving toward ALI cell exposures is comparable to the rationale behind the increase in mouth/nose controlled breathing exposure versus intratracheal instillation in animal studies: particle deposition and distribution patterns differ greatly when delivered by a bolus suspension versus by aerosol. Although suspensions deliver a defined dose instantaneously, inhalation is a

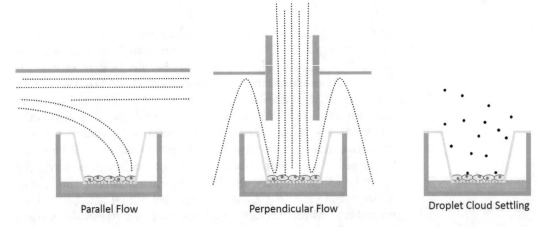

Parallel Flow          Perpendicular Flow          Droplet Cloud Settling

**Fig. 1** Pictorial representation of the aerosol flow profiles of the MEC (*left*), Vitrocell or NACIVT (*middle*), and ALICE (*right*) exposure chambers with aerosol in *blue*. The MEC, NACIVT, and Vitrocell system work with aerosols of a wide size range, while the ALICE system relies on gravitational sedimentation, which requires aerosol droplets in the μm range

dynamic not static process, and the dose deposited in the lungs is related to the amount of air inhaled and the breathing rate. In addition, particle concentration, size distribution, and morphology are all influenced by nanoparticle agglomeration, which occurs differently in liquid and gas phases. An additional benefit of ALI models is that cells cultured at the ALI develop more differentiated cell monolayers that more closely resemble the in vivo lung epithelium when compared to cells cultured under media [7–9].

Several ALI exposure systems have been developed, which allow for nanoparticle exposure of cells cultured on transwell membranes. The essential components for ALI exposure systems are as follows: "(1) complex pulmonary cell systems, which can be cultivated for at least several hours at the ALI, (2) direct contact between the cultivated cells and the inhalable substances without interfering medium, (3) uniform exposure of the entire cell layer, (4) temperature and humidity conditioning of the air to maintain cell integrity (T ~ 37 °C; relative humidity >85%), and (5) precise control of the substance concentration and purity for accurate dosimetry" [10]. Equally important are the methods of test aerosol generation and characterization. For particles dispersed in liquid, nebulizers are used to generate aerosols and a variety of studies test nebulized nanoparticle suspensions with ALI cell exposure systems [11–13]. The three referenced systems rely on methods of diffusion and/or sedimentation as deposition mechanisms (Fig. 1). An important difference is the direction of aerosol flow toward the cells, which is either parallel, perpendicular, or without flow (droplet cloud sedimentation), and has implications on the deposition efficiency for different particle sizes and the stress to the cells from passing air. The systems also differ regarding the maximal number of transwell samples that can be exposed per experiment.

This chapter describes an air–liquid interface (ALI) cell exposure assay for nanoparticle aerosols using the multiculture exposure chamber (MEC) system [11]. We selected the MEC system for this protocol for its ease of use (well plates containing transwells can be placed directly in the chamber) and versatility (up to 144 transwells can be exposed at once); however, the methods could easily be adapted for other ALI exposure systems. First, we describe how to culture human lung cells at the air–liquid interface and perform in vitro exposure to nanoparticle aerosols using the MEC system. Then we explain how to analyze the toxicological effects of the nanoparticle aerosol using a reactive oxygen species (ROS) assay, which indicates oxidative stress due to redox imbalance, and a lactate dehydrogenase (LDH) release assay, which indicates cellular necrosis as LDH is released from cells upon loss of plasma membrane integrity. We describe these assays using ferumoxytol iron oxide nanoparticles as model nanoparticles, and we also provide optional methods to analyze iron content in cells following aerosol exposure. The reader could adapt these methods for use with other types of nanoparticles as desired.

## 2  Materials

### 2.1  Culture of Lung Cells at the Air–Liquid Interface

1. A549 human alveolar type II-like lung epithelial cell line (*see* **Note 1**).

2. 1× Hank's balanced salt solution (HBSS), pH 6.7–7.8.

3. 1× phosphate-buffered saline (PBS), pH 7.2.

4. Complete cell culture medium: 500 mL of Dulbecco's Modified Eagle Medium (DMEM) with Glutamax 5 g/L D-glucose. Supplemented with 50 mL of Fetal Bovine Serum (FBS) and 5 mL of Penicillin–Streptomycin (10,000 units/mL of penicillin and 10,000 μg/mL of streptomycin).

5. 1× 0.25% Trypsin–EDTA.

6. 6- or 24-well plates and polyethylene transwell inserts with 3 μm pore size (*see* **Note 2**).

7. Forceps.

8. Biological safety cabinet.

9. $CO_2$ incubator.

10. Light microscope.

11. Hemocytometer.

12. Micropipettes.

13. Pipetman.

14. Aspiration system.

15. Volt–Ohm meter with chopstick electrodes (*optional*).

### 2.2  In Vitro Aerosol Exposure

1. In vitro air–liquid interface (ALI) exposure system: This protocol describes the use of a laboratory developed multiculture exposure chamber (MEC) [11] (*see* **Note 3**).

2. Nanoparticle suspension in ultrapure water: This protocol describes the use of ferumoxytol (*see* **Note 4**).

3. Air supply (respirable air cylinder and/or technical house air).

4. 1-jet Collison nebulizer.

5. Conductive tubing (inner diameter 8 mm).

6. Digital thermal mass flow controller.

7. Thermal mass flow meter.

8. Aerosol particle counters: This protocol describes the use of the Scanning Mobility Particle Sizer (SMPS) (*see* **Note 5**).

### 2.3  Toxicity Assays

1. Microplate reader.

2. Flat-bottom 96-well plates.

3. LDH Assay Kit: This protocol describes the use of the CytoTox-ONE Homogenous Membrane Integrity Assay Kit.

Other commercial kits that are comparable include lactate, NAD$^+$, diaphorase, and resazurin as the components of the assay mixture.

4. DCFH-DA working solution: Dissolve 24.4 mg of 2,7-dichlorodihydrofluorescein diacetate (DCFH-DA) in 50 mL methanol to make 1 mM DCFH-DA concentrated stock solution. This solution can be stored at $-20\,°C$ for up to 4 months. Dilute 1 mM DCFH-DA stock solution 100×s, or mix 0.1 mL of DCFH-DA stock solution with 9.9 mL HBSS for each 96-well plate, to make 10 μM DCFH-DA in HBSS.

5. Hydrogen peroxide ($H_2O_2$) stock solution: In a 50 mL volumetric flask, add 10 mL of Milli-Q water. Then add 0.5 mL $H_2O_2$ and fill up to the mark with Milli-Q water to achieve a final concentration of 0.1 M $H_2O_2$. This solution can be stored at 4 °C.

6. $H_2O_2$ working solution: Dilute 0.1 M $H_2O_2$ stock solution to 200 μM by adding 20 μL to a 10 mL volumetric flask and diluting with Milli-Q water.

7. Prussian blue iron stain kit (*optional*): This protocol describes the use of the Iron Stain Kit. Other commercial kits that are comparable include solutions of 4% w/v potassium ferrocyanide, 1.2 mM hydrochloric acid, and 1% w/v pararosaniline.

8. 50 mL conical tubes.

# 3 Methods

*Safety note*: This is a pressurized system with the potential to break and release hazardous materials. Have your safety and health staff verify the setup is safe for operation. All cell culture work should be conducted in a biological cabinet with standard personal protection equipment (lab coat, gloves, goggles). During testing and operation of the aerosol system, respiratory protection (N100 or FFP3 filter masks) is needed whenever accidental exposure cannot be excluded.

## 3.1 Culture of Lung Cells at the Air–Liquid Interface

*Note*: Cell cultures should be maintained in an incubator at 37 °C, 5% $CO_2$, and >80% relative humidity. The reader is assumed to be familiar with aseptic culture technique.

1. Harvest A549 cells cultured under liquid cover from culture flask. Remove medium and rinse cells with 3 mL PBS. Remove PBS and add 3 mL of trypsin. Incubate cells at 37 °C for 5–10 min. Examine flask using a light microscope to confirm cells have detached. Add 7 mL of complete medium to neutralize trypsin. Collect cells using a serological pipette with pipetman and transfer to a 15 mL conical tube.

2. Count cells using a hemocytometer to determine cell concentration. Remove 10 μL of cell suspension from the 15 mL conical tube using a micropipette and add to hemocytometer. Count cells in four $1 \times 1$ mm squares. Average and multiply by 10,000 to calculate number of cells per mL in harvested cell suspension. Repeat this step two additional times for higher statistical accuracy.

3. Prepare well plates with culture medium. For 6-well plates, add 1 mL to each well. For 24-well plates, add 0.3 mL to each well.

4. Seed transwell membrane inserts on apical side. For A549 cells at a seeding density of $1 \times 10^5$ cells/cm$^2$, add $3.5 \times 10^4$ cells to 24-well transwell or $5 \times 10^5$ cells to 6-well transwell.

5. Add complete medium to apical side of transwell to reach the recommended volume of 2 mL for 6-well plates and 0.5 mL for 24-well plates.

6. Culture cells under liquid cover for 7 days, replacing the medium every 1–2 days. (*Optional*: Monitor transepithelial electrical resistance (TEER) using a volt–ohm meter (*see* **Note 6**). Measure TEER of a cell-free transwell for the baseline.)

7. After 7 days, remove apical medium and culture cells for at least 1 day in ALI conditions, replacing basolateral medium every 1–2 days. Replace medium before exposing cells to aerosol. (*Optional*: Continue to monitor TEER in ALI conditions. Add 0.1 mL (6 well) or 0.02 mL (24 well) of prewarmed medium to apical side and incubate for 30 min before TEER measurement. Measuring TEER before conducting oxidative stress assay is not recommended (*see* **Note 7**).)

### 3.2 In Vitro Aerosol Exposure

1. Prepare nanoparticle suspensions at desired concentration in a minimum of 10 mL and a maximum of 200 mL ultrapure water.

2. Load nanoparticle suspension in nebulizer and ensure that nozzle tip is adequately submerged in suspension without blocking jet.

3. Connect nebulizer to aerosolization system. *See* Fig. 2 for an example system configuration. Ensure valve to SMPS is open and value to MEC is closed.

4. Load the exposure chamber with lung cell samples. Inside a biological safety cabinet, first disinfect MEC by wiping interior surfaces with 70% ethanol. After allowing ethanol to dry, line interior with aluminum foil to reduce electrostatic deposition of aerosol onto the MEC inner surface. Place the well plates containing the lung cells on transwells grown at the ALI for 7 days into the well plate holders with lids removed. Close MEC and remove from biological safety cabinet.

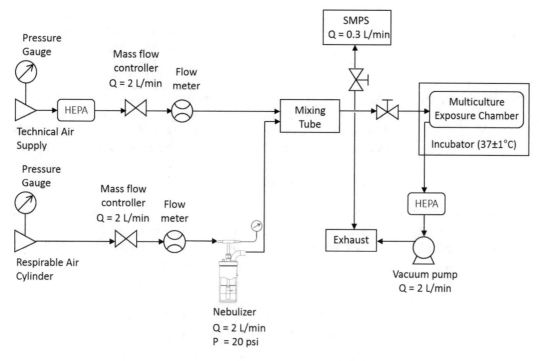

**Fig. 2** Example aerosolization system configuration for in vitro air–liquid interface exposures

5. Place MEC inside incubator preheated to 37 °C. Connect all fittings and tubing. Ensure valve to the MEC inlet is opened.

6. Turn on the air supply, set the pressure using the gas regulator, and set the flow rate using the mass flow controller. The 1-jet Collison nebulizer should be supplied with air at a flow rate of 2 L/min and pressure of 20 psig (*see* **Note 8**). Turn on the vacuum pump to the exposure chamber to start the exposure. Nebulize the nanoparticle suspension for a 60 min period.

7. Physicochemical characterization of the generated aerosol should be conducted during the experiment. At a minimum, particle size should be measured using an aerosol particle counter, such as a scanning mobility particle sizer (*see* **Note 5**). For the physicochemical characterization of aerosolized ferumoxytol [14], particle size was determined using transmission electron microscopy, scanning mobility particle sizing, particle correlation spectroscopy, and nanoparticle tracking analysis. Metal analysis was conducted using atomic absorption spectroscopy and inductively coupled plasma optical emission spectroscopy.

8. After the 60 min exposure duration, turn off the vacuum pump to the exposure chamber and then shut the aerosol inlet valve to stop the exposure.

9. Disconnect the MEC from the aerosolization system setup. Remove the MEC from the incubator and transfer it to a biological safety cabinet. Open the MEC, immediately remove lung cell culture samples, and begin toxicity assays.

### 3.3 Toxicity Assays

This protocol describes two different toxicity assays. The first describes how to measure reactive oxygen species generation, which is indicative of oxidative stress, using DCFH-DA (*see* **Note 9**). The second describes how to measure lactate dehydrogenase (LDH) release from cells, which indicates disruption in cell membrane integrity and, indirectly, cell death.

### 3.3.1 Measuring Reactive Oxygen Species Generation Using DCFH-DA

1. Prepare new well plates containing 10 μM DCFH-DA working solution in each well. For 6-well plates, add 1 mL of DCFH-DA working solution per well. For 24-well plates, add 0.3 mL of DCFH-DA working solution per well.

2. Using sterile forceps, transfer transwells containing nanoparticle-exposed lung cells to the prepared well plates containing DCFH-DA dye. Save basolateral medium for the lactate dehydrogenase (LDH) release assay.

3. Cover the plate with aluminum foil to prevent photoactivation and incubate the cells with the DCFH-DA working solution at 37 °C in a humidified $CO_2$ incubator for 30 min.

4. Following incubation, load the well plate into the microplate reader and measure the fluorescence of DCF using excitation/emission wavelengths of 485/530 nm (*see* **Note 9**).

5. Report ROS production as the fluorescence intensity increase of treated cells relative to the baseline measurements, as shown in Fig. 3.

### 3.3.2 Measuring Lactate Dehydrogenase (LDH) Release

1. From the well plate containing the saved basolateral medium, transfer 100 μL of supernatant to a 96-well plate.

2. Equilibrate this plate to room temperature for approximately 20–30 min.

3. Add 100 μL of CytoTox-ONE Reagent to each well and mix or shake gently by hand for 30 s.

4. Incubate at room temperature for 10 min.

5. To each well, add 50 μL of stop solution (3% w/v sodium dodecyl sulfate) provided in the CytoTox ONE assay kit.

6. Shake the plate for 10 s and record the fluorescence within 1 h with an excitation wavelength between 560 nm and an emission wavelength between 590 nm (*see* **Note 10**).

7. Report the LDH release as a percentage based on the negative control after background correction (Fig. 4). This sets the

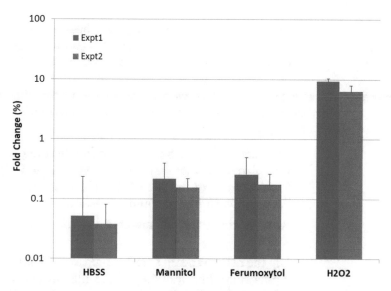

**Fig. 3** Example result for DCFH-DA assay after 1 h aerosol exposure. Here, cells were analyzed after different aerosol exposure conditions, including HBSS (1×), mannitol (0.59 mg/mL), and ferumoxytol (0.4 mg Fe/mL). $H_2O_2$ (1 mM) was added to the cells to serve as a positive control and to ensure dye activation

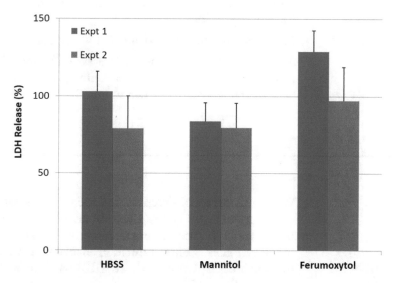

**Fig. 4** Example result for LDH assay after 1 h aerosol exposure. Here, cells were analyzed after different aerosol exposure conditions, including HBSS (1×), mannitol (0.59 mg/mL), and ferumoxytol (0.4 mg Fe/mL)

untreated cells (negative control) at 100% LDH release, which is used as the reference for no leakage. Alternatively, cytotoxicity can be reported as a percentage based on the positive control (e.g., adding 1% Triton-X 100 solution to lyse cells for maximum LDH release) after background correction (*see* **Note 11**). This sets the lysed cells (positive control) at 100% cytotoxicity, which is used as the reference for 100% cell death.

*3.3.3 Iron Visualization (Optional)*

This protocol can be used to visualize iron-containing nanoparticles, such as those in ferumoxytol, deposited on the cell monolayer after exposure.

1. After measuring ROS production (Subheading 3.3.1), transfer well plate to a chemical fume hood then add 100 μL of 4% formaldehyde to the top of the membrane to cover the cells. Fix cells for 15 min at room temperature.

2. Remove the transwell membranes from the inserts using a scalpel.

3. Dip the membranes in a freshly prepared 1:1 solution of 4% w/v potassium ferrocyanide and 1.2 mM hydrochloric acid and incubate for 10 min at room temperature. This stains the membranes for iron-containing compounds.

4. Rinse the membranes in deionized water and incubate in freshly prepared 2% v/v pararosaniline–water solution for 5 min at room temperature. This counterstains the membranes for polysaccharides.

5. Rinse the membranes in deionized water again.

6. Mount the membranes on coverglass and image with a bright-field microscope. A representative image showing the appearance of stained cells is shown in Fig. 5.

# 4    Notes

1. While this protocol describes the use of A549 lung epithelial cells, several types of lung cell monocultures and cocultures have been reported. Reported monocultures include established cell lines (Calu-3 human bronchial epithelial, BEAS-2B human bronchial epithelial, 16HBE14o- human bronchiolar epithelial-like, A549 human alveolar type II-like epithelial, LK004 human lung fibroblast) and primary lung cells (human bronchial epithelial—normal or diseased). Reported cocultures include laboratory developed [15–17] and commercially available (MucilAir, EpiAirway) tissues. Any of these model systems could be utilized with this protocol.

2. Transwell manufacturers provide excellent technical guidance on material and pore size selection for different applications. In general, in vitro lung models using transwells with pore sizes ranging from 0.4, 1, and 3 μm have been reported. Transwells

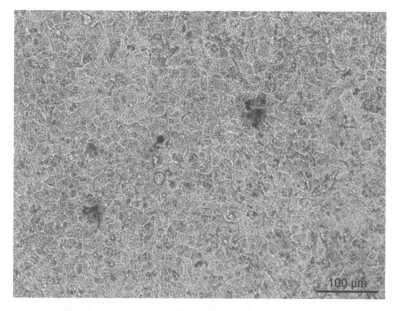

**Fig. 5** Iron stained A549 cells exposed to ferumoxytol aerosol for 1 h. The *dark blue/black stain* indicates the presence of the iron oxide nanoparticles, while the pink counterstain reveals polysaccharides

with 0.4 μm pore size have been demonstrated to result in the least cell translocation and tracer dye leakage [16, 18].

3. Over 30 different ALI cell exposure systems have been developed. These include two commercially available systems (Vitro-Cell, Cultex). These could be utilized instead of the MEC system described in this protocol.

4. While this protocol describes the use of ferumoxytol, it may be adapted for use with other types of nanoparticles that are of interest to the reader.

5. Nanoparticle aerosol concentrations are commonly measured using size mobility particle sizer/condensation particle counters (SMPS/CPC). Other online particle counters that can be used include fast mobility particle sizers (FMPS) and diffusion size classifiers (e.g., DiscMini).

6. TEER measurements indicate cell monolayer integrity and permeability. TEER values for A549 cells can be measured with values ranging between 20–60 $\Omega{\cdot}cm^2$ [19] and 140–180 $\Omega{\cdot}cm^2$ after 7 days of culture [17]. Note that TEER values for other cell lines (e.g., Calu-3) or primary lung cell models (e.g., MucilAir) that form functional tight junctions and a "tight" monolayer can be 3–10 times larger.

7. After cells are placed under ALI culture conditions, addition of liquid to the apical side can induce temporary oxidative stress.

As oxidative stress is an endpoint measured in this protocol, it is recommended to take TEER measurements after conducting the oxidative stress assay.

8. The 1-jet Collison nebulizer can be operated at pressures between 20 and 100 psig and flow rates of 2–7 L/min. According to the manufacturer, at 20 psig the droplets produced have a mass median diameter of 2 μm, and the liquid use rate is 1.5 mL/h.

9. DCFH-DA is a popular fluorescent probe for the detection of oxidative stress in cells. Since the probe can be prone to auto-oxidation, the use of carboxy-2′,7′-dichlorodihydrofluorescein-diacetate (carboxy-$H_2$DCF-DA), which is more stable, is also reported. Upon crossing the cell membrane, esterases hydrolyze DCFH-DA to DCFH, which remains trapped within cells. The oxidation of DCFH yields DCF, a fluorescent, fluorescein-based compound that can be measured using excitation/emission wavelengths of 485–495/520–530 nm.

10. The absorbance peak of resorufin is broad allowing the use of excitation/emission wavelengths in the ranges of 530–570/580–620 nm.

11. It is recommended to optimize the microplate reader settings to the positive control in order to ensure that the microplate reader's maximum readable value is not exceeded.

## Acknowledgements

This work was supported by the Whitaker International Program through a postdoctoral grant to NAL, the Wagoner Foreign Study Scholarship through a summer research grant to NJL, the Institute for Work and Health in Lausanne, Switzerland, and Virginia Commonwealth University.

## References

1. OECD. Test No. 403: Acute Inhalation Toxicity. OECD Publishing

2. OECD. Test No. 436: Acute Inhalation Toxicity—Acute Toxic Class Method. OECD Publishing

3. Srinivas A, Rao PJ, Selvam G, Goparaju A, Murthy BP, Reddy NP (2012) Oxidative stress and inflammatory responses of rat following acute inhalation exposure to iron oxide nanoparticles. Hum Exp Toxicol 31(11):1113–1131. doi:10.1177/0960327112446515

4. Srinivas A, Rao PJ, Selvam G, Murthy PB, Reddy PN (2011) Acute inhalation toxicity of cerium oxide nanoparticles in rats. Toxicol Lett 205(2):105–115. doi:10.1016/j.toxlet.2011.05.1027

5. Sung JH, Ji JH, Song KS, Lee JH, Choi KH, Lee SH, Yu IJ (2011) Acute inhalation toxicity of silver nanoparticles. Toxicol Ind Health 27(2):149–154. doi:10.1177/0748233710382540

6. Wang L, Wang L, Ding W, Zhang F (2010) Acute toxicity of ferric oxide and zinc oxide nanoparticles in rats. J Nanosci Nanotechnol 10(12):8617–8624. doi:10.1166/jnn.2010.2483

7. Grainger CI, Greenwell LL, Lockley DJ, Martin GP, Forbes B (2006) Culture of Calu-3 cells

at the air interface provides a representative model of the airway epithelial barrier. Pharm Res 23(7):1482–1490. doi:10.1007/s11095-006-0255-0

8. Lehmann AD, Daum N, Bur M, Lehr C-M, Gehr P, Rothen-Rutishauser BM (2011) An in vitro triple cell co-culture model with primary cells mimicking the human alveolar epithelial barrier. Eur J Pharm Biopharm 77 (3):398–406. doi:10.1016/j.ejpb.2010.10.014

9. Steimer A, Haltner E, Lehr CM (2005) Cell culture models of the respiratory tract relevant to pulmonary drug delivery. J Aerosol Med 18 (2):137–182. doi:10.1089/jam.2005.18.137

10. Paur H-R, Cassee FR, Teeguarden J, Fissan H, Diabate S, Aufderheide M, Kreyling WG, Hänninen O, Kasper G, Riediker M, Rothen-Rutishauser B, Schmid O (2011) In-vitro cell exposure studies for the assessment of nanoparticle toxicity in the lung—A dialog between aerosol science and biology. J Aerosol Sci 42 (10):668–692. doi:10.1016/j.jaerosci.2011.06.005

11. Akrivi A, Emmanouil D, Nastassja L, Michael R, Eleni P, Athanasios GK (2013) Development of a dose-controlled multiculture cell exposure chamber for efficient delivery of airborne and engineered nanoparticles. J Phys Conf Ser 429(1):012023

12. Deschl U, Vogel J, Aufderheide M (2011) Development of an in vitro exposure model for investigating the biological effects of therapeutic aerosols on human cells from the respiratory tract. Exp Toxicol Pathol 63 (6):593–598. doi:10.1016/j.etp.2010.04.013

13. Lenz AG, Karg E, Lentner B, Dittrich V, Brandenberger C, Rothen-Rutishauser B, Schulz H, Ferron GA, Schmid O (2009) A dose-controlled system for air-liquid interface cell exposure and application to zinc oxide nanoparticles. Part Fibre Toxicol 6(1):1–17. doi:10.1186/1743-8977-6-32

14. Graczyk H, Bryan LC, Lewinski N, Suarez G, Coullerez G, Bowen P, Riediker M (2014) Physicochemical characterization of nebulized Superparamagnetic Iron Oxide Nanoparticles (SPIONs). J Aerosol Med Pulm Drug Deliv 28(1):43–51. doi:10.1089/jamp.2013.1117

15. Kasper JY, Hermanns MI, Unger RE, Kirkpatrick CJ (2015) A responsive human triple-culture model of the air–blood barrier: incorporation of different macrophage phenotypes. J Tissue Eng Regen Med. doi:10.1002/term.2032

16. Klein SG, Serchi T, Hoffmann L, Blömeke B, Gutleb AC (2013) An improved 3D tetraculture system mimicking the cellular organisation at the alveolar barrier to study the potential toxic effects of particles on the lung. Part Fibre Toxicol 10(1):1–18. doi:10.1186/1743-8977-10-31

17. Rothen-Rutishauser BM, Kiama SG, Gehr P (2005) A three-dimensional cellular model of the human respiratory tract to study the interaction with particles. Am J Respir Cell Mol Biol 32(4):281–289. doi:10.1165/rcmb.2004-0187OC

18. Ye D, Dawson KA, Lynch I (2015) A TEM protocol for quality assurance of in vitro cellular barrier models and its application to the assessment of nanoparticle transport mechanisms across barriers. Analyst 140(1):83–97. doi:10.1039/C4AN01276C

19. Sporty JL, Horálková L, Ehrhardt C (2008) In vitro cell culture models for the assessment of pulmonary drug disposition. Expert Opin Drug Metab Toxicol 4(4):333–345. doi:10.1517/17425255.4.4.333

# Chapter 22

# Returning to the Patent Landscapes for Nanotechnology: Assessing the Garden that It Has Grown Into

## Diana M. Bowman, Douglas J. Sylvester, and Anthony D. Marino

### Abstract

The patent landscape, like a garden, can tell you much about its designers and users; their motivations, biases, and general interests. While both patent landscapes and gardens may appear to the casual observer as refined and ordered, an in-depth exploration of the terrain is likely to reveal unforeseen challenges including, for example, alien species, thickets, and trolls. As this Chapter illustrates, patent landscapes are dynamic and have been forced to continually evolve in response to technological innovation. While emerging technologies, such as biotechnology and information communication technology have challenged the traditional patent landscape, resulting in the pruning of certain elements here and there, the overarching framework and design has largely remained intact. But will this always be the case? As the field of nanotechnology continues to evolve and mature, the aim of this Chapter is to map how the technology has evolved and grown within the confines of existing structures and underlying foundation of the patent landscape and the implications thereof for the technology, industry, and the public more generally. The Chapter concludes by asking the question whether the current patent landscape will be able to withstand the ubiquitous nature of the technology, or whether nanotechnology, in combination with other emerging technologies, will be a catalyst for governments and policy makers to completely redesign the patent landscape.

**Key words** Intellectual property, TRIPS Agreement, Patent thickets, Patent pools, Trolls, Technology innovation

## 1 Introduction

One is tempted to think of the patent landscape as a refined English garden. Views of gently rolling lawns spotted by outcroppings of majestic trees, a few revival buildings, and inundated by hundreds of floral and shrub varieties might leave the casual observer with the view that it is entirely organic and naturalistic. For those who look closer, however, one sees the tenders' efforts. The lack of straight lines, walls, or delineated beds masks the perfect visual delineation of the form—a form evolving over decades (if not centuries) and one that is largely in balance. Organic contours hide the hundreds of small decisions that are continually being made to retain the

Sarah Hurst Petrosko and Emily S. Day (eds.), *Biomedical Nanotechnology: Methods and Protocols*, Methods in Molecular Biology, vol. 1570, DOI 10.1007/978-1-4939-6840-4_22, © Springer Science+Business Media LLC 2017

appropriate balance. But those decisions are nevertheless made. Although such gardens may give the impression that they arise "just so," they hide the enormous and complicated efforts of their tenders to organize, weed, and design them.

Since the 1970s, patent gardens have been under continued attack, and their carefree style seems under threat as alien species invade these once tranquil spaces. The dual shocks of software [1] and much more importantly, biotechnology [2], wreaked havoc on these once tranquil and, seemingly, unchanging spaces. Well-tended beds, ancient perennials, and majestic arbors were threatened by rapidly growing and unanticipated thickets and brambles [3–6]. Rolling meadows were quickly dotted with pitfalls, and once-languid pools [7], now choked with unforeseen infestations, threatened to become unsightly and unrecoverable bogs and quagmires [8]. Worst of all, these gardens (and especially their beautiful marble bridges) were invaded by trolls [9, 10]!

In these dark days, the garden's tenders created new tools, brought in help from abroad (as discussed in Subheading 2.3) and, although forced to make certain concessions to these alien species (*see*, generally, the *Uruguay Round Agreements Act of 1994*), finally succeeded by the end of the millennia to return much of the garden to its former apparent placidity. Sure some ancient varietals were replaced by new and, for a time, foreign blossoms. A few hedges, long put up to keep out unwanted visitors, or at least to make the entry difficult, were, if not fully removed, trimmed a tad. Finally, it appears that furtive efforts were made to gather pitchforks and torches to drive off those pesky trolls—although some are obviously lurking underneath some of the murkier bridges [10]. In the end, the threat, although not entirely gone, seemed largely under control and much of what we had always expected in our patent garden remained familiar and friendly.

However, new threats are looming on the edges of our serene plot. Tendrils of invasive species can be seen sprouting all over the garden and destructive vines threaten the integrity of the garden's walls. In short, nanotechnology brings with it the potential to upset not only some aspects of the patent garden, but may force a complete rethinking of its function and form (*see* **Note 1**). Biotechnology thickets may have grown over some beloved blooms, but nanotechnology's brambles have threatened to take down the entire field [6, 8]. Those seeking refuge in the cooling waters of the garden's pools, now clear after a decade of invasion, may find them once again choked to a vivid green. And, horror, the trolls appear to be breeding again! Our patent garden, so perfect in its form to handle the challenges of past patent revolutions, seems particularly unable to handle what nanotechnology, as well as other emerging technologies such as additive manufacturing and genome-editing technologies, may be bringing.

Leaving aside the metaphor of the garden, the increasing pace, complexity, and importance of technological revolutions have put real pressure on the patent system and the policy makers, and regulators, charged with overseeing their development and operational activities. Revered doctrines, designed for pre- and early industrial innovations, seem quaint (if not dangerous) for these times [6]. Patent institutions, organized around silos of knowledge and focused on local inventorship, may not be able to stand in the face of massively complex and global innovations. Finally, the pace of patenting, both in terms of process and conceptual foundations, seems dangerously ill suited to technological advances that have the ability to challenge existing national and international legal frameworks, including those relating to patents, in the blink of an eye. In numerous other publications, we have examined the many challenges nanotechnology poses to traditional regulatory structures related to environmental and human health and safety (*see*, for example, [11–18]). In this Chapter, we examine the challenges that nanotechnology has, and will, pose for patent frameworks and institutions.

We already know that biotechnology radically shifted the patent landscape both in terms of patenting practices of researchers [19], and the scope and breadth of patentable subject matter [20–22]. In just a few short years, biotechnology and rapidly advancing pharmaceutical patenting forced a fundamental rethinking of what was and should be patentable [22], as well as substantial hand-wringing about why we allow patenting of socially beneficial inventions at all [5]. Although biotechnology, software, and pharmaceutical patenting may have spurred a substantial "rethinking [of] intellectual property rights" [22], much of the prior system remains in place. Patents are still, largely, national affairs [23] with massive bureaucratic costs that view patents as arising from traditional scientific disciplines [24, 25]. In addition, patents are still issued without great oversight and on increasingly early stage technologies [26]. These were, similarly, issues in the biotechnology and pharmaceutical revolutions of the 1980s, but the system was able to accommodate these issues without fundamentally altering the process and purpose of patenting.

Returning to our metaphor—it may be time to decide whether the garden as it has been known for centuries should be redesigned. The garden, whether we are conscious of it or not, picks winners and losers. It gives preference to those that bloom first—crowding out latecomers and preventing variations. Hedges and walls, meant to keep out unwelcome visitors and maintain tranquility, reduce hybridization and competition and, arguably, reduce overall social utility. Finally, those darn trolls really need to be run out!

In this Chapter, we explore how nanotechnology has sought to challenge the traditional patent landscape. In particular, we wish to address more of the garden metaphors (although, mercifully, these are not our creations) that speak of potential thickets, bogs, brambles, quagmires, pitfalls, pools, and other problems that new technologies may pose for our little patent garden. To this end, this Chapter proceeds in several Subheadings. In Subheading 2.1, we set out some of the patent basics that have arisen over the course of the past few centuries. In so doing, our analysis focuses mainly on the United States but we do include some important comparative and multinational aspects of the current landscape. In Subheading 2.2, we discuss some of the unique aspects of nanotechnology's development, current regulation, and potential that many have predicted will pose real problems for traditional patent systems. Further, we provide an overview of research into whether some of those predictions have started to come true. Subheading 2.3 sets out some of the remaining challenges that nanotechnology may pose for patent frameworks. Finally, in Subheading 2.4 we set forth some thoughts about how we may avoid some of the potential problems with nanotechnology and how the patent system may need to reform itself to better accommodate technological invasions that will inevitably occur again. And do so with increasing frequency.

## 2   Discussion

### 2.1   Patent Basics

For centuries, the patent system has been predicated on a few unchanging and nearly universal precepts. First, patents are national in scope (and often favor national inventors) [27, 28]. Second, patents reward innovation by granting rights to inventors [27]. Third, patents only apply to inventions and not discoveries [27]. And, fourth, patents are ultimately intended to benefit society by encouraging technological innovation and must, therefore, seek the balance between encouraging invention and ensuring social benefits through access and use [23, 28]. To achieve these various principles, patent systems around the world have created both institutions and doctrinal frameworks dedicated to ensuring their fulfillment.

These precepts were built up during periods of relatively low patenting and invention. They made sense in that era but their relevance today is increasingly coming under question. First, the national scope of patents produces real inefficiencies in that inventors must seek approval in each nation, greatly increasing the cost on both inventors and societies to manage that system. In addition, in an era of increasing global research and patenting, it makes little sense to continue to favor one citizen inventor over another merely because of accident of birth.

As a result, there have been some tentative efforts to streamline this process. First, the United States Patent and Trademark Office (USPTO) and Japanese Patent Offices (along with many others) have begun to take some of the administrative burden off of inventors by allowing for streamlined patenting through individual treaty arrangements or application of the Patent Cooperation Treaty (*see* **Note 2**). A discussion of how these procedures have streamlined the process is outside the scope of this Chapter, but the real point is that it has not been wildly successful. It is still an extraordinarily complex and expensive process to engage in global patenting and calls for reform continue to grow [29, 30].

A secondary, and closely related, framework for achieving patent's precepts is through doctrinal limitations on patentability. At their base, these rules seek to ensure that only those patents that *add something* to the world of knowledge in a given area are patentable [28]. The doctrines of novelty, nonobviousness, and utility are all, in some respects, attempts to ensure that only those patents worthy of intellectual property protection are granted a bundle of legal rights [23, 31]. Yet, given patent law's overall goal of providing social goods through appropriate incentivized innovation, a fundamental practical principle of nearly every patent system has been to "grant the patent and let the market figure it out!" In low-innovation periods, this approach makes perfect sense [26]. In addition, it may even have net benefits in periods of explosive innovation in technological applications. Where it is deeply problematic is in times—and you guessed that the nanotechnology revolution may be one of those times—of immense patenting of basic research and fundamental research tools [6]. Again, this is an issue we take up later in the Chapter.

One area of genuine progress in patent law, at least in terms of overall efficiency and systemic fairness, has been in the area of doctrinal harmonization. In particular, the World Trade Organization's (WTO) *Trade-Related Intellectual Property Rights (TRIPS) Agreement* has provided an opportunity for tremendous progress in the harmonization of basic standards of patentability. Pursuant to Article 27(1) of the *TRIPS Agreement*,

> patents shall be available for any invention, whether products or processes, in all fields of technology provided they are new, involve an inventive step and are capable of industrial application.

While conditions for the granting of a patent and nature of the exclusive rights will vary between jurisdictions, Mandel [31] notes that these requirements, and therefore the bundle of rights granted, are "largely harmonized throughout the world" (*see* **Note 3**). Unfortunately, as hinted earlier, substantive or doctrinal harmonization has not been accompanied with procedural efficiencies [28].

Despite these efforts at substantive harmonization, it is important to note that not all subject matter may be the subject of a patent grant and that, at its outer limits, the question of what is patentable is still an open discussion. As highlighted by Article 27 of the *TRIPS Agreement*, patents may be granted for "any inventions" but the Agreement does not define what constitutes an "invention." This area of potential disunity has been largely avoided by the practice of most nations to adopt broad approaches to patentability [32] thereby ensuring continuation and global propagation of the "patent now, sue later" mentality noted earlier. To be clear, we do not view this practice as necessarily problematic. Indeed, as we note later in this Chapter, an equally thorny problem is making patent grants too difficult and/or expensive to obtain because they create disincentives to innovate, rob inventors of the fruits of their labor, or unduly delay beneficial applications. Nobody said tending the patent garden was easy.

As a result of the efforts of the WTO and, indeed, the United States through numerous bilateral agreements [33], there is wide consensus on what is *not patentable* as well as buy-in for the general principle that most things should be. According to Mandel [31],

> laws of nature, natural phenomena, abstract ideas, aesthetic creations, and information and data per se generally are not patent eligible. Almost everything else is.

In most jurisdictions, "discoveries" or "products of nature" fall outside the traditional scope of patentable subject matter. *Prime facie*, the boundary of what does and does not constitute a patentable subject should be simple. However, the reality is somewhat more complex with Eisenberg [2], for example, having observed a "shifting landscape of discovery in genetics and genomics research" that presents moral and conceptual difficulties about what is or should be patentable. In particular, biotechnology (and, to a much lesser extent, software) forced national patent systems to reevaluate the scope of what they consider to be patentable subject matter. For example, in 1976 the Australian Patent Office adopted a fairly liberal approach to the patenting of living subject matter when it held that,

> living organisms were determined to be patentable provided they were not in a naturally occurring state and they had improved or altered useful properties, and not merely changed morphological characteristics which had no effect on the working of the organism [34].

While the United States has adopted a similarly liberal approach to what it considers to be patentable within this area, the position of these two jurisdictions may be contrasted to that of, for example, the European Union (EU). Pursuant to the *European Parliament and Council Directive 98/44/EC of 6 July 1998 on the legal protection of biotechnological inventions*, the European Union places specific limitations on the patenting of, for example, plants and animal varieties (*see*, for example, Article 4(1)) (*see* **Note 4**).

Nanotechnology, and other emerging technologies—such as synthetic biology and gene editing technologies such as clustered regularly interspaced short palindromic repeats, or 'CRISPR' (*see* **Note 5**)—would appear to have the potential to further impact on, and potentially blur, this landscape. Where biotechnology created fundamental challenges for many on the moral nature of what is patentable [35, 36], areas of nanotechnology seem less controversial although the potential for ethical challenges remains. As a result, the real challenges for the patenting of nanotechnology will flow from doctrinal conceptions of novelty and nonobviousness, and the fact that nanotechnology does not fall neatly into any one silo.

## 2.2 Nanotechnology Background

Nanotechnology is a field of technological effort that holds tremendous promise as well as potential peril [37, 38]. Despite the concerns of many groups and governments, the regulatory response to nanotechnology has been, largely, one of research and development. Nanotechnology-specific safety or environmental legislation has been slow to develop [39], with only a handful of jurisdictions having, as of January 2016, pass and/or implemented legislative instrument containing nano-specific provisions. The majority of these measures have been focused on the provision of information to consumers in relation to, for example, cosmetic products and foodstuffs, and the creation of mandatory nano-registries (*see* **Note 6**). Such action has not deterred governments from investing in the fundamental research and development of nanotechnology, and the development of their workforce needed to further the commercialization of the technology. In the United States, for example, two (of the four) goals of the National Nanotechnology Initiative (NNI) are to "[a]dvance a world-class nanotechnology research and development program" and "[fo]ster the transfer of new technologies into products for commercial and public benefit" [40]. To that end, the United States federal government has ponied up more than $US22 billion between 2001 and 2016 to foster basic and applied research into the technology under the NNI [41]; an investment of $1.6 billion in 2016 alone [42]. Other countries, including China and Russia, have been equally quick to promote their nascent nanotechnology efforts with government funding [43–47].

One consequence of this approach is that nanotechnology may be the most multijurisdictional and multinational technology to have emerged so far [48]. The result of this is that nanotechnology patents are likely to have numerous inventors from more than one institution and, just as likely, from more than one country [49, 50].

For example, the atomic force microscope (AFM), one of the most basic research tools necessary to do almost any work in nanotechnology [51], was patented in 1988 (*see* later for a discussion of how patenting of basic research tools may be a problem) and awarded to IBM and, in particular, its Swiss research center (*see* **Note 7**). The initial patented invention was not multinational; however, a recent survey of patents arising out the original AFM patent shows that more than 3000 patents now relate to (either in improvements, modifications, or processes) to the original AFM patent [8].

In writing the 2011 version of this Chapter, the authors conducted a rather cursory survey of just a few dozen of those complementary patents and discovered related filings from Japan, German, the United States, China, Canada, France, and many others. Among these patents, a select few showed researchers from different jurisdictions collaborating on inventions (*see* **Note 8**). The authors repeated this brief survey in January 2016 for the purposes of this updated Chapter. The results were, not surprising, comparable.

In 2011, we leafed through a few of the more than 2000 pending nano-patent applications at that time and found numerous multinational collaborations on inventions including inventors from the United States collaborating with inventors from (1) India, (2) Great Britain, (3) the Netherlands, (4) Poland, (5) Belgium, and (6) Japan. In January 2016, we repeated this process. Using the USPTO's Full-Text and Image Database, we searched the specifications of patents for the word "nano." Out of 57,580 hits, we searched through the first five pages, selecting nanotechnology patents that were filed after 2010. We observed inventors from Korea, the United States, Germany, Japan, Taiwan, China, Canada, France, and Russia collaborating together on patents involving nanotechnology. A random sample of 50 of these patents was then used to test the level of cross-country collaboration involving nanotechnology. The results showed high cross-country collaboration and supported our earlier findings.

Perhaps more important, as collisions between patent holders will inevitably grow so will calls for patent pools (*see* Subheading 2.4 for a broader discussion of these) that will require inventors from numerous jurisdictions to collaborate on and share potential inventions. As inventors and inventions will become increasingly multinational, so the pressure on national patent systems to increase efficiency and harmonize doctrines that discriminate against foreign inventors will also increase.

Massive government funding often means direct funding to universities for early stage research. In this way, nanotechnology is nothing new under the sun—governments have a long history of funding basic research into potential applied sciences. What *is* new, however, is that the outgrowth of this funding has come in an era of hyper-patenting on all points of the research curve. Indeed, this is one area that may separate nanotechnology from all other prior technological revolutions—every aspect of this technology may be patented [6].

All other technological innovations, from biotechnology to software, initially developed during a time when those who received federal funding (universities mainly) were unable to effectively commercialize or patent inventions [6]. Software, arising in the 1960s and 1970s, not only arose in an era where basic research could not often be patented as a result of its federal funding, but also because it arose in a culture of publication over patenting. Biotechnology, although eventually reaching the hyper-patenting stage, was also developed in an era and culture that dampened much of the urge to patent. The only two historical parallels to nanotechnology are the patent wars of the radio and the airplane [6, 52, 53]. In each of those cases, market-based attempts to override patent thickets failed. Government action was required in each case to break the patent logjam and allow research to transform to public goods.

Nanotechnology is not only global in its funding and commercial reach, but in its ability to patent not only the applications of the technology but the basic research tools necessary to conduct the research. The ability of inventors to patent basic research tools is not, doctrinally, novel [54]. What is new is, as already noted, the willingness and desire of universities, the origin of most basic research in industrialized countries, to patent such tools and, more important, the erosion of traditional experimental use exceptions [55]. Indeed, there are very few basic nanotechnology building blocks and research tools that are not, already, patented [6, 56, 57]. There has been much hand-wringing about this fact and real concern that nanotechnology's real potential will be swallowed up in a morass of litigation [58]. This is an issue we explore in more detail in Subheadings 2.3 and 2.4.

Another characteristic, as described by Maynard [59], is its multidisciplinarity "which crosses established boundaries of scientific inquiry and agency jurisdiction." Based on the complexity and multidisciplinarity of most nanotechnology patents, concern was expressed early on by several commentators regarding the capacity of national patent offices to review such patent requests, and to do so in a timely manner [60]. Central to these concerns was the fear that poor review at patent offices [6, 61] would result in over-patenting of inventions that do not meet minimum requirements of patentability in all areas [62]. As Tegart [62] noted, "Inventors need fast and unbureaucratic help to realize an idea with importance for the future."

The creation of a specific nanotechnology class for nanotechnology-based patents by the USPTO in 2004—*Class 977-Nanotechnology*—would appear to have been a move by the patent office to alleviate and/or avoid some of these concerns [63]. Accordingly to the USPTO, the creation of the class, and the 250 cross-reference art collection subclasses within Class 977, will "improve the ability to search and examine nanotechnology-related patent documents" [64]. Patent offices around the world, under the auspices of the International Patent Classification system (or IPC) (*see* **Note 2**), have similarly taken steps to specifically distinguish and 'tag' nanotechnology-related patent submissions. As of 1 January 2011, all nanotechnology-related patents have been assigned a special 'tag'—B82Y—to assist in the review process.

Eight subcategories (i.e., B82Y5, B82Y10, etc.) further assist the patent office with the retrieval and review process (*see* **Note 9**).

Finally, as evidenced by their financial commitments, the potential of nanotechnology is obviously not lost on governments. Much of this potential is in public health and, in particular, in cancer research and other disease areas [37, 65]. Although there is real concern that patenting practices may hinder the commercial viability of nanotechnology applications, a larger concern is that legal wrangling will slow the very real health benefits promised by nanotechnologies. Slowing down economic growth may be a cost more easily born by nations with strong patenting regimes—but as debates over pharmaceuticals have made clear—issues involving health applications of the technology include nano-based cancer treatments such as Abraxane [66] (*see* **Note 10**) and dendrimer-based drug delivery systems such as Starpharma's docetaxel (US Patent 8,837,823; US Patent 8,703,116; US Patent 8,420,067) have the potential to rework patent systems in much more substantial ways.

## 2.3 Mapping the Current Nanotechnology Patent Landscape

Given the general background of nanotechnology patenting discussed earlier, the question that many have been asked is whether the potential disaster many feared is starting, or perhaps, has come to pass. The short answer is that it looks like things are not exactly going to plan at the patent offices.

With Lux Research [57] having stated that "corporations, start-ups, and labs depend on patents to protect their nanotech innovation–and turn them into cash," the importance of securing patents for nanotechnology-based inventions—including both product and processes—is arguably best highlighted by reference to the levels of patent activity within key patent offices such as the USPTO and the European Patent Office (EPO). Efforts to report on this activity, including performance data relating to jurisdictions, institutions, and individuals, have included work by, for example, Marinova and McAleer [67], Huang et al. [68–70], Bawa [71], Koppikar et al. [72], Heinze [73], Lux Research [57, 60], and Chen and Roco [74]. All have shown a deluge of nanotechnology patents and have lamented the emergence of a thicket surrounding the technology as a whole. The patent law landscape has changed radically since these reports were created, though. In 2013, Congress enacted the *America Invents Act*, Public Law 112-29 (*see* **Note 3**), which, among other things, created Inter Partes Review (IPR) proceedings at the Patent Trial and Appeals Board (PTAB). In an IPR, plaintiffs are allowed to challenge the validity of an issued patent on certain grounds without risking an infringement trial and at a much lower cost. From its enactment in 2013 to the end of 2014, PTAB has invalidated close to three quarters of all challenged claims in IPRs [75]. PTAB, then, presents a novel and easily accessible way to invalidate patents in the heart of the thicket.

In one of the first published studies examining longitudinal patent activity for nanoscale science and engineering activities within the USPTO, Huang et al. [69] reported rapid growth in patenting activity over the period 1976–2002. By using a "full text" key-word based approach (see **Note 11**), and subsequent filtering process, the authors found that the USPTO processed approximately 8600 "nano-based" patents over this period; patenting activity was found to be steep after 1997 and 2001. These periods of growth in patent activity occurred, it was observed by the authors, around periods of program growth and other institutional activities within, for example, the United States. It is perhaps unsurprising then that the authors [68] found that "the [nanoscale science and engineering] patents grew significantly faster than the USPTO database as a whole, especially beginning with 1997."

Other key findings reported by Huang et al. [68] included the diversity of countries and institutions involved in the patenting activity (albeit still dominated by the United States) and the strength of patenting activity within particular technological fields including chemicals, catalysts, and pharmaceuticals. The observed growth in patenting activity would appear to highlight the importance of patent law, and the protections therefore afforded to patentees under the legal framework, this is despite the costs associated with securing patent protection for an invention (see **Note 12**).

In a 2005 study of patenting activity within the USPTO, Lux Research [60] similarly reported a "ramp-up" in the number of patents being issued by the national patent office, with steep growth continuing in the post-2003 period. According to their analysis,

> the number of nanotech patents issued ha[d] risen steadily from a base of 125 in 1985 to 4,995 today ...Nanotech patents far outpace other areas of innovation, with a compound growth rate of 20 % versus just 2 % for patents overall [60].

Their analysis supported the findings of Huang et al. [68], with Lux Research noting that patentees were more likely to be from the United States than any other jurisdiction but with a growing percentage from other jurisdictions. In addition, patents were likely to be assigned to a patentee in the private sector than any other sector (for example, university, government, and/or research organization). Along with significant growth in patent activity, the authors found that the average number of claims within each patent had also increased. In their words:

> inventors are authoring more sophisticated patents that cover more nuanced variations of the same theme in a single filing. The average nanotech patent issued in 2005 has 23.5 claims, compared to only 15.8 in 1985 [60].

As will be discussed later, this observed trend has significant implications for not only patent examiners, who must be able to deal with the complexities associated with the applications, but also patent growth more generally.

In addition to overall patenting activity within the USPTO, Lux Research [60] looked at patenting activity (applications and grants) for eight specific nanomaterials, each of which has the ability to be utilized across five different applications areas. They included carbon nanotubes, metal nanoparticles, ceramic nanoparticles, dendrimers, quantum dots, fullerenes, and nanowires. The purpose of the report was to provide an in-depth analysis of patent density for each of these platform materials, determine the breadth of the patent claims, and identify areas of potential entanglement; vulnerability to potential challenges (conflict) and market potential of patents were also considered [60].

Based on this examination, the authors found that significant growth occurred in relation to patenting activity for all eight materials; ceramic nanoparticles and carbon nanotubes, which have broad applications across numerous fields, were found to have experienced particularly steep growth over the time period examined, resulting in high patent density [60]. This, they suggested, had the potential to create an unfavorable patent environment for inventors/patentees in relation to, for example, carbon nanotubes within the electronics field and ceramic nanoparticles within personal health care and cosmetics applications. However, the authors [60] went on to suggest that there may still be hidden opportunities in relation to these two materials, and that given their potential breadth as structural materials, "it [was] likely that these nanomaterials will emerge as battles worth fighting by 2008." Varying trends in patent filing and density, as well as future potential based on market opportunities, were observed for the other six materials [60].

The continued increase in patenting activity in key national patent offices suggests that industry, research organizations, universities, and other key bodies remain positive about the market opportunities and associated economic benefits for nanotechnology-based inventions. This is despite the costs associated with technological innovation and the increasingly vocal debates occurring within jurisdictions over potential, yet unquantified, risks associated with the manufacturing of certain families of nanomaterials (such as carbon nanotubes) [76, 77] and/or the use of specific types of nanomaterials in consumer products (including fullerenes and insoluble metal oxide nanoparticles) [78–80]. Yet, as the ETC Group [56] have sought to remind us, the successful granting of a patent, albeit for a nanotechnology-based invention or any other type, is not enough in itself to ensure the commercial success of that invention. Success or failure, as witnessed in the European Union in relation to, for example, genetically modified foods is dependent on a far broader range of criteria, including consumer acceptance of the invention and/or technology [12, 81, 82].

While much of the literature relating to intellectual property rights and nanotechnology has canvassed the patent landscape and paints a detailed picture thereof, there is an increasing body of work that has focused on the potential challenges and barriers that the technology and its inventors may face in the coming years. Concerns have been expressed, for example, in relation to the breadth of patent claims for platform or structural materials, patent thickets, overlapping patent claims, and the institutional capacity of patent offices to assess nanotechnology-based applications in a timely manner. Many of these issues are not in themselves unique to nanotechnology; rather, as highlighted later, they are common to other emerging technologies, including synthetic biology, and reflect many of the experiences of prior technologies. What is important to note is that these challenges can drastically impact time to market of a product, which in turn may delay, for example, in the consumers accessing life-saving nano-based applications (such as a nanodrug or delivery system).

In addition to these economic inefficiencies, the studies we have discussed earlier, as well as countless others, have shown that feared patent thickets, brambles, quagmires, and other natural disaster themed descriptions have apparently taken over the landscape [83–86]. In basic research tools, fundamental materials, building-block structures, and numerous other crucially important aspects of nanotechnology, vast numbers of overlapping and broadly written patents, held by varied institutions and competitors, have already issued. We have already seen, in the United States, a series of patent infringement lawsuits filed among competitors. Although no study has yet compared the level of infringement suit activity compared to prior technologies, there are many reasons to believe that nanotechnology's future may be threatened, or at least made less bright, by these looming controversies.

Recent suits surrounding nanotechnology patents are microcosms of issues facing the entire industry. The court in *Cephalon, Inc. v. Abrazis Biosciences, LLC, 618 F. Appx. 663 (Fed. Cir. 2015)* dealt with one of these cornerstone questions when it grappled with creating a definition for the word "nanoparticle." The patent at issue, which dealt primarily with formulations of taxane drugs that allowed for bolus injection, claimed both "nanoparticles and microparticles of a taxane" with "a mean diameter between about 0.01 and 5 μm." The court, then, was faced with the question of: What is the actual difference between a nanoparticle and microparticle when the range given includes both without differentiation? This would be like asking our hypothetical gardener to identify the difference between a shrub and a bush when they are told that both range from having one branch to having 1000 branches. Both questions, nano vs. micro and shrub vs. bush, are subject to the subjective interpretation of the person answering the question, but thankfully the federal courts have developed rules for this type of situation. Ambiguous technical terms in the claims of a patent are given their "widely accepted meaning" within the scientific community (*see* **Note 13**). Here, the court, after listening to expert testimony from both sides, defined "'microparticles' as particles that have a diameter between 1 and 1000 micrometers and greater than that of nanoparticles and 'nanoparticles' as particles that have a diameter of between about 1 and 1000 nanometers and less than that of microparticles." In our garden, then, the difference between a shrub and bush would be that a shrub is smaller than a bush and has 500 or fewer branches while a bush is larger than a shrub and has 500 or more branches. What happens if the plant has exactly 500 branches? The court in *Cephalon* left that question unanswered. One can only imagine how courts will be able to sort out cases involving more complex nanomaterials in the future.

In *Collins v. Nissan N. Am., Inc., 2013 U.S. Dist. LEXIS 15749, (E.D. Tex. 2013)*, we see, yet again, the federal courts grappling with what actually constitutes nanotechnology. The patent here claimed "a nanophase diamond film comprising nodules of carbon" that can be used in various applications from razor blades to hard-disk drives to add hardness, wear resistance, and a host of other nanoscale advantages. Here the court had to ask: What is a "nodule of carbon?" The court imagined a nanoscale cobblestone street made of carbon when it defined "nodules" as "discrete clusters of carbon atoms of rounded or irregular shape." Back to that garden. In *Collins*, we see the court presented with three shrubs and saying that this definitely isn't a bush.

*In re Kumar, 418 F.3d 1361 (Fed. Cir. 2005)*, on the other hand, is a case that deals with the PTO's difficulties with the ephemeral nature of the "nanotechnologist." The patent application in *Kumar* was for "aluminum oxide particles having a specified size, range, and distribution" used to polish electronics. One of the problems that courts face is that, when interpreting patent claims, they are viewed from the perspective of "a person having ordinary skill in the art" (PHOSITA). The problem arises in nanotechnology because there is no "normal nanotechnologist" that can be placed on the witness stand to say what is obvious to a PHOSITA and the inventor gets to testify to this regard instead. This means the inventor is defining what is "nanoparticle" or a "nodule," and this can lead to overly broad patents due to the inventor's self-interest. Our gardener is now attempting to say that four bushes in a row are called a hedge and no one else gets any input in the decision.

### 2.4 Moral and Ethical Implications of Nanotechnology Patenting

It is arguably not surprising when considered against the backdrop of the patenting of human genes debate and associated concerns over the breadth of patents being granted on human genes, that this issue has also become a topic of debate in regards to the patenting of nanotechnology. This has been the case with platform or structural materials, such as the eight considered by Lux Research [57] in their report. The concern here, as articulated by the ETC Group, is primarily in relation to the issues of concentrated ownership and therefore control over patents, and the subsequent implication of this in terms of economics, innovation, and access—especially for developing economics [56]. Of course such concerns are not new, nor unique to nanotechnology. However, it would appear that nanotechnology does create additional challenges here, with the ETC Group [56] have suggested that, for example,

> breathtakingly broad nanotech patents are being granted that span multiple industrial sectors and include sweeping claims on entire classes of the Periodic Table.

They go on to suggest that [56],

> it's not just the opportunity to patent the most basic enabling tools, but the ability to patent the nanomaterials themselves, the products they are used in and the methods of making them.

The ETC Group's concern is that the dense concentration of patents, which are held by a small number of patent holders, combined with the ability to patent basic nanoscale materials, "could mean monopolizing the basic elements that make life possible" [56]. Given their concern over patent concentration and breadth of claims being made, the nongovernmental organization went on to highlight the emergence of so-called patent thickets within the nanotechnology patent landscape. This refers to, as explained by Shapiro [3], "an overlapping set of patent rights requiring that those seeking to commercialize new technology obtain licenses from multiple patentees." Patent thickets therefore have the ability to hinder technological innovation and therefore the commercialization of new technologies.

In their examination of patenting activity for four nanomaterials and one tool within the USPTO and the EPO (carbon nanotubes, inorganic nanostructures, quantum dots, dendrimers, and Scanning Probe Microscopes), the ETC Group were able to paint a picture of the emerging patent thicket for some nanomaterials; this was illustrated primarily by reference to the number of patents relating to each of the applications currently held by different institutions, and the so-called patent density for each applications. By way of example, the ETC Group reported that the USPTO had issued 227 patents for carbon nanotubes between 1999 and 2004. Between January 2004 and March 2016, there have been 2569 patents filed drawn to carbon nanotubes. While they found that the patents for the material were held by a number of different parties, across a range of different sectors, they nevertheless came to the conclusion that a patent thicket for the material had already occurred and that,

> a swarm of existing patents, whose claims are often broad, overlapping and conflicting, means that researchers hoping to develop new technology based on carbon nanotubes must first negotiate licenses from multiple patent owners [56].

The ETC Group were not the only commentators to have voiced their concern about the potential implications of overlapping patent claims and the emergence of "patent thickets" or "nano-thickets," with a number of commentators having expressed concern over the potential creation, and the implications thereof, for nanotechnology [52, 57, 60, 83–87].

Having observed the problems associated with patent thickets in other areas of technological innovation, including biotechnology and information and communication technologies, Clarkson and DeKorte [83] noted that patent thickets have the potential to give rise to a range of issues, including the unintentional infringement of patents and the subsequently liability created as a consequence of the said infringement, the problem of anticommons, the creation of barriers to entry, and the need for licensing. While the issues are not unique to any one area, they [83] noted that, "the nanotechnology patent space experiences an even greater level of these problems because it is much more complicated than other technology areas."

In order to determine the extent of the growing patent thickets for nanotechnology, Clarkson and DeKorte [83] undertook a mapping exercise of patent space and density within the USPTO. The authors then used network analytic techniques as a way to "visualize" the growth at three time points—2000, 2002, and 2004. By plotting individual patents and then references between patents, the authors demonstrated not only the growth in nanotechnology patents within the USPTO during that time period, but also the increasing interconnectivity—or network—between the patents. This visualization process enabled the authors to map potential patent thickets.

In February 2016, we sought to test these concerns regarding patent thickets in relation to three types of nanomaterials (buckeyballs, carbon nanotubes, and fullerenes). Using Google Patents, we searched for United States patents assigned to three large chemical companies, Dow, BASF, and DuPont, that contained the words "carbon nanotube," "fullerene," and "dendrimer." Our results, unsurprisingly, showed a deluge of patents on all three nanostructures (*see* **Note 14**). This suggests that the fears surrounding the existence of a nanotechnology-based thicket are not unfounded. There is hope, though. Patents, unlike diamonds, are not forever and expire after their 20-year term is over. Many of the patents covering basic nanostructures, such as US Patents 5,424,054 and 5,747,161, which cover early carbon nanotube technology, are set to expire soon or have expired already. This may lead to an explosion of commercial uses for the original embodiments of these technologies, but also may lead to an even bramblier thicket depending on the strategies used by the patent owners' lawyers.

If patent thickets cannot be avoided, what strategies may therefore be employed in order to protect patents while also promoting innovation? Traditionally cross-licensing arrangements—which

Shapiro [3] has eloquently defined as 'an agreement between two companies that grants each the right to practice the other's patents—are one way in which this may be achieved. However, while such arrangements are relatively straightforward when involving only two parties, Clarkson and DeKorte [83] note a number of limitations, including high transaction costs, which can make such arrangements prohibitive when more than two parties are involved in the contractual negotiations. In light of these limitations, Clarkson and DeKorte [83] proposed a second alternative for avoiding validity challenges and potential patent litigation: "patent pools." These contractual undertakings are, as summarized by Clark et al. [88] "an agreement between two or more patent owners to license one or more of their patents to one another or third parties." Such arrangements have been an important tool for providing parties with access to proprietary information for over a century. It has been suggested that the ability for parties to readily access patent information through such pool arrangements promote innovation within areas which may have otherwise become the subject of patenting blocking and legal challenges and at a lower transactional cost than cross-licensing [3, 83].

But establishing and relying on patent pools as a mechanism to access patent information would appear to be only one potential approach to addressing the challenges presented by the thickets. Another, arguably somewhat more radical, approach would be to promote an "open source" approach. The open source "movement" has been widely adopted in relation to software development [89] and to varying degrees within the field of medicine and drug development [90, 91]. The movement's potential application to the nanotechnology patents landscape has therefore raised some level of discussion among commentators. One of the earliest contributions was from Bruns [92], who looked at the applicability of the open source movement to molecular nanotechnology. In his view, "open source approaches might offer advantages for faster, more reliable and more accessible research and development" [92]. He advocated the adoption of an open source approach to the technology where public money had been used to generate the intellectual property in question. Bruns' [92] argument was that such an approach would not only encourage innovation but also assist in diffusing the technology, and its associated benefits, to developing economies.

While the open source movement has continued to gain traction within, for example, the field of software development, "there is [still] not yet an "open source nanotechnology" movement" [93]. This may in part be explained by the fact that software development is process based, where developers of nanotechnology are at this time largely focused on product generation. Moreover, with open source software, the primary 'cost' is the programmer's time, which they give freely in order to further develop and refine the code lines. The same cannot be said with the development of nanotechnology, which requires not only human resources but also infrastructure and consumables. Commentators such as Prisco [94], Peterson [95], and Pearce [96] have, however, begun to further explore open source for nanotechnology. As such, we would suggest that the nanotechnology patent landscape is likely to evolve over the short to medium term, with the open source movement just one way in which individuals and organizations attempt to circumnavigate the emerging patent thickets and promote technology innovation.

As any individual with a green thumb will know, tending to a garden—albeit a refined English garden or a small herb garden—requires constant care and attention. Any such garden is dynamic by its very nature and will evolve over time. A constant state of vigilance is needed to ward off pests and other challenges, and the more proactive, educated, and vigilant the gardener is, the better the outcome will be.

As this Chapter has sought to highlight, there are many similarities between the needs and challenges of a garden and that of a patent landscape. As with our garden, the patent landscape has evolved and been refined over centuries in response to new species and the introduction of new technologies. Sometimes the landscape has been better prepared to handle the attacks than others. Nanotechnology is one of the more recent species to strain the fundamental features of the landscape, pushing up against historical walls and threatening traditionally well-tended fields. This is due to a number of factors: its multidisciplinary character, its transjurisdictional nature, the ability for inventors to apply for and be granted patents not only to products but also the basic building blocks, and claims which relate to a diverse number of areas and/or applications. It is also in part due to the immense public and private sector interests in nanotechnology, and the rush to secure legal rights over their inventions.

But the question is: will nanotechnology be permitted to devastate that which has taken centuries to build up? Or will the gardeners—primarily national governments in this instance—see the emergence and growth of nanotechnology as an opportunity to reconsider the borders and features of the current landscape and revamp it accordingly? This would of course involve significant time and energy, but with other equally complex and multifaceted technologies already in the research and development pipeline, including, for example, additive manufacturing and CRISPR technologies, it would appear that policy makers need to "stop and smell the roses" in order to ensure the economic and social benefits of the technology are released. Perhaps it is time to modernize the landscape to meet the needs of the current climate—a more global approach to patenting is one obvious option—and provide the gardeners with the tools that they need to do their job in a timely and efficient manner.

But those who use and enjoy the garden must also take some responsibility for its future in order to ensure that the benefits of the technology are realized. Rather than, for example, relying on costly and time-consuming litigation, beneficiaries of the patent system should be encouraged to explore arrangements such as cross-licensing and patent pools early on, or where appropriate, be encouraged to look to open source approaches. Governments can also play a role here by, for example, creating a framework that encourages and/or rewards these approaches.

## 3   Notes

1. In this Chapter, we do not provide background or definitional sections on what we consider to be nanotechnology.

2. The Patent Cooperation Treaty (PTC) was concluded on 19 June 1970 and was most recently modified on 3 October 2001. The PTC, as explained by World Intellectual Property Organization, "makes it possible to seek patent protection for an invention simultaneously in each of a large number of countries by filing an "international" patent application. Such an application may be filed by anyone who is a national or resident of a PCT Contracting State" [97]. As of January 2016, 158 State parties were members of the International Patent Cooperation Union (the Assembly established by Article 1 of the PTC) [97]. The IPC, which was established under the auspices of the *Strasbourg Agreement Concerning the International Patent Classification of March 24, 1971* (as amended on September 28, 1979), is a hierarchical classification system for patents. Sections, which are categorized on the basis of technology, sit at the top of the hierarchy, of which there are eight. There are then ~70,000 subdivisions that fall under these eight sections.

3. On September 16, 2011, the President of the United States signed into law the Leahy-Smith America Invents Act (or AIA) (Public Law 112-29). The passage of the AIA, as noted by Rantanen and Petherbridge [98], "represent[ed] the most significant legislative event affecting patent law and practice in more than half a century" within the United States. A cornerstone of the AIA was to shift the US patent system from a "first-to-invent" to a "first-inventor-to-file" system. This provision of the AIA came into effect on March 16, 2013, at which time it brought the US patent system into alignment with the patent systems of the vast majority of other countries [99, 100].

4. *Directive 98/44/EC* also prohibits the patenting of an invention where, pursuant to Article 6(1), "their commercial exploitation would be contrary to *ordre public* or morality;...." Expressed captured by this prohibition are the 2(a) processes

for cloning human beings; (b) the process for modifying the germ line genetic of human beings; and (c) use of human embryos for industrial or commercial purposes;. . .."

5. The first patent for a CRISPR-related invention was filed on October 15, 2013 with the USPTO and assigned to the Broad Institute, Inc. and Massachusetts Institute of Technology on April 15, 2014. As stated in the abstract of US patent no. 8697359 (CRISPR-Cas systems and methods for altering expression of gene products), the invention "provides for systems, methods, and compositions for altering expression of target gene sequences and related gene products."

6. Regulation (EC) No 1223/2009 of the European Parliament and of the Council of 30 November 2009 on cosmetic products [101] was the first piece of national and/or supranational legislative instrument to specifically regulate nanomaterials. For the purposes of the Regulation, a "nanomaterial" is defined as "an insoluble or biopersistant and intentionally manufactured material with one or more external dimensions, or an internal structure, on the scale from 1 to 100 nm;. . ." Among other things, Regulation (EC) No 1223/2009 requires cosmetic containing nanomaterials indicate the presence of the nanomaterials by listing the nanoscale ingredient in the list of ingredients on its label immediately followed by: (nano) (*see* Article 19). Pursuant to Article 16, the responsible party for placing a new cosmetic products containing nanomaterials into the European market is required to notify the Commissions within six months of that action occurring (*see* Article 16(3)). Nano-specific provisions may also be found in, for example, Regulation (EU) No 1169/2011 of the European Parliament and of the Council of 25 October 2011 on the provision of food information to consumers and New Zealand's Cosmetic Group Standard. Countries such as France and Belgium have also established mandatory registries for nanomaterials within their borders.

7. See United States Patent Number: 4 724 318.

8. See, for example, United States Patent Number: 9 252 208, 9 263 551, 9 246 015. Other citations available upon request.

9. The B82Y tag was introduced in 2011, superseding the Y01N tag that had been used previously for nanotechnology-related patents prior to this. The B82Y tag builds on, and extends, the categorization of nanotechnology patents under the Y01N system [102].

10. Abraxane/Taxol (generic name: paclitaxel) was the first nanotechnology-based drug to be approved by the United States Food and Drug Administration in January 2005. As explained by Miele et al. [66], the drug consists of a "novel, albumin-bound, 130-nm particle formulation of paclitaxel, free

from any kind of solvent." The formulation of the drug results in increased bioavailability of the chemotherapeutic agents. Initially approved to treat metastatic pancreatic cancer, the drug is today used to treat a wide range of cancers including breast cancer and lung cancers.

11. As noted by Huang et al., there were "seven basic keywords with several variations" [68].

12. For a more recent longitudinal study of nanotechnology-patenting activity within the USPTO, the European Patent Office (EPO) and the Japan Patent Office (JPO). See also Chen and Roco [70].

13. *Phillips v. AWH Corp.*, 415 F.3d 1303, 1314 (Fed. Cir. 2005).

14. Our Google Patents search yielded the following results:

    (a) US patents held by Dow Chemical: 151 dendrimer; 64 quantum dots; 3 CNTs,

    (b) US patents held by BASF: 373 dendrimer; 31 quantum dots; 1781 CNTs, and

    (c) US patents held by DuPont: 10 dendrimer; 6 quantum dots; 167 CNTs.

## References

1. Graham S, Mowery DC (2003) Intellectual property protection in the U.S. software industry. In: Cohen W, Merrill D (eds) Patents in the knowledge-based economy. Board on Science, Technology and Economic Policy (STEP). The National Academies, Washington, DC

2. Eisenberg RS (2002) How can you patent genes? Am J Bioeth 2(3):3–11

3. Shapiro C (2000) Navigating the patent thicket: cross licenses, patent pools, and standard setting. Innov Policy Econ 1:119–150

4. Burk DL, Lemley MA (2003) Policy levers in patent law. Va Law Rev 89:1575–1696

5. Heller MA, Eisenberg RS (1998) Can patents deter innovation? The anticommons in biomedical research. Science 280 (5364):698–701

6. Lemley M (2005) Patenting nanotechnology. Stanf Law Rev 58(2):601–630

7. Bessen J (n.d.) Patent thickets: strategic patenting of complex technologies. http://www.researchoninnovation.org/thicket.pdf. Accessed 26 Dec 2009

8. D'Silva J (2009) Pools, thickets and open source nanotechnology. http://ssrn.com/abstract=1368389. Accessed 13 Dec 2009

9. Rantanen J (2006) Slaying the troll: litigation as an effective strategy against patent threats.

Santa Clara Comput High Technol Law J 23 (1):159–210

10. Magliocca GN (2007) Blackberries and barnyards: patent trolls and the perils of innovation. Notre Dame Law Rev 82(5):1809–1838

11. Abbott KW, Sylvester DJ, Marchant GE (2010) Transnational regulation of nanotechnology: reality or romanticism? In: Hodge GA, Bowman DM, Maynard AD (eds) International handbook on regulating nanotechnologies. Edward Elgar, Cheltenham, pp 525–543

12. Marchant GE, Sylvester DJ (2006) Transnational models for regulation of nanotechnology. J Law Med Ethics 34(4):714–725

13. Bowman DM, van Calster G (2007) Does REACH go too far? Nat Nanotechnol 1:525–526

14. Marchant GE, Sylvester DJ, Abbott KA, Gaudet LM (2012) International harmonization of nanotechnology oversight. In: Dana DA (ed) The nanotechnology challenge: creating law and legal institutions for uncertain risks. Cambridge University Press, Cambridge, pp 179–201

15. Maynard AD, Bowman DM, Hodge GA (2011) The wicked problem of regulating sophisticated materials. Nat Mater 10:554–557

16. Bowman DM, Gatof J (2015) Reviewing the regulatory barriers for nanomedicine: global questions and challenges. Nanomedicine 10 (21):3275–3286

17. Foss Hansen S, Maynard AD, Baun A, Tickner JA, Bowman DM (2014) What are the warning signs that we should be looking for? In: Hull M, Bowman DM (eds) Nanotechnology risk management: perspectives and progress, 2nd edn. Elsevier, London

18. Bowman DM, Ludlow K (2013) Assessing the impact of a 'for government' review on the nanotechnology regulatory landscape. Monash Law J 38(3):168–212

19. Rai AK, Eisenberg RS (2003) Bayh-Dole reform and the progress of biomedicine. Law Contemp Probl 66(1–2):289–314

20. Caulfield T, Cook-Deegan RM, Kieff FS, Walsh JP (2006) Evidence and anecdotes: an analysis of human gene patenting controversies. Nat Biotechnol 24(9):1091–1094

21. Klein RD (2007) Gene patents and genetic testing in the United States. Nat Biotechnol 25(9):989–990

22. Andrews LB (2002) Genes and patent policy: rethinking intellectual property rights. Nat Rev Genet 3(10):803–808

23. Dinwoodie GB, Hennessey WO, Perlmutter S (2001) International intellectual property law and policy. LexisNexis, Newark

24. Eisenberg RS (1989) Patents and the progress of science: exclusive rights and experimental use. Univ Chic Law Rev 56(3):1017–1086

25. Rai AK (1999) Regulating scientific research: intellectual property rights and the norms of science. Northwest Univ Law Rev 94 (1):77–152

26. Masur JS (2008) Process as purpose: costly screens, value asymmetries, and examination at the patent office. http://ssrn.com/abstract=1105184. Accessed 26 Dec 2009

27. United States Patent and Trademark Office (2005) General information concerning patents. http://www.uspto.gov/patents/basics/index.html#patent. Accessed 15 Dec 2009

28. Webber PM (2003) Protecting your inventions: the patent system. Nat Rev Drug Discov 2(10):823–830

29. Maskus KE (2000) Intellectual property rights in the global economy. Institute for International Economics, Washington, DC

30. Grossman GM, Lai EC (2004) International protection of intellectual property. Am Econ Rev 94(5):1635–1653

31. Mandel G (2010) Regulating nanotechnology through Intellectual Property Rights. In: Hodge GA, Bowman DM, Maynard AD (eds) International handbook on regulating nanotechnologies. Edward Elgar, Cheltenham, pp 388–407

32. Caulfield T, Gold ER, Cho MK (2000) Patenting human genetic material: refocusing the debate. Nat Rev Genet 1(3):27–231

33. Abbott FM (2006) Intellectual property provisions of bilateral and regional trade agreements in light of U.S. federal law. http://www.unctad.org/en/docs/iteipc20064_en.pdf. Accessed 13 Dec 2009

34. Australian Government (2008) Patentable subject matter—issues paper. Advisory Council on Intellectual Property, Canberra

35. Drahos P (1999) Biotechnology patents, markets and morality. Euro Intell Prop Rev 21 (9):441–449

36. Bagley MA (2003) Patent first, ask questions later: morality and biotechnology in patent law. William Mary Law Rev 45:469

37. Royal Society and Royal Academy of Engineering (2004) Nanoscience and nanotechnologies: opportunities and uncertainties. RS-RAE, London

38. Maynard AD (2007) Nanotechnology: the next big thing, or much ado about nothing? Annu Occup Hyg 51(1):1–12

39. Lux Research (2009) Nanotech's evolving environmental, health, and safety landscape: the regulations are coming. Lux Research, New York

40. National Nanotechnology Initiative (n.d.) About the NNI-Home. http://www.nano.gov/html/about/home_about.html. Accessed 15 Dec 2009

41. National Nanotechnology Initiative (2016) Funding. http://www.nano.gov/about-nni/what/funding#content#content#content. Accessed 16 Jan 2016

42. The White House (2016) Budget of the United States government, fiscal year 2016. U.S. Government Printing Office, Washington, DC

43. Roco MC (2005) International perspectives on government nanotechnology funding in 2005. J Nanopar Res 7:707–712

44. European Commission (2005) Nanosciences and nanotechnologies: an action plan for Europe 2005–2009. European Parliament, Brussels

45. Gao Y, Jin B, Shen W, Sinko PJ, Xie X, Zhang H, Jia L (2016) China and the United States—global partners, competitors and

collaborators in nanotechnology development. Nanomedicine 12(1):13–19

46. Gokhberg L, Fursov K, Karasev O (2012) Nanotechnology development and regulatory framework: the case of Russia. Technovation 32(3):161–162

47. Liu L, Van de Voorde, M., Werner, M., & Fecht, H. J. (Eds.) (2015) Overview on nanotechnology R&D and commercialization in the Asia Pacific region. In: The nano-micro interface: bridging the micro and nano worlds. John Wiley & Sons, New York, pp 37–54

48. Hullmann A, Meyer M (2003) Publications and patents in nanotechnology: an overview of previous studies and the state of the art. Scientometrics 58(3):507–527

49. Zucker LG, Darby MR (2005) Socio-economic impact of nanoscale science: initial results and nanobank. (working paper 11181). http://www.nber.org/papers/w11181. Accessed 13 Dec 2009

50. Zucker LG, Darby M, Furner J, Lieu R, Ma H (2007) Minerva unbound: knowledge stocks, knowledge flows, and new knowledge production. Res Policy 36:850–863

51. Binnig G, Quate CF, Gerber C (1986) Atomic force microscope. Phy Rev Lett 56 (9):930–934

52. Sabety T (2004) Nanotechnology innovation and the patent thicket: which IP policies promote growth? Albany Law J Sci Technol 15:477–516

53. Johnson HA (2004) Wright patent wars and early American aviation. J Air Law Commer 69(1):21–64

54. Mueller JM (2001) No dilettante affair: rethinking the experimental use exception to patent infringement for biomedical research tools. Wash Law Rev 76(1):1–66

55. Sylvester DJ, Menkhus E, Granville KJ (2005) Innovation law handbook. Available at SSRN. http://ssrn.com/abstract=999451. Accessed 26 Dec 2009

56. ETC Group (2005) Nanotech's "second nature" patents: implications for the Global South. ETC Group, Ottawa

57. Lux Research (2006) Nanotech battles worth fighting. Lux Research, New York

58. Harris DL, Hermann K, Bawa R et al (2004) Strategies for resolving patent disputes over nanoparticle drug delivery systems. Nanotechnol Law Bus 1:372–390

59. Maynard AD (2006) Nanotechnology: a research strategy for addressing risk. Project on Emerging Nanotechnologies, Washington, DC

60. Lux Research (2005) The nanotech intellectual property landscape. Lux Research, New York

61. Thomas JR (2001) Collusion and collective action in the patent system: a proposal for patent bounties. Univ Ill Law Rev 1:305–353

62. Tegart G (2004) Nanotechnology: the technology for the twenty-first century. Foresight 6(6):364–370

63. Guston G (ed) (2010) Encyclopedia of nanoscience and society. Sage, Thousand Oaks

64. United States Patent and Trademark Office (2012) Class 977 nanotechnology cross-reference art collection. http://www.uspto.gov/patents/resources/classification/class_977_nanotechnology_cross-ref_art_collec tion.jsp. Accessed 11 Jan 2016

65. Ferrari M (2005) Cancer nanotechnology: opportunities and challenges. Nat Rev Cancer 5:161–171

66. Miele E, Spinelli GP, Miele E, Tomao F, Tomao S (2009) Albumin-bound formulation of paclitaxel (Abraxane® ABI-007) in the treatment of breast cancer. Int J Nanomed 4:99–105

67. Marinova D, McAleer M (2003) Nanotechnology strength indicators: international rankings based on US patents. Nanotechnology 14:R1–R7

68. Huang Z, Hu R, Pray C (2003) Longitudinal patent analysis for nanoscale science and engineering: country, institution and technology field. J Nanopart Res 5:333–363

69. Huang Z, Chen H, Chen ZK, Roco MC (2004) International nanotechnology development in 2003: country, institution, and technology field analysis based on USPTO patent database. J Nanopart Res 6:325–354

70. Huang Z, Chen H, Li X, Roco MC (2006) Connecting NSF funding to patent innovation in nanotechnology (2001–2004). J Nanopart Res 8:859–879

71. Bawa R (2004) Nanotechnology patenting in the US. Nanotechnol Law Bus 1(1):31–51

72. Koppikar V, Maebius SB, Rutt JS (2004) Current trends in nanotech patents: a view from inside the patent office. Nanotechnol Law Bus 1:24–30

73. Heinze T (2004) Nanoscience and nanotechnology in Europe: analysis of publications and patent applications including comparisons with the United States. Nanotechnol Law Bus 1(4):1–19

74. Chen H, Roco MC (2008) Mapping nanotechnology innovations and knowledge. Springer, New York

75. Fitzpatrick, Cella, Harper & Scinto (2014) 2014 findings on USPTO contested proceedings. http://www.postgranthq.com/wp-content/uploads/2014/10/PostgrantHQ_Reporter.pdf. Accessed 6 Feb 2016

76. Mullins S (2009) Are we willing to heed the lessons of the past? Nanomaterials and Australia's asbestos legacy. In: Hull M, Bowman DM (eds) Nanotechnology environmental health and safety: risks, regulation and management. Elseiver, New York, pp 49–69

77. Poland CA, Duffin R, Kinloch I, Maynard AD, Wallace W et al (2008) Carbon nanotubes introduced into the abdominal cavity of mice show asbestos-like pathogenicity in a pilot study. Nat Nanotechnol 3(7):423–428

78. Scientific Committee on Consumer Products (2007) Opinion on safety of nanomaterials in cosmetic products, SCCP/1147/07. European Commission, Brussels

79. Mu L, Sprando RL (2010) Application of nanotechnology in cosmetics. Pharm Res 27(8):1746–1749

80. Gulson B, McCall MJ, Bowman DM, Pinheiro T (2015) A review of critical factors for assessing the dermal absorption of metal oxide nanoparticles from sunscreens applied to humans, and a research strategy to address current deficiencies. Arch Toxicol 89(11):1909–1930

81. Bauer MW, Gaskell G (eds) (2002) Biotechnology: the making of a global controversy. Cambridge University Press, London

82. Jasanoff S (2005) Designs on nature: science and democracy in Europe and the United States. Princeton University Press, Princeton

83. Clarkson G, DeKorte D (2006) The problem of patent thickets in convergent technologies. Ann N Y Acad Sci 1093:180–200

84. Lee A (2006) Examining the viability of patent pools for the growing nanotechnology patent thicket. Nanotechnol Law Bus 3:317–328

85. Harris DL (2008) Carbon nanotube patent thickets. In: Allhoff F, Lin P (eds) Nanotechnology & society: current and emerging ethical issues. Springer, New York, pp 163–186

86. Miller J, Serrato R, Represas-Cardenas JM, Kundahl G (2005) The handbook of nanotechnology: business, policy, and intellectual property law. John Wiley & Sons, New York

87. Bastani B, Fernandez D (2004) Intellectual property rights in nanotechnology. Fernandez & Associates, Menlo Park

88. Clark J, Piccolo J, Stanton B, Tyson K (2000) Patent pools: a solution to the problem of access in biotechnology patents? USPTO, Washington, DC

89. Lakhani KR, von Hippel E (2003) How open source software works: "free" user-to-user assistance. Res Policy 32(6):923–943

90. Economist (2004) An open-source shot in the arm? Economist 10 June, 17

91. Munos B (2006) Can open-source R&D reinvigorate drug research? Nat Rev Drug Discov 5(9):723–729

92. Bruns B (2001) Open sourcing nanotechnology research and development: issues and opportunities. Nanotechnology 12:198–210

93. Kelty C, Lounsbury M, Yavuz CT, Colvin VL (n.d.) Towards open source nanotechnology: arsenic removal and alternative models of technology transfer. http://opensourcenano.net/images/GRC-Poster2.pdf. Accessed 15 Apr 2009

94. Prisco G (2006) Globalization and open source nano economy. kurzweilai.net. http://www.kurzweilai.net/meme/frame.html?main=/articles/art0659.html. Accessed 12 Apr 2009

95. Peterson CL (2008) Citizen-controlled sensing: using open source and nanotechnology to reduce surveillance and head off Iraq-style wars. http://www.opensourcesensing.org/proposal.pdf. Accessed 12 Apr 2009

96. Pearce JM (2013) Open-source nanotechnology: solutions to a modern intellectual property tragedy. Nano Today 8(4):339–341

97. World Intellectual Property Organization (2016) Patent Cooperation Treaty (PTC). http://www.wipo.int/treaties/en/registration/pct/. Accessed 7 Jan 2016

98. Rantanen J, Petherbridge L (2011) Toward a system of invention registration: the Leahy-Smith America Invents Act. Mich Law Rev 110:2012–2101

99. Sedia AJ (2007) Storming the last bastion: The Patent Reform Act of 2007 and its assault on the superior first-to-invent rule. DePaul J Art Technol Intell Prop Law 18:79–107

100. Abrams D, Wagner RP (2013) Poisoning the next apple? The America Invents Act and individual inventors. Stanf Law Rev 65:517

101. EurActiv.com (2009) Germany opposed to 'nano' label for cosmetics, 24 November. http://www.euractiv.com/en/enterprise-jobs/germany-opposed-nano-label-cosmetics/article-187583. Accessed 15 Dec 2009

102. European Patent Office (2013) Nanotechnology and patents. http://documents.epo.org/projects/babylon/eponet.nsf/0/623ECBB1A0FC13E1C12575AD0035EFE6/$File/nanotech_brochure_en.pdf. Accessed 16 Jan 2016

# Erratum to: Biomedical Nanotechnology

## Sarah Hurst Petrosko and Emily S. Day

Sarah Hurst Petrosko and Emily S. Day (eds.), *Biomedical Nanotechnology: Methods and Protocols*, Methods in Molecular Biology, vol. 1570, DOI 10.1007/978-1-4939-6840-4, © Springer Science+Business Media LLC 2017

DOI 10.1007/978-1-4939-6840-4_23

In Chapter 16, *NanoScript: A Versatile Nanoparticle-Based Synthetic Transcription Factor for Innovative Gene Manipulation*, the co-author Ki-Bum Lee's name was originally published as KiBum Lee. This has also been corrected in the table of contents and list of contributors.

The updated online version of the original chapter can be found under
http://dx.doi.org/10.1007/978-1-4939-6840-4_16

Sarah Hurst Petrosko and Emily S. Day (eds.), *Biomedical Nanotechnology: Methods and Protocols*, Methods in Molecular Biology, vol. 1570, DOI 10.1007/978-1-4939-6840-4_23, © Springer Science+Business Media LLC 2017

# INDEX

Sarah Hurst Petrosko and Emily S. Day (eds.), *Biomedical Nanotechnology: Methods and Protocols*, Methods in Molecular Biology, vol. 1570, DOI 10.1007/978-1-4939-6840-4, © Springer Science+Business Media LLC 2017

Printed in the United States
By Bookmasters